TECHNOLOGIES AND APPROACHES TO REDUCING THE FUEL CONSUMPTION OF MEDIUM- AND HEAVY-DUTY VEHICLES

Committee to Assess Fuel Economy Technologies for
Medium- and Heavy-Duty Vehicles

Board on Energy and Environmental Systems
Division on Engineering and Physical Sciences

Transportation Research Board

NATIONAL RESEARCH COUNCIL
OF THE NATIONAL ACADEMIES

THE NATIONAL ACADEMIES PRESS
Washington, D.C.
www.nap.edu

THE NATIONAL ACADEMIES PRESS **500 Fifth Street, N.W.** **Washington, DC 20001**

NOTICE: The project that is the subject of this report was approved by the Governing Board of the National Research Council, whose members are drawn from the councils of the National Academy of Sciences, the National Academy of Engineering, and the Institute of Medicine. The members of the committee responsible for the report were chosen for their special competences and with regard for appropriate balance.

This study was supported by Contract DTNH22-08-H-00222 between the National Academy of Sciences and the U.S. Department of Transportation, National Highway Traffic Safety Administration. Any opinions, findings, conclusions, or recommendations expressed in this publication are those of the author(s) and do not necessarily reflect the views of the organizations or agencies that provided support for the project.

International Standard Book Number-13: 978-0-309-14982-2
International Standard Book Number-10: 0-309-14982-7

Copies of this report are available in limited supply, free of charge, from:

Board on Energy and Environmental Systems
National Research Council
500 Fifth Street, N.W.
Keck W934
Washington, DC 20001
202-334-3344

Additional copies of this report are available for sale from:

The National Academies Press
500 Fifth Street, N.W.
Lockbox 285
Washington, DC 20055
(800) 624-6242 or (202) 334-3313
(in the Washington metropolitan area)
Internet: http://www.nap.edu

Printed in the United States of America

THE NATIONAL ACADEMIES

Advisers to the Nation on Science, Engineering, and Medicine

The **National Academy of Sciences** is a private, nonprofit, self-perpetuating society of distinguished scholars engaged in scientific and engineering research, dedicated to the furtherance of science and technology and to their use for the general welfare. Upon the authority of the charter granted to it by the Congress in 1863, the Academy has a mandate that requires it to advise the federal government on scientific and technical matters. Dr. Ralph J. Cicerone is president of the National Academy of Sciences.

The **National Academy of Engineering** was established in 1964, under the charter of the National Academy of Sciences, as a parallel organization of outstanding engineers. It is autonomous in its administration and in the selection of its members, sharing with the National Academy of Sciences the responsibility for advising the federal government. The National Academy of Engineering also sponsors engineering programs aimed at meeting national needs, encourages education and research, and recognizes the superior achievements of engineers. Dr. Charles M. Vest is president of the National Academy of Engineering.

The **Institute of Medicine** was established in 1970 by the National Academy of Sciences to secure the services of eminent members of appropriate professions in the examination of policy matters pertaining to the health of the public. The Institute acts under the responsibility given to the National Academy of Sciences by its congressional charter to be an adviser to the federal government and, upon its own initiative, to identify issues of medical care, research, and education. Dr. Harvey V. Fineberg is president of the Institute of Medicine.

The **National Research Council** was organized by the National Academy of Sciences in 1916 to associate the broad community of science and technology with the Academy's purposes of furthering knowledge and advising the federal government. Functioning in accordance with general policies determined by the Academy, the Council has become the principal operating agency of both the National Academy of Sciences and the National Academy of Engineering in providing services to the government, the public, and the scientific and engineering communities. The Council is administered jointly by both Academies and the Institute of Medicine. Dr. Ralph J. Cicerone and Dr. Charles M. Vest are chair and vice chair, respectively, of the National Research Council.

www.national-academies.org

COMMITTEE TO ASSESS FUEL ECONOMY TECHNOLOGIES FOR MEDIUM- AND HEAVY-DUTY VEHICLES

ANDREW BROWN, JR., *Chair*, NAE, Delphi Corporation
DENNIS N. ASSANIS, NAE, University of Michigan
ROGER BEZDEK, Management Information Services, Inc.
NIGEL N. CLARK, West Virginia University
THOMAS M. CORSI, University of Maryland
DUKE DRINKARD, Southeastern Freight Lines
DAVID E. FOSTER, University of Wisconsin
ROGER D. FRUECHTE, Consultant
RON GRAVES, Oak Ridge National Laboratory
GARRICK HU, Consultant
JOHN H. JOHNSON, Michigan Technological University
DREW KODJAK, International Council on Clean Transportation
DAVID F. MERRION, Detroit Diesel (retired)
THOMAS E. REINHART, Southwest Research Institute
AYMERIC P. ROUSSEAU, Argonne National Laboratory
CHARLES K. SALTER, Consultant
JAMES J. WINEBRAKE, Rochester Institute of Technology
JOHN WOODROOFFE, University of Michigan Transportation Research Institute
MARTIN B. ZIMMERMAN, University of Michigan

Staff

DUNCAN BROWN, Study Director
DANA CAINES, Financial Associate
LANITA JONES, Administrative Coordinator
JOSEPH MORRIS, Senior Program Officer, Transportation Research Board
JASON ORTEGO, Senior Program Assistant (until December 2009)
MADELINE WOODRUFF, Senior Program Officer
E. JONATHAN YANGER, Senior Project Assistant
JAMES J. ZUCCHETTO, Director, Board on Energy and Environmental Systems

BOARD ON ENERGY AND ENVIRONMENTAL SYSTEMS

DOUGLAS CHAPIN, *Chair*, NAE,[1] MPR Associates, Inc., Alexandria, Virginia
RAKESH AGRAWAL, NAE, Purdue University, West Lafayette, Indiana
WILLIAM BANHOLZER, NAE, The Dow Chemical Company, Midland, Michigan
ANDREW BROWN, JR., NAE, Delphi Technologies, Troy, Michigan
MARILYN BROWN, Georgia Institute of Technology, Atlanta, Georgia
MICHAEL CORRADINI, NAE, University of Wisconsin, Madison, Wisconsin
PAUL DECOTIS, Long Island Power Authority, Long Island, NY
E. LINN DRAPER, JR., NAE, American Electric Power, Lampasas, Texas
CHRISTINE EHLIG-ECONOMIDES, NAE, Texas A&M University, College Station, Texas
WILLIAM FRIEND, NAE, University of California Presidents Council on National Laboratories, Washington, DC
SHERRI GOODMAN, CNA, Alexandria, Virginia
NARAIN HINGORANI, NAE, Independent Consultant, Los Altos Hills, California
MICHAEL OPPENHEIMER, Princeton University, Princeton, New Jersey
MICHAEL RAMAGE, NAE, ExxonMobil Research and Engineering Company (retired), Moorestown, New Jersey
DAN REICHER, Google.org, Warren, Vermont
BERNARD ROBERTSON, NAE, Daimler-Chrysler (retired), Bloomfield Hills, Michigan
MAXINE SAVITZ, NAE, Honeywell, Inc. (retired), Los Angeles, California
MARK THIEMENS, NAS,[2] University of California, San Diego
RICHARD WHITE, Oppenheimer's Private Equity & Special Products, New York, NY

Staff

JAMES J. ZUCCHETTO, Director, Board on Energy and Environmental Systems
DUNCAN BROWN, Senior Program Officer
DANA CAINES, Financial Associate
ALAN CRANE, Senior Program Officer
K. JOHN HOLMES, Senior Program Officer
LANITA JONES, Administrative Coordinator
MADELINE WOODRUFF, Senior Program Officer
E. JONATHAN YANGER, Senior Project Assistant

[1]National Academy of Engineering.
[2]National Acaedemy of Science.

Acknowledgments

The Committee to Assess Fuel Economy Technologies for Medium- and Heavy-Duty Vehicles is grateful to all of the company, agency, industry, association, and national laboratory representatives who contributed significantly of their time and efforts to this National Research Council (NRC) study, either by giving presentations at meetings or by responding to committee requests for information.

We acknowledge the valuable contributions of individuals and organizations that provided information and made presentations at our meetings, as listed in Appendix B. We especially recognize the organizations that hosted site visits for the committee's work as outlined in Chapter 1.

The committee was aided by consultants in various roles who provided analyses to the committee, which it used in addition to other sources of information. Special recognition is afforded the TIAX team of Michael Jackson, Matthew Kromer, and Wendy Bockholt; and the Argonne National Laboratory team of Aymeric Rousseau, Antoine Delorme, Dominik Karbowski, and Ram Vijayagopal.

We wish to recognize the committee members for taking on this daunting charter and accomplishing it on schedule within tight budget requirements. The staff of the NRC Board on Energy and Environmental Systems has been exceptional in organizing and planning meetings, gathering information, and drafting sections of the report. Duncan Brown, Dana Caines, LaNita Jones, Joseph Morris, Jason Ortego, Jonathan Yanger, and James Zucchetto have done an outstanding job of facilitating the work of the committee and providing their knowledge and experience to help the committee in its deliberations. Lastly, the committee chair expresses his personal appreciation to Lori Motley, Delphi executive assistant, for her administrative support provided to this overall effort.

This report has been reviewed in draft form by individuals chosen for their diverse perspectives and technical expertise, in accordance with procedures approved by the NRC's Report Review Committee. The purpose of this independent review is to provide candid and critical comments that will assist the institution in making its published report as sound as possible and to ensure that the report meets institutional standards for objectivity, evidence, and responsiveness to the study charge. The review comments and draft manuscript remain confidential to protect the integrity of the deliberative process.

We wish to thank the following individuals for their review of this report:

Paul Blumberg, Consultant
Fred Browand, University of Southern California
Douglas Chapin, MPR Associates, Inc.
Robert Clarke, Truck Manufacturers Association
Coralie Cooper, Northeast States for Coordinated Air Management
Joe Fleming, Consultant
Winston Harrington, Resources for the Future
John Heywood, Massachusetts Institute of Technology
Larry Howell, General Motors (retired)
Thomas Jahns, University of Wisconsin
James Kirtley, Massachusetts Institute of Technology
Priyaranjan Prasad, Ford Motor Company (retired)
Mike Roeth, Consultant
Russell Truemner, AVL Powertrain Engineering, Inc.

Although the reviewers listed above have provided many constructive comments and suggestions, they were not asked to endorse the conclusions or recommendations, nor did they see the final draft of the report before its release. The review of this report was overseen by Elisabeth Drake, NAE, Massachusetts Institute of Technology (retired). Appointed by the NRC, she was responsible for making certain that an independent examination of this report was carried out in accordance with institutional procedures and that all review comments were carefully considered. Responsibility for the final content of this report rests entirely with the authoring committee and the institution.

Andrew Brown, Jr., *Chair*
Committee to Assess Fuel Economy Technologies
for Medium- and Heavy-Duty Vehicles

Contents

Tables and Figures

TABLES

FIGURES

Summary

Liquid fuel consumption by medium- and heavy-duty vehicles (MHDVs) represents 26 percent of all U.S. liquid transportation fuels consumed and has increased more rapidly—in both absolute and percentage terms—than consumption by other sectors. In early recognition of these trends, which are forecast to continue until 2035 (DOE, EIA, 2009), the Energy Independence and Security Act of 2007 (EISA; Public Law 110-140, Dec. 19, 2007), Section 108, was passed, requiring the U.S. Department of Transportation (DOT), for the first time in history, to establish fuel economy standards for MHDVs. In December 2009 the U.S. Environmental Protection Agency (EPA) formally declared that greenhouse gas (GHG) emissions endanger public health and the environment within the meaning of the Clean Air Act, a decision that compels EPA to consider establishing first-ever GHG emission standards for new motor vehicles, including MHDVs. If the United States is to reduce its reliance on foreign sources of oil, and reduce GHG emissions from the transportation sector, it is important to consider how the fuel consumption of MHDVs can be reduced.

Following the passage of EISA, the National Research Council appointed the Committee to Assess Fuel Economy Technologies for Medium- and Heavy-Duty Vehicles. The committee considered approaches to measuring fuel economy (the committee uses fuel consumption), assessed current and future technologies for reducing fuel consumption, addressed how such technologies may be practically implemented in vehicles, discussed the pros and cons of approaches to improving the fuel efficiency of moving goods as opposed to setting vehicle fuel consumption standards, and identified potential costs and other impacts on the operation of MHDVs (see Chapter 1 and Appendix A for the complete statement of task).

The legislation also requires DOT's National Highway Traffic Safety Administration (NHTSA) to conduct its own study on the fuel consumption of commercial medium- and heavy-duty highway vehicles and work trucks and then to establish a rulemaking to implement a commercial medium- and heavy-duty on-highway and work-truck fuel efficiency improvement program.

The organization of this Summary follows that of the report's chapters: Chapter 1 provides background; Chapter 2 provides vehicle fundamentals; Chapter 3 surveys the current U.S., European, and Asian approaches to fuel economy and regulations; Chapters 4 and 5 review and assess technologies to reduce fuel consumption; Chapter 6 assesses direct and indirect costs and benefits of integrating fuel consumption reduction technologies into vehicles; Chapter 7 presents a review of potential unintended consequences and the alternative nontechnology approaches to reducing fuel consumption; and Chapter 8 reviews options for regulatory design. The Summary presents the committee's major findings and recommendations from each chapter; fuller discussion and additional findings are found in the report.

VEHICLE FUNDAMENTALS, FUEL CONSUMPTION, AND EMISSIONS

Medium- and heavy-duty trucks, motor coaches, and transit buses, Class 2b through Class 8, are used in every sector of the economy. The purposes of these vehicles range from carrying passengers to moving goods. For some vehicles and driving cycles this simple relationship breaks down (as with a bucket truck, which carries one or two passengers but delivers no freight). It brings services and capability (the bucket, tools, and spare parts) to a job site. This results in a broad range of varying duty cycles, from high-speed operation on highways with few stops to lower-speed urban operation with many stops per mile. For the purposes of estimating fuel consumption benefits of various technologies in this report, the committee examined seven different types of vehicles and made assumptions about the duty cycles that would characterize their operations: (1) tractor trailer, (2) Class 6 box truck, (3) Class 6 bucket truck, (4) refuse truck, (5) transit bus, (6) motor coach, and (7) pickup/van. When DOT promulgates standards for fuel consumption, it will have to address the duty cycles that characterize different types of vehicles and their wide range of applications.

The fundamental engineering metric for measuring the fuel efficiency of a vehicle is fuel consumption, the amount of fuel used, assuming some standard duty or driving cycle, to deliver a given transportation service, for example, the amount of fuel a vehicle needs to go a mile or the amount of fuel needed to transport a ton of goods a mile. For light-duty vehicles (cars and light trucks), the corporate average fuel economy (CAFE) program uses miles per gallon (mpg). This measure, although derived from measurements of fuel consumption in gallons/mile, is not the appropriate measure for MHDVs, since these vehicles are designed to carry loads in an efficient and timely manner. A partially loaded tractor trailer would consume less fuel per mile than a fully loaded truck, but this would not be an accurate measure of the fuel efficiency of moving goods. However, normalizing fuel consumption by the payload and using the calculation of gallon/ton-mile—the load-specific fuel consumption (LSFC)—the fully loaded truck would have a much lower LSFC number than the partially loaded truck, reflecting the ability of the truck to accomplish the task of delivering goods.

Major Findings and Recommendations— Chapters 1 and 2: Introduction and Fundamentals

Finding 2-1. Fuel consumption (fuel used per distance traveled; e.g., gallons per mile) has been shown to be the fundamental metric to properly judge fuel efficiency improvements from both engineering and regulatory viewpoints, including yearly fuel savings for different technology vehicles.

Finding 2-2. The relationship between the percent improvement in fuel economy (FE) and the percent reduction in fuel consumption (FC) is nonlinear; e.g., a 10 percent increase in FE (miles per gallon) corresponds to a 9.1 percent decrease in FC, whereas a 100 percent increase in FE corresponds to a 50 percent decrease in FC. This nonlinearity leads to widespread consumer confusion as to the fuel-savings potential of the various technologies, especially at low absolute values of FE.

Finding 2-3. MHDVs are designed as load-carrying vehicles, and consequently their most meaningful metric of fuel efficiency will be in relation to the work performed, such as fuel consumption per unit payload carried, which is load-specific fuel consumption (LSFC). Methods to increase payload may be combined with technology to reduce fuel consumption to improve LSFC. Future standards might require different values to accurately reflect the applications of the various vehicle classes (e.g., buses, utility, line haul, pickup, and delivery).

Recommendation 2-1. Any regulation of medium- and heavy-duty vehicle fuel consumption should use LSFC as the metric and be based on using an average (or typical) payload based on national data representative of the classes and duty cycle of the vehicle. Standards might require different values of LSFC due to the various functions of the vehicle classes e.g., buses, utility, line haul, pickup, and delivery. Regulators need to use a common procedure to develop baseline LSFC data for various applications, to determine if separate standards are required for different vehicles that have a common function. Any data reporting or labeling should state an LSFC value at specified tons of payload.

COMPARING THE REGULATORY APPROACHES OF THE UNITED STATES, JAPAN, AND EUROPEAN COMMUNITY

Although a CAFE regulatory program has been implemented for light-duty vehicles, where the responsibility for the manufacture and certification of vehicles is well defined and the configurations of cars and light trucks for sale are well defined and of limited number, the MHDV world is much more complicated. There are literally thousands of different configurations for vehicles, including bucket trucks, pickup trucks, garbage trucks, delivery vehicles, and long-haul tractor trailers. Their duty cycles vary greatly. Some stop and go every few seconds; others spend most of their time at highway speeds. Furthermore, the party responsible for the final truck configuration is often not well defined. For example, a body builder (vehicle integrator) may be the manufacturer of record, but the body builder may not design or even specify the chassis and power train. For tractor-trailer combinations, the tractor and trailer are always made and often owned by different companies, and a given tractor may pull hundreds of different trailers of different configurations over its life. Many trucks are custom made, literally one of a kind.

Even though the regulation of such vehicles will be much more complicated than it is for light-duty vehicles, the barriers are not insurmountable. Safety and emission regulations have been implemented, and regulations for fuel consumption in medium- and heavy-duty trucks already exist in Japan and are under development by the European Commission. California is building on the EPA's SmartWay Partnership to implement its own approach to regulating truck fuel consumption.

Major Findings and Recommendations— Chapter 3: Current Regulatory Approaches

Finding 3-1. Although it took years of development and substantial effort, regulators have dealt effectively with the diversity and complexity of the vehicle industry for current laws on fuel consumption and emissions for light-duty vehicles. Engine-based certification procedures have been applied to address emissions from heavy-duty vehicles and the myriad of nontransportation engines.

Finding 3-2. The heavy-duty-truck fuel consumption regulations in Japan, and those under consideration and study by the European Commission, provide valuable input and experience to the U.S. plans. In Japan the complexity of MHDV configurations and duty cycles was determined to lend itself to the use of computer simulation as a cost-effectives means to calculate fuel efficiency, and Japan is not using extensive full-vehicle testing in the certification process.

TECHNOLOGIES AND COSTS OF REDUCING FUEL CONSUMPTION

The committee has evaluated a wide range of fuel-saving technologies for medium- and heavy-duty vehicles. Some technologies, such as certain aerodynamic features, automated manual transmissions, and wide-base single low-rolling-resistance tires, are already available in production. Some of the technologies are in varying stages of development, while others have only been studied using simulation models. Reliable, peer-reviewed data on fuel-saving performance is available only for a few technologies in a few applications. As a result, the committee had to rely on information from a wide range of sources, (e.g., information gathered from vehicle manufacturers, component suppliers, research labs, and major fleets during site visits by the committee), including many results that have not been duplicated by other researchers or verified over a range of duty cycles.

There is a tendency among researchers to evaluate technologies under conditions which are best suited to that specific technology. This can be a serious issue in situations where performance is strongly dependent on duty cycle, as is the case for many of the technologies evaluated in this report. One result is that the reported performance of a specific technology may be better than what would be achieved by the overall vehicle fleet in actual operation. Another issue with technologies that are not fully developed is a tendency to underestimate the problems that could emerge as the technology matures to commercial application. Such issues often result in implementation delays as well as a loss of performance compared to initial projections. As a result of these issues, some of the technologies evaluated in this report may be available later than expected, or at a lower level of performance than expected. Extensive additional research would be needed to quantify these issues, and regulators will need to allow for the fact that some technologies may not mature as expected.

The fuel-saving technologies that are already available on the market generally result in increased vehicle cost, and purchasers must weigh the additional cost against the fuel savings that will accrue. In most cases, market penetration is low at this time. Most fuel-saving technologies that are under development will also result in increased vehicle cost, and in some cases, the cost increases will be substantial. As a result, many technologies may struggle to achieve market acceptance, despite the sometimes substantial fuel savings, unless driven by regulation or by higher fuel prices. Powertrain technologies (for diesel engines, gasoline engines, transmissions, and hybrids) as well as vehicle technologies (for aerodynamics, rolling resistance, mass/weight reduction, idle reduction, and intelligent vehicles) are analyzed in Chapters 4 and 5. Tables S-1 and S-2 provide the committee's estimate of the range of fuel consumption reduction that is potentially achievable with new technologies in the period 2015 to 2020, compared to a 2008 baseline.[1] Figure S-1 provides estimates for potential fuel consumption reductions for typical new vehicles in the 2015 to 2020 time frame.

The technologies were grouped into time periods based on the committee's estimate of when the technologies would be proven and available. In practice, the timing of their introduction will vary by manufacturer, based in large part on individual company product development cycles. In order to manage product development costs, manufacturers must consider the overall product life cycle and the timing of new product introductions. As a result, widespread availability of some technologies may not occur in the time frames shown.

The percent fuel consumption reduction (% FCR) numbers shown for individual technologies and other options are not additive. For each vehicle class, the % FCR associated with combined options is as follows:

$$\%\ FCR_{package} = 100\ [1 - (1 - \{\%\ FCR_{tech1}/100\})\ (1 - \{\%\ FCR_{tech2}/100\}) \ldots \{(1 - \{\%\ FCR_{techN}/100\})]$$

where $\%\ FCR_{techx}$ is the percent benefit of an individual technology.

The major enabling technologies necessary to achieve these reductions are hybridization, advanced diesel engines, and aerodynamics. Hybridization is particularly important in those applications with the stop-and-go duty cycles characteristic of many MHDVs, such as refuse trucks and transit buses, as well as bucket trucks. Diesel and gasoline engine advancements are helpful in all applications and will include continuing improvements to fuel injection systems, emissions control, and air handling systems, in addition to commercialization of waste heat recovery systems. Essentially all Class 8 vehicles will continue with diesel engines as the prime mover. The third major technology improvement is total vehicle aerodynamics, especially in over-the-road applications like tractor trailers and motor coaches. Other technologies that will play a role in reducing fuel consumption in all vehicle segments include low-rolling-resistance tires, improved transmissions, idle-reduction technologies, weight reduction, and driver management and coaching.

The applications of these technologies can be put into packages and then applied to the seven types of MHDVs analyzed. The resulting fuel consumption reduction for each

[1] More information on the baseline can be found in Chapter 6 and in TIAX (2009).

TABLE S-1 Range of Fuel Consumption Reduction Potential, 2015-2020, for Power Train Technologies

Technology	Fuel Consumption Reduction (%)
Diesel engines	15 to 21
Gasoline engines	Up to 24
Diesel over gasoline engines	6 to 24
Improved transmissions	4 to 8
Hybrid power trains	5 to 50

NOTE: Potential fuel reductions are not additive. For each vehicle class, the fuel consumption benefit of the combined technology packages is calculated as follows: [% $FCR_{package}$ = 100 [1 – (1 – {% FCR_{tech1} /100}) (1 – {% FCR_{tech2}/100)} … (1 – {% FCR_{techN} /100})]. Values shown are for one set of input assumptions. Results will vary depending on these assumptions.

TABLE S-2 Range of Fuel Consumption Reduction Potential, 2015-2020, for Vehicle Technologies

Technology	Fuel Consumption Reduction (%)
Aerodynamics	3 to 15
Auxiliary loads	1 to 2.5
Rolling resistance	4.5 to 9
Mass (weight) reduction	2 to 5
Idle reduction	5 to 9
Intelligent vehicle	8 to 15

NOTE: Potential fuel reductions are not additive. For each vehicle class, the fuel consumption benefit of the combined technology packages is calculated as follows: [% $FCR_{package}$ = 100 [1 – (1 – {% FCR_{tech1}/100 }) (1 – {% FCR_{tech2}/100)} … (1 – {% FCR_{techN}/100})]. Values shown are for one set of input assumptions. Results will vary depending on these assumptions.
SOURCE: Adapted from TIAX (2009).

vehicle type will be dependent on the typical vehicle application and the typical duty cycle. The results of the packages on fuel consumption reduction from a 2008 baseline are shown for the 2015 to 2020 time frame in Figure S-1.

The technology packages that result in the fuel consumption reduction for each application also have projected costs. The costs are estimated assuming the technologies will be produced at large enough volumes to achieve economies of scale in the 2015 to 2020 time frame. The committee has also determined several ways to measure costs versus benefits. The first measure, dollars per percent fuel saved, is the cost of the technology package divided by the percent reduction in fuel consumption. The second measure, dollars per gallon saved per year, accounts for the fact that some vehicles are normally driven more miles than others. The measure calculates how much it costs to save one gallon of fuel each year for the life of the vehicle by adopting the relevant technology. The third measure, "breakeven" fuel price, represents the fuel price that would make the present discounted value

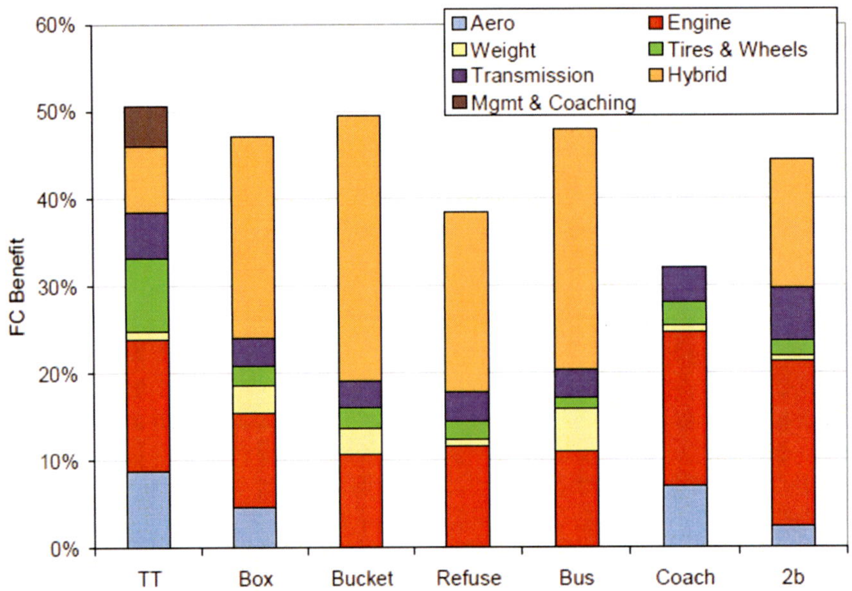

FIGURE S-1 Comparison of 2015-2020 new-vehicle potential fuel-saving technologies for seven vehicle types: tractor trailer (TT), Class 3-6 box (box), Class 3-6 bucket (bucket), Class 8 refuse (refuse), transit bus (bus), motor coach (coach), and Class 2b pickups and vans (2b).
NOTE: TIAX (2009) only evaluated the potential benefits of driver management and coaching for the tractor-trailer class of vehicles. It is clear to the committee that other vehicle classes would also benefit from driver management and coaching, but studies showing the benefits for specific vehicle classes are not available. For more information, see the subsection "Driver Training and Behavior" in Chapter 7. Also, potential fuel reductions are not additive. For each vehicle class, the fuel consumption benefit of the combined technology packages is calculated as follows: [% $FCR_{package}$ = 100 [1 – (1 – {% FCR_{tech1}/100}) (1 – {% FCR_{tech2}/100)} … (1 – {% FCR_{techN}/100})]. Values shown are for one set of input assumptions. Results will vary depending on these assumptions. SOURCE: TIAX (2009).

of the fuel savings equal to the total costs of the technology package applied to the vehicle class.

The breakeven fuel price shown in Table S-3 does not necessarily reflect how vehicle buyers would evaluate technologies, because they often do not plan to own a vehicle for its full life, they may use a different discount rate, and they would need to consider operation and maintenance costs, which are excluded from the calculation. However, a lifetime breakeven price is a useful metric for considering both the private and the societal costs and benefits of regulation. Although incomplete, the measures shown in Table S-3 are suggestive of the differences in economic viability of the various technology options for the indicated vehicle classes. It is important to remember, however, that these breakeven prices are calculated assuming that all the technologies are applied as a package. In fact, individual fuel-saving technologies applied in a given vehicle class may face much lower or much higher breakeven values than the aggregate figures listed in Table S-3. For more detailed information on the values summarized in Table S-3, see Tables 6-18 and 6-19 in Chapter 6.

The findings and recommendations below combine material from Chapters 4 through 6 and therefore do not match the numbering in those chapters but are presented instead as "Finding 4/5/6-X."

TABLE S-3 Fuel Consumption Reduction Potential for Typical New Vehicles, 2015-2020, and Cost-Effectiveness Comparisons for Seven Vehicle Configurations

			Cost-Effectiveness Metric		
Vehicle Class	Fuel Consumption Reduction (%)	Capital Cost ($)	Dollars per Percent Fuel Saved	Dollars per Gallon Saved per Year	Breakeven Fuel Price[a] ($/gal)
Tractor-trailer	51	84,600	1,670	7.70	1.10
Class 6 box truck	47	43,120	920	29.30	4.20
Class 6 bucket truck	50	49,870	1,010	37.80	5.40
Class 2b pickup	45	14,710	330	33.70	4.80
Refuse truck	38	50,800	1,320	18.90	2.70
Transit bus	48	250,400	5,230	48.00	6.80
Motor coach	32	36,350	1,140	11.60	1.70

NOTE: Numbers in last three columns are rounded. Also, these point estimates will vary depending on input assumptions. For each vehicle class, the fuel consumption benefit of the combined technology packages is calculated as follows: [$\% \text{ FCR}_{package} = 100 [1 - (1 - \{\% \text{ FCR}_{tech1}/100\}) (1 - \{\% \text{ FCR}_{tech2}/100)\} \ldots (1 - \{\% \text{ FCR}_{techN}/100\})]$. Values shown are for one set of input assumptions. Results will vary depending on these assumptions.

[a]Calculated assuming a 7 percent discount rate and a 10-year life, excluding incremental operating and maintenance costs associated with the technologies.

SOURCE: Adapted from TIAX (2009).

Major Findings and Recommendations—Chapters 4, 5, and 6: Technologies and Direct Impacts

Finding 4/5/6-1. The fuel consumption reduction potential of specific power train and vehicle technologies is extremely dependent on application (pickup vs. tractor trailer) and duty cycle (start-stop vs. steady state, variations in load, etc.).

Finding 4/5/6-2. Technologies vary significantly in the cost-benefit evaluation. Some technologies are economically viable at today's fuel prices. Others examined require significantly higher fuel prices or correspondingly high valuations of environmental and security externalities to justify their application.

Finding 4/5/6-3. Cost per percent fuel saved is a widely used metric for evaluating the cost/benefit of fuel-saving technologies, and this metric is also used here. Unfortunately, this metric can be very misleading, because it leaves out the critical component of total annual vehicle fuel consumption. Table S-3 shows great discrepancies between cost per percent fuel saved and cost per gallon saved.

Recommendation 4/5/6-1. The federal government should continue to support programs in industries, national laboratories, private companies, and universities to develop MHDV technologies for reducing fuel consumption.

INDIRECT EFFECTS AND EXTERNALITIES

In addition to the direct costs and benefits associated with the application of new technologies, there are also *indirect* costs, benefits, and externalities (impacts that are not expressed in market terms) that should be discussed and addressed. Some of these indirect effects represent unintended consequences associated with technologies or policies designed to spur greater fuel efficiency in MHDVs. Although it recognizes that it did not address an exhaustive list of indirect effects, the committee emphasizes the importance of assessment of such effects during policy development to help avoid or mitigate negative unintended consequences.

Major Findings and Recommendations—Chapter 6: Indirect Effects and Externalities

Finding 6-9. A number of indirect effects and unintended consequences associated with regulations aimed at reducing fuel consumption in the trucking sector can be important. In particular, regulators should consider the following effects in the development of any regulatory proposals: rate of replacement of older vehicles (fleet turnover impacts), increased ton-miles shipped due to the lower cost of shipping (rebound effect), purchasing one class of vehicle rather than another in response to a regulatory change (vehicle class shifting), environmental co-benefits and costs, congestion, safety, and incremental weight impacts.

Finding 6-10. Consumer buying in anticipation of new regulations (pre-buy) and retention of older vehicles can slow the rate of fleet turnover and the rate at which regulatory standards can affect fleet-wide fuel consumption.

Finding 6-11. Elasticity estimates vary over a wide range, and it is not possible to calculate with a great deal of confidence what the magnitude of the "rebound" effect is for heavy-duty trucks. The rebound effect measures the increase in ton-miles shipped resulting from a reduction in the cost of shipping. Estimates of fuel savings from regulatory standards will be somewhat misestimated if the "rebound" effect is not considered.

Finding 6-12. Standards that differentially affect the capital and operating costs of individual vehicle classes can cause purchase of vehicles that are not optimized for particular operating conditions. The complexity of truck use and the variability of duty cycles increase the probability of these unintended consequences.

Finding 6-16. Some fuel-efficiency-improving technologies will add weight to vehicles and push those vehicles over federal threshold weights, thereby triggering new operational conditions and affecting, in turn, vehicle purchase decisions. More research is needed to assess the significance of this potential impact.

Finding 6-17. Some fuel-efficiency-improving technologies will reduce cargo capacity for trucks that are currently "weighed-out" and will therefore force additional trucks onto the road. More research is needed to assess the significance of this potential impact.

Recommendation 6-1. NHTSA, in its study, should do an economic/payback analysis based on fuel usage by application and different fuel price scenarios. Operating and maintenance costs should be part of any study.

ALTERNATIVE APPROACHES

There may be more effective, less costly, and complementary approaches than vehicle fuel efficiency standards for reducing fuel consumption of MHDVs, such as training truck drivers on best practices, adjusting size and weight restrictions on trucks, implementing market-based instruments (e.g., fuel taxes), providing incentives for mode shifting, or developing intelligent vehicle and highway systems. As DOT/NHTSA conduct regulatory analyses of fuel efficiency options, indirect costs and alternative approaches will have to be identified.

Major Findings and Recommendations—Chapter 7

Finding 7-1. The committee examined a number of approaches for reducing fuel consumption in the trucking sector and found suggestive evidence that several approaches—particularly driver training and longer combination vehicles (LCVs)—offer potential fuel savings for the trucking sector that rival the savings available from technology adoption for certain vehicle classes and/or types. Any government action taken to reduce fuel consumption in the trucking sector should consider these alternatives.

Finding 7-2. Fuel taxes offer a transparent and efficient method for internalizing the potential societal costs of climate change and oil imports (e.g., energy security) and reducing fuel consumption in road transport. Fuel taxes operate to make fuel-saving technologies more attractive and provide incentives for saving fuel in operations, while involving fewer unintended consequences than standards.

Recommendation 7-1. Although the committee recognizes the political difficulty associated with increasing fuel taxes, it strongly recommends that Congress consider fuel taxes as an alternative to mandating fuel efficiency standards for medium- and heavy-duty trucks.

Finding 7-5. A cap-and-trade system, such as is being considered by Congress that would limit total carbon dioxide (CO_2) emissions by primary energy producers, would have implications for the trucking sector. Regulators would then not need to develop standards for CO_2 emissions that apply to specific trucks and trucking operations, avoiding the complexity of different classes and duty cycles of trucks. On the other hand, the cap-and-trade system would likely involve new administrative burdens for monitoring emissions from the primary producers and policing the system.

Finding 7-7. When there are several fuel-saving options and complex truck operating conditions, performance standards are likely to be superior to specific technology requirements.

Finding 7-8. Increasing vehicle size and weight limits offers potentially significant fuel savings for the entire tractor-trailer combination truck fleet. This approach would need to be weighed against increased costs of road repair. Example case studies explored in this report demonstrate fuel savings of up to 15 percent or more. These savings are similar in size but independent and accumulative of other actions that may be taken to improve fuel consumption of vehicles; therefore the net potential benefit is substantial. To achieve these savings would require the federal government to:

- Change regulatory limits that currently restrict vehicle weight to 80,000 lb and that freeze LCV operations on the Federal Interstate System.
- Establish a regulatory structure that assures safety and compatibility with the infrastructure. One possible regulatory structure has been proposed by the Transportation Research Board in *Regulation of Weights,*

Lengths, and Widths of Commercial Motor Vehicles, Special Report 267 (TRB, 2002).
- Consider the necessary changes that would be required to permit reasonable access of LCVs to vehicle breakdown yards and major shipping facilities in close proximity to the interstate.

Recommendation 7-2. Congress should give serious consideration to liberalizing weight and size restrictions and should consider how the potential fuel savings and other benefits of such liberalization can be realized in a way that maintains safety and minimizes the cost of potential infrastructure changes.

Finding 7-10. Intelligent transportation systems enable more efficient use of the existing roadway system by improving traffic flow and reducing or avoiding congestion.

Finding 7-12. There are significant opportunities for savings in fuel, equipment, maintenance, and labor when drivers are trained properly. Indications are that this could be one of the most cost-effective and best ways to reduce fuel consumption and improve the productivity of the trucking sector. For example, cases evaluated herein demonstrate potential fuel savings of ~2 to 17 percent with appropriately trained drivers.

Recommendation 7-3. The federal government should encourage and incentivize the dissemination of information related to the relationship between driving behavior and fuel savings. For example, one step in this direction could be to establish a curriculum and process for certifying fuel-saving driving techniques as part of commercial driver license certification and to regularly evaluate the effects of such a curriculum.

APPROACHES TO FUEL CONSUMPTION REDUCTION AND REGULATIONS

This is an important juncture for the nation. The choices that will be made over the course of the next few years will establish the regulatory design for MHDV fuel consumption standards for the next several decades at least. While the stringency of the standards themselves may be revisited from time to time, the regulatory design elements (regulated parties, certification tests and procedures, compliance methods)—once established—are far more difficult to modify.

In many cases, the commercial vehicle market is sophisticated, driven by knowledgeable purchasers who focus on the efficiency of their operations, including the fuel costs associated with accomplishing their tasks. Thus, one of the most important challenges facing NHTSA is how to enhance and improve upon the commercial trucking industry's existing desire to maximize the fuel economy of its trucks and fleets.

At the same time, there are commonly acknowledged characteristics in the commercial truck and buses marketplace that may be improved by a regulatory approach, such as split incentives between owners and operators (e.g., trailers) and the short payback period of 18 months to 2 years, that create barriers to the adoption of efficiency technologies for many purchasers, suggesting that a well-designed regulatory program may yield important benefits.

Due to the complexity of the vehicle market, the committee was not able to give adequate consideration to the noncommercial markets such as personal pickup trucks, school buses, and personal motor homes. NHTSA should consider these applications in its regulatory proposal.

A fundamental concern raised by the committee and those who testified during its public sessions was the tension between the need to set a uniform test cycle for regulatory purposes and existing industry practices of seeking to minimize fuel consumption of medium- and heavy-duty vehicles designed for specific routes that may include grades, loads, work tasks, or speeds inconsistent with the regulatory test cycle. This concern emphasizes the critical importance of achieving fidelity between certification values and real-world results, in order to avoid driving decisions that hurt rather than help real-world fuel consumption.

Because regulations can lead to unintended consequences, either because the variability of tasks within a vehicle class is not adequately dealt with or because regulations may lead to distortions between classes in the costs of accomplishing similar tasks, the committee urges NHTSA to carefully consider all factors when developing its regulatory proposal.

Major Finding and Recommendations—Chapter 8

Finding 8-1. While it may seem expedient to focus initially on those classes of vehicles with the largest fuel consumption (i.e., Class 8, Class 6, and Class 2b, which together account for approximately 90 percent of fuel consumption of MHDVs), the committee believes that selectively regulating only certain vehicle classes would lead to very serious unintended consequences and would compromise the intent of the regulation. Within vehicle classes, there may be certain subclasses of vehicles (e.g., fire trucks) that could be exempt from the regulation without creating market distortions.

Finding 8-2. Large original equipment manufacturers (OEMs), which have significant engineering capability, design and manufacture almost all Class 2b, 3, and 8b vehicles. Small companies with limited engineering resources make a significant percentage of vehicles in Classes 4 through 8a, although in many cases they buy the complete chassis from larger OEMs. Regulators will need to take the limitations of these smaller companies into account.

Finding 8-3. Commercial trailers are produced by a separate group of manufacturers that are not associated with truck manufacturers. Trailers, which present an important opportunity for fuel consumption reduction, can benefit from improvements in aerodynamics and tires.

Recommendation 8-1. When NHTSA regulates, it should regulate the final-stage vehicle manufacturers since they have the greatest control over the design of the vehicle and its major subsystems that affect fuel consumption. Component manufacturers will have to provide consistent component performance data. As the components are generally tested at this time, there is a need for a standardized test protocol and safeguards for the confidentiality of the data and information. It may be necessary for the vehicle manufacturers to provide the same level of data to the tier suppliers of the engines, transmissions, and after-treatment and hybrid systems.

Recommendation 8-3. NHTSA should establish fuel consumption metrics tied to the task associated with a particular type of MHDV and set targets based on potential improvements in vehicle efficiency and vehicle or trailer changes to increase cargo-carrying capacity. NHTSA should determine whether a system of standards for full but lightly loaded (cubed-out) vehicles can be developed using only the LSFC metric or whether these vehicles need a different metric to properly measure fuel efficiency without compromising the design of the vehicles.

Finding 8-7. Some certification and compliance methods seem more practical than others, and the committee acknowledges that there may be other options or variations that have yet to be identified. Regulating total vehicle fuel consumption of MHDVs will be a formidable task due to the complexity of the fleet, the various work tasks performed, and the variations in fuel-consumption-related technologies within given classes, including vehicles of the same model and manufacturer.

Finding 8-9. Using the process and results from existing engine dynamometer testing for criteria emissions to certify fuel economy standards for MHDVs would build on proven, accurate, and repeatable methods and put less additional administrative burden on the industry. However, to account for the fuel consumption benefits of hybrid power trains and transmission technology, the present engine-only tests for emissions certification will need to be augmented with other power train components added to the engine test cell, either as real hardware or as simulated components. Similarly, the vehicle attributes (aerodynamics, tires, mass) will need to be accounted for, one approach being to use vehicle-specific prescribed loads (via models) in the test cycle. This will require close cooperation among component manufacturers and vehicle manufacturers.

Recommendation 8-4. Simulation modeling should be used with component test data and additional tested inputs from power train tests, which could lower the cost and administrative burden yet achieve the needed accuracy of results. This is similar to the approach taken in Japan, but with the important clarification that the program would represent all of the parameters of the vehicle (power train, aerodynamics, and tires) and relate fuel consumption to the vehicle task.

Finding 8-13. There is an immediate need to take the findings and recommendations in this report and begin the development of a regulatory approach. Significant engineering work is needed to produce an approach that results in fuel efficiency standards that are cost-effective and that accurately represent the effects of fuel-consumption-reducing technologies. The regulations should fit into the engineering and development cycle of the industry and provide meaningful data to vehicle purchasers.

Recommendation 8-5. Congress should appropriate money for and NHTSA should implement as soon as possible a major engineering contract that would analyze several actual vehicles covering several applications and develop an approach to component testing and related data collection in conjunction with vehicle simulation modeling to arrive at LSFC data for these vehicles. The actual vehicles should also be tested by appropriate full-scale test procedures to confirm the actual LSFC values and the reductions measured with fuel consumption reduction technologies in order to validate the evaluation method.

Recommendation 8-6. NHTSA should conduct a pilot program to "test drive" the certification process and validate the regulatory instrument proof of concept. It should have these elements:

- Gain experience with certification testing, data gathering, compiling, and reporting. There needs to be a concerted effort to determine the accuracy and repeatability of all the test methods and simulation strategies that will be used with any proposed regulatory standards and a willingness to fix issues that are found.
- Gather data on fuel consumption from several representative fleets of vehicles. This should continue to provide a real-world check on the effectiveness of the regulatory design on the fuel consumption of trucking fleets in various parts of the marketplace and in various regions of the country.

REFERENCES

DOE, EIA. 2009. Annual Energy Outlook 2010 (Preliminary). Washington, D.C., December.

TIAX, LLC. 2009. Assessment of Fuel Economy Technologies for Medium- and Heavy-Duty Vehicles. Final Report. Report to the National Academy of Sciences. Cambridge, Mass. September.

TRB (Transportation Research Board). 2002. Special Report 267: Regulation of Weights, Lengths, and Widths of Commercial Motor Vehicles. Washington, D.C.: TRB.

1

Introduction

In the United States, medium- and heavy-duty vehicles (MHDVs) consume a significant and increasing amount of fuel. In 2008 these vehicles consumed 26 percent of all U.S. transportation liquid fuels and 20 percent of all U.S. liquid fuels. MHDVs consumed 3.9 million barrels per day (mbpd), compared to total 2008 U.S. liquid fuel consumption of 19.5 mbpd.

Liquid fuel consumption by MHDVs has increased more rapidly—in both absolute and percentage terms—than consumption by other sectors, and the Energy Information Administration (EIA) forecasts that this will continue. EIA projects that in 2035 these classes of vehicles will consume 30 percent of all U.S. transportation liquid fuels and 23 percent of all U.S. liquid fuels. That total will represent 5.1 mbpd, compared with total projected 2035 U.S. liquid fuel consumption of 22.1 mbpd. Thus, the fuel efficiency of these classes of vehicles is of high and increasing importance (DOE, EIA, 2009c). Furthermore, in December 2009 the U.S. Environmental Protection Agency (EPA) formally declared that greenhouse gas (GHG) emissions endanger public health and the environment within the meaning of the Clean Air Act, a decision that compels EPA to consider establishing first-ever GHG emission standards for new motor vehicles, including MHDVs. If the United States is to reduce its reliance on foreign sources of oil, and reduce GHG emissions from the transportation sector, it is important to consider how the fuel consumption of MHDVs can be reduced.

ORIGIN OF STUDY AND STATEMENT OF TASK

The National Research Council (NRC) Committee to Assess Fuel Economy Technologies for Medium- and Heavy-Duty Vehicles was formed in response to a congressional mandate to the National Highway Traffic Safety Administration (NHTSA), an agency of the U.S. Department of Transportation (DOT), under Section 108 of the Energy Independence and Security Act (EISA) of 2007. NHTSA was directed to contract with the National Academies to undertake a study and develop a report that evaluates medium- and heavy-duty truck fuel economy. The legislation also (1) mandates that NHTSA itself conduct a study on the fuel efficiency of commercial medium- and heavy-duty on-highway vehicles and work trucks and (2) mandates that NHTSA then conduct a rulemaking to implement a commercial medium- and heavy-duty on-highway and work-truck fuel efficiency improvement program.[1]

The language in Section 108 directs the National Academy of Sciences to address the following items in its report:

(1) an assessment of technologies and costs to evaluate fuel economy for medium-duty and heavy-duty trucks;
(2) an analysis of existing and potential technologies that may be used practically to improve medium-duty and heavy-duty truck fuel economy;
(3) an analysis of how such technologies may be practically integrated into the medium-duty and heavy-duty truck manufacturing process;
(4) an assessment of how such technologies may be used to meet fuel economy standards to be prescribed under section 32902(k) of title 49, United States Code, as amended by this subtitle; and
(5) associated costs and other impacts on the operation of medium-duty and heavy-duty trucks, including congestion.

In response to that language, the NRC developed a statement of task for the committee that directs it to:

- Consider approaches to measuring fuel economy for medium- and heavy-duty vehicles that would be required for setting standards;

[1]The legislation uses both the terms "fuel economy" and "fuel efficiency." Fuel economy, generally miles per gallon or kilometers per liter, is commonly used in comparing the efficiency of light-duty vehicles, which have similar size and driving cycles. In comparing the fuel consumption of trucks and buses, its usefulness is limited, given there is a wide difference in mass and driving cycles. In particular, a metric is needed that reflects the work done by the vehicle. The committee discusses the appropriate metric for trucks and buses later in Chapter 1 and in Chapter 2.

- Assess current and potential technologies and estimate improvements in fuel economy for medium-duty and heavy-duty trucks that might be achieved;
- Address how the technologies identified in Task 2 above may be used practically to improve medium-duty and heavy-duty truck fuel economy;
- Address how such technologies may be practically integrated into the medium-duty and heavy-duty truck manufacturing process;
- Assess how such technologies may be used to meet fuel economy standards;
- Discuss the pros and cons of approaches to improving the fuel efficiency of moving goods as opposed to setting vehicle fuel economy standards; and
- Identify the potential costs and other impacts on the operation of medium-duty and heavy-duty trucks. (See Appendix A for the full statement of task.)

The committee discussed these tasks with the DOT/NHTSA representatives, as well as relevant congressional staff, prior to and at the committee's first meeting. The purpose of these discussions was to explore what information and data could be made available to the committee and to take advantage of the expertise available on the committee to determine the extent to which the tasks could be addressed. It should be noted that the study does not address the use of alternative fuels to substitute for fossil-fuel-based diesel or gasoline. Domestic production of alternative fuels such as biodiesel or natural gas could help to reduce demand for imports of petroleum or reduce emissions of greenhouse gases, but these technologies and/or strategies are not addressed. The committee provides some insights in Chapter 6 into the unintended consequences that could arise from various approaches that might be used to reduce the fuel consumption of vehicles. In addition, Chapter 7 explores the advantages and disadvantages of alternative approaches to reducing fuel consumption, since many of these alternatives involve regulatory changes, and Chapter 8 discusses fuel consumption regulatory approaches.

POLICY MOTIVATION

The President and Congress have placed among the highest national objectives that of reducing petroleum imports. Despite efforts to wean the United States away from oil toward more acceptable fuels, it has become increasingly dependent on oil (Figure 1-1).

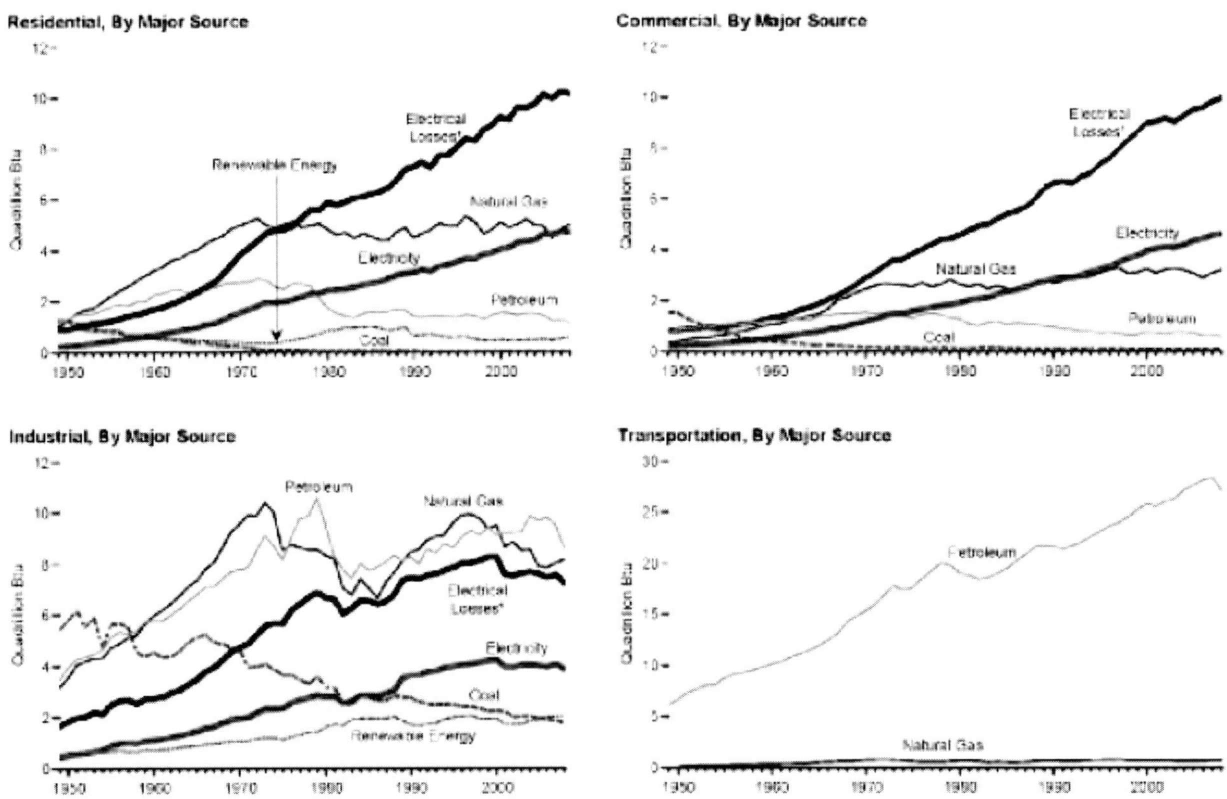

FIGURE 1-1 Energy consumption by major source end-use sector, 1949-2008.
SOURCE: DOE, EIA (2009b, p. 39).

While fuel consumed per mile by light-duty vehicles improved substantially between 1966 and 2007, fuel consumption of the average heavy-duty vehicle remained nearly constant (Figure 1-2). However, this trend hides an important factor regarding trucking. The mission is not just to move the truck and driver from one place to another but to deliver cargo. If total fuel consumed, total miles traveled, and total tons shipped are considered for the United States as a whole, a U.S. average payload specific fuel consumption for the entire medium- and heavy-duty fleet can be calculated for this sector. Figure 1-3 shows the results of dividing the total fuel consumed by the miles traveled and tons moved each year, to produce a fuel consumption per ton shipped and per mile driven (gallon/ton-mile) from 1975 to 2005. The amount of fuel required to move a given amount of freight a given distance has been reduced by more than half over this time period. This is a result of many factors, including:

- Improved efficiency of engines and drivelines
- Improved vehicle aerodynamics
- Improved tire rolling resistance
- Widespread implementation of electronic control features such as road speed governors
- Regulatory changes that allowed the use of longer, wider, and taller trailers and higher maximum weight limits
- Operational efficiency improvements by trucking companies to reduce the amount of distance traveled with little or no load

The improvement trend in this U.S. average payload specific fuel consumption for trucks has slowed in the past several years, at least in part due to the requirement to introduce new pollution controls for EPA-regulated air pollutants such as nitrogen oxides (NO_x) and particulate matter (PM). The resulting changes to diesel engines have tended to degrade their thermal efficiency. Gasoline engines also suffered degradation in performance when first required to meet regulated emission standards. The development of the three-way catalyst has allowed the recapture of much of the

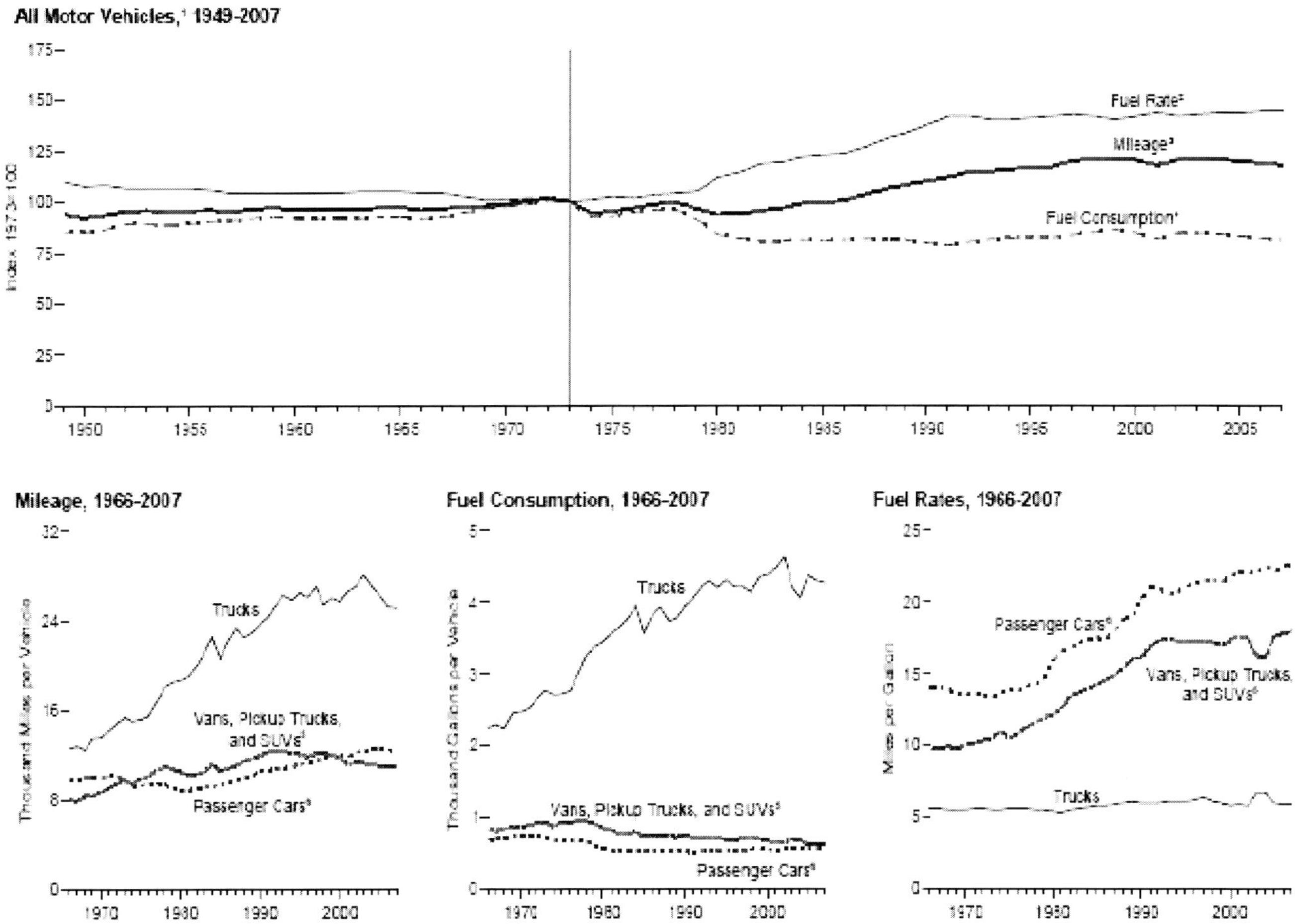

FIGURE 1-2 Motor vehicle mileage, fuel consumption, and fuel rates. SOURCE: DOE, EIA (2009a, Figure 2.8).

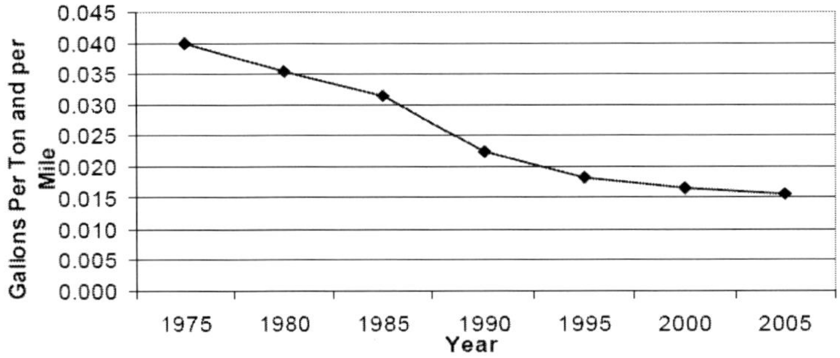

FIGURE 1-3 U.S. average payload-specific fuel consumption. SOURCES: Data from Federal Highway Administration, Highway Statistics Summary to 1995, Table VM-201A, and Highway Statistics (annual releases), Table VM-1, Washington D.C., available at http://www.fhwa.dot.gov/ohim/summary95/vm201a.xlw, accessed Feb. 25, 2010; total tons hauled from Bob Costello, American Trucking Association.

lost performance; however, the three-way catalyst mandates that current gasoline engines must operate at stoichiometric air/fuel ratios. If effective lean air/fuel ratio aftertreatment systems could be developed, further reductions in gasoline engine fuel consumption would be readily achievable. Gasoline-powered medium- and heavy-duty vehicles have followed the same historical fuel consumption and emission trends as the light-duty gasoline-powered vehicles.

The improved efficiency of both light- and heavy-duty vehicles has been overwhelmed by an increase in annual vehicle miles traveled (VMT). VMT has grown more quickly in the trucking sector than in the light-duty sector, resulting in medium- and heavy-duty vehicles taking up a growing share of total transportation-related petroleum consumption.

In fact, the U.S. transportation system relies nearly exclusively on petroleum, as shown in Figure 1-1 (DOE, EIA, 2009a). That dependence grows more each year, despite attempts to substitute other fuels and energy sources. NHTSA's programs to improve fuel consumption are generally consistent with the EISA of 2007. The law also requires the DOT, for the first time in history, to establish fuel economy standards for medium- and heavy-duty vehicles. The gross vehicle weight ratings (GVWRs) for these vehicles range from 8,500 to more than 80,000 lb. (GVW and gross combined weight [GCW] refer to gross vehicle weight, which is limited by regulation. GVWR is the manufacturer's stated maximum GVW rating for a vehicle. The legal weight limit may be lower than the manufacturer's rating in some cases. GVW and GVWR apply to single-unit vehicles and to the tractor in a tractor-trailer combination.)

In addition, the use of fossil fuels for transportation produces carbon dioxide, an important greenhouse gas that contributes to climate change; governments around the world have taken action to reduce the use of fossil energy in their economies. The United States in particular is pursuing alternative sources of fuel and attempting to increase efficiency in oil usage, which will lower oil consumption and reduce greenhouse gas emissions.

As a result of these initiatives, vehicle manufacturers are required to reduce both fuel consumption and exhaust emissions. Light-duty vehicle manufacturers have already made significant improvements in reducing fuel consumption and even more progress in reducing vehicle emissions. The improvements in light-duty vehicle (cars and light trucks) fuel economy have been spurred in part by corporate average fuel economy (CAFE) standards. For medium- and heavy-duty vehicles greater than 8,500 pounds GVW, no such standards currently exist. Emissions of NO_x and PM from heavy-duty vehicles will be significantly reduced by regulations that have gone into effect. However, reductions in fuel consumption of the large medium- and heavy-duty vehicle fleet have not been as impressive, partly because of the growth in the number of miles driven by large trucks during the past decade. If current trends continue, heavy vehicles will consume an important fraction of the fuel used for on-the-road vehicles. Therefore, if the United States is to reduce its reliance on foreign sources of oil, it will be necessary to reduce the fuel consumption of medium- and heavy-duty vehicles.

The recession has interrupted the constant growth in demand. The trucking industry and manufacturers continue to lay off workers on a vast scale (something that does not show in the 2007 data used throughout this report), and it is difficult to accurately extrapolate demand.

WEIGHT CLASSES AND USE CATEGORIES

Figure 1-4 gives the reader an idea of the diversity of medium- and heavy-duty vehicles. It is based on the DOT classification system using a truck's GVWR. This information was developed by Davis and Diegel of Oak Ridge National Laboratory for the U.S. Department of Energy (DOE) Transportation Databook (Davis et al., 2009) and used extensively by the NESCCAF/ICCT (2009). The committee refers to that material (Table 5.7) for the following observations:

- Class 1 and 2 vehicles lighter than 10,000 lb are considered light trucks, such as pickups, small vans, and sport utility vehicles. They generally have spark-

Light-Duty		*Medium Heavy-Duty*				*Heavy-Duty*	
Class 1	Class 2	Class 3	Class 4	Class 5	Class 6	Class 7	Class 8
Less than 6,000 lb	6,000 to 10,000 lb	10,000 to 14,000 lb	14,000 to 16,000 lb	16,000 to 19,500 lb	19,500 to 26,000 lb	26,000 to 33,000 lb	Greater than 33,000 lb

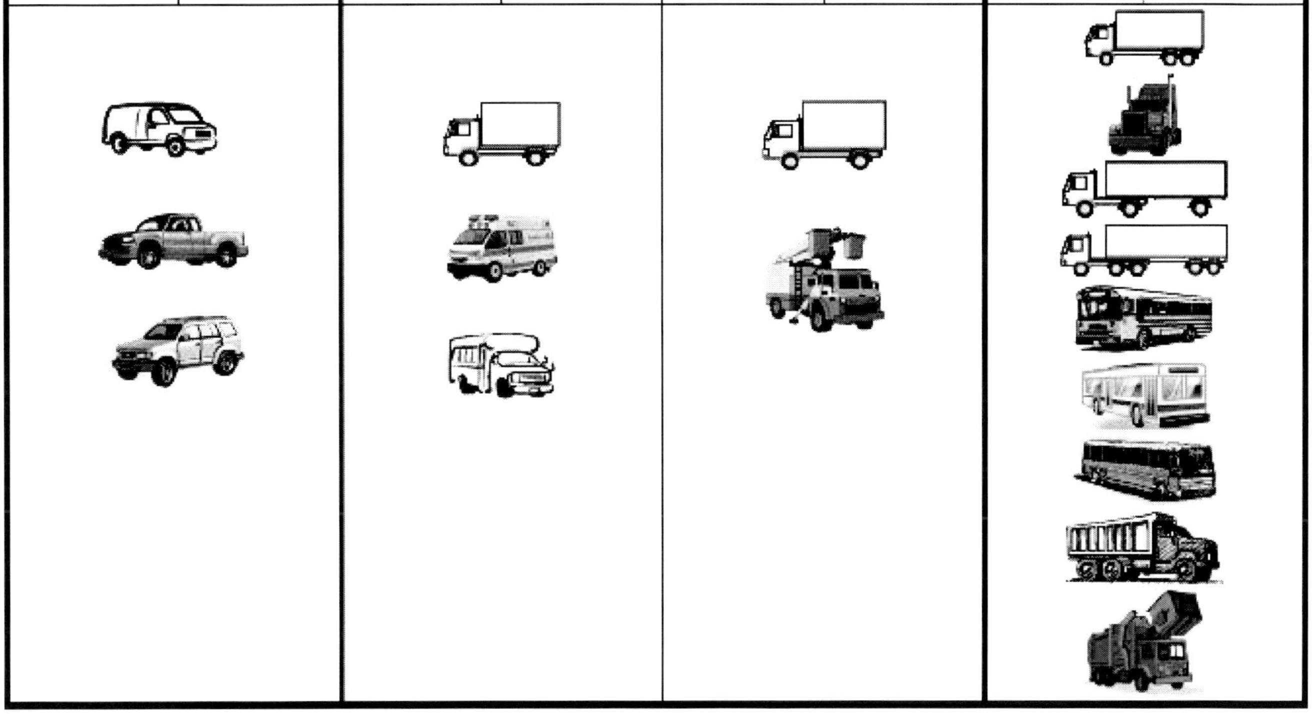

FIGURE 1-4 Illustrations of typical vehicle weight classes. SOURCE: Davis et al. (2009, pp. 5-6).

ignited gasoline-fueled internal combustion engines, and more than 80 percent are for personal use. This class of vehicle up to about 8,500 lb comes under CAFE requirements for cars. Class 2 trucks with GVWR above 8,500 lb are similar to Class 3 trucks.

- Class 3 and above are primarily commercial vehicles. A mix of gasoline and diesel engines is used in Classes 3 through 7, and diesel engines are almost exclusively used in Class 8.
- Classes 3 through 6 are medium- and heavy-duty vehicles with single rear axles.
- Classes 7 and 8 are heavy-duty vehicles with two or more rear axles.
- Class 8 combination trucks have a tractor and one or more trailers and a GCW of up to 80,000 lb, with higher weights allowed in specific circumstances.

ENERGY CONSUMPTION TRENDS AND TRUCKING INDUSTRY ACTIVITY

The number of medium- and heavy-duty trucks has increased substantially as the U.S. economy has grown. Over the period from 1970 to 2003, energy consumption by lightweight trucks grew 4.7 percent annually, while that of passenger cars grew only 0.3 percent. Meanwhile, energy consumption by heavy trucks increased 3.7 percent annually. Figure 1-3 displays this divergence in growth. It also displays the underlying pattern that it is not so much the change in fuel economy as a dramatic increase in annual miles driven by heavy vehicles. The continuation of these trends is documented in recent data through 2008 (NRC, 2008).

Trucks and trucking are important contributors to the national income, of course. According to the Economic Census of 2002 (the latest available), the truck transportation industry consisted of more than 112,698 separate establishments, with total revenues of $165 billion (DOC, Census Bureau, 2005). These establishments employ 1.4 million workers, who take home an annual payroll of $47 billion. Truck and bus manufacturing also accounts for a significant share of national income. According to the same census, light truck and utility vehicle manufacturers have total shipments of $137 billion. Heavy-duty truck manufacturing had sales of $16 billion. Another way to look at the trucking industry's economic contribution is to compare it with other industries in the transportation sector, in which it accounts for about one-fourth of the sector's total revenues (see Figure 1-5).

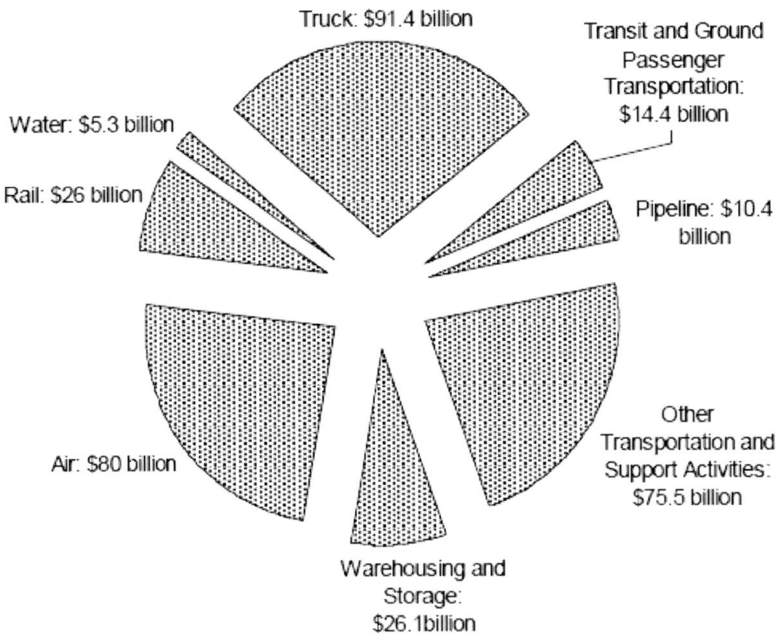

FIGURE 1-5 Total revenue of for-hire transportation services compared with revenue of other sectors of the transportation industry, 2002. SOURCE: DOC, Census Bureau (2005).

FACTORS AFFECTING IMPROVEMENTS IN FUEL CONSUMPTION

Medium- and heavy-duty trucks and buses are load-carrying vehicles that are designed to perform work in an efficient and timely manner. This makes them different from light-duty vehicles. In the U.S. EPA light-duty fuel economy tests, the vehicle is tested at its empty weight plus the equivalent weight of a couple of passengers.

For light-duty vehicles, load has only a modest impact on fuel consumption. For example, a car with four passengers and luggage will use only slightly more fuel than the same car with only a driver. For most light-duty vehicles, the loaded weight is approximately 25 to 35 percent more than the empty weight. However, for a heavy-duty vehicle, a loaded vehicle weighs more than double the empty weight.

Light-duty vehicles use the sales-weighted average of the fuel consumption (e.g., gallons/mile) for the urban and highway schedule converted into fuel economy (e.g., miles/gallons) to compare to the CAFE standards. Therefore, from a regulatory viewpoint, fuel consumption is also the fundamental measure (or "metric").

Because trucks and buses are designed to carry payloads, and the loaded weight of a truck may be more than double the empty weight, the way to represent an appropriate attribute-based fuel consumption metric is to normalize the fuel consumption to the payload the vehicle hauls. As noted previously, this metric is called load-specific fuel consumption (LSFC) and is measured in gallons of fuel per payload tons per 100 miles. The lower the fuel consumption (FC) of the vehicle and the higher the payload the vehicle carries, the lower the LSFC. The payload has an important effect on the fuel consumption, and the average value used for potential standards should be based on national data representative of the class and duty cycle of the vehicle.

This report uses FC or LSFC data throughout except where fuel economy (FE) data are specified in the literature. Later in the report, metrics for the fuel efficiency of medium- and heavy-duty vehicles sold and used in the United States will be discussed. How best to quantify this will be addressed consistent with U.S. national objectives to reduce the nation's reliance on oil. More detailed discussion of FE, FC, and LSFC is presented in Chapter 2.

TASK ORGANIZATION AND EXECUTION

Recognizing the challenge and complexity of its work, the committee organized its members in working groups focused on the individual tasks outlined in its charge. There were four such groups, each with its own leader responsible for task work, coordination, and scheduling under the umbrella of the broader committee.

Given the constrained time and legislative deadline it faced, the committee used specialized consultants to execute various portions of the study directed by a committee working group. The consultants and their assignments were:

TIAX, LLC—Developing Detailed Forecasts of Fuel Consumption Reducing Technologies:

- Supported the committee's evaluation of medium- and heavy-duty vehicle technologies by researching the technologies and their costs through intensive interviews of manufacturers, fleet owners and, others to produce a detailed matrix relating technologies and vehicle types over time.
- Developed a detailed matrix of fuel-saving technologies, their fuel consumption benefits, and their costs.
- Focused on a 10-year time frame.
- Arranged specific site visits for the committee.

Argonne National Laboratory—Modeling and Simulation:

- Provided quantitative data to support the committee in its task of establishing a report to support rulemaking on medium- and heavy-duty vehicle fuel consumption.
- Provided modeling and simulation analyses of technologies now and into the future for eight vehicle applications: pickup truck, van, delivery straight truck, bucket truck, combination tractor trailer, refuse hauler, urban bus, and intercity highway bus.

Cambridge Systematics and ERG:

- Examined possible consequences or side effects of fuel economy standards or of technologies to improve the fuel efficiency of medium- and heavy-duty vehicles.
- Examined alternative approaches to improving fuel efficiency.

The four consultants' reports are available in the National Academies Public Access File associated with this study.

In addition, the committee distributed a questionnaire to aerodynamics technology stakeholders in the private and academic sectors from which it received useful input. The committee's efforts were also aided by several industry organizations that hosted site visits providing relevant information, which proved especially helpful for the work of TIAX. They included the following:

- The University of Michigan Transportation Research Institute (UMTRI), Ann Arbor
- Ford Motor Co.
- Azure Dynamics Corp.
- Arvin Meritor, Inc.
- Navistar International Corp.
- ISE Corp.
- Allison Transmission, Inc.
- Peterbilt Trucks (PACCAR Co.)
- Auto Research Center, Inc., Indianapolis, Ind.
- U.S. Environmental Protection Agency (EPA), Ann Arbor, Mich.
- Southwest Research Institute (SwRI), San Antonio, Tex.
- Detroit Diesel Corp.
- Eaton Corp.
- Tank and Automotive Research, Development and Engineering Center (TARDEC), U.S. Army
- PACCAR, Inc.
- Volvo Trucks North America
- Cummins, Inc.
- Great Dane Trailers (Great Dane LLC)
- Walmart
- Transportation Research Center, Inc.
- National Highway Transportation Safety Administration (NHTSA), Washington, D.C.

All of these consultants and industry partners provided invaluable assistance to the committee in its efforts.

REPORT STRUCTURE

This report begins with a summary of the key findings and recommendations. Chapter 1, the introduction, lays the factual background for the reader. Next, Chapter 2 provides vehicle fundamentals necessary for a thorough understanding of the topics addressed in the report. Chapter 3 surveys the current U.S., European, and Asian approaches to fuel economy and regulations. Chapter 4 reviews and assesses power train technologies for reducing load-specific fuel consumption. Chapter 5 covers vehicle technologies for reducing load-specific fuel consumption. The direct and indirect costs and benefits of integrating fuel economy technologies medium- and heavy-duty vehicles are addressed in Chapter 6. Chapter 7 discusses alternative approaches to be considered for reducing the fuel consumption of such vehicles. Chapter 8 discusses approaches to measurement and regulation of fuel consumption.

BIBLIOGRAPHY

DOC (U.S. Department of Commerce), Census Bureau. 2005. 2002 Economic Census, Industry Product Analysis. Washington, D.C. March.

DOE, EIA (U.S. Department of Energy, Energy Information Administration). 2009a. Annual Energy Review 2008. Report No. DOE/EIA-0384(2008). Washington, D.C.

DOE, EIA. 2009b. Annual Energy Outlook 2009. Report No. DOE/EIA-DOE/EIA-0383(2009). Washington, D.C. March.

DOE, EIA. 2009c. Annual Energy Outlook 2010 (Preliminary). Washington, D.C. December.

Davis, S., S. Diegel, and R.G. Boundy. 2009. Transportation Energy Data Book, Edition 28, Table 5.7. Knoxville, Tenn.: U.S. Department of Energy, Energy Efficiency and Renewable Energy. Report No. ORNL-6984.

Federal Highway Administration, Highway Statistics Summary to 1995, Table VM-201A, and Highway Statistics (annual releases), Table VM-1. Washington D.C.

ICCT (Northeast States Center for a Clean Air Future/International Council on Clean Transportation). 2009. Setting the Stage for Regulation of Heavy-Duty Vehicle Fuel Economy and GHG Emissions: Issues and Opportunities. February.

NRC (National Research Council). 2008. Review of the 21st Century Truck Partnership. Washington, D.C.: The National Academies Press.

2

Vehicle Fundamentals, Fuel Consumption, and Emissions

This chapter addresses the makeup of the trucking industry and the complexity of the trucking sector. It also discusses measures of vehicle fuel economy and consumption, and their measurement, as well as the importance and diversity of vehicle duty cycles for different vehicle applications.

Trucks and buses are classified by weight, based on the gross vehicle weight rating (usually abbreviated as GVW, but sometimes GVWR), which is the maximum in-service weight set by the manufacturer. The GVW includes the empty weight of the vehicle plus the maximum allowed cargo load. For vehicles that pull trailers, the maximum weight rating is the gross combination weight (GCW). Note that the vehicle structure and especially axle and suspension components are specifically designed and manufactured in adherence to the target GVW or GCW. The use categories of vehicles are not as well defined as weight classes and depend on widely varying industry usage. For example, the same vehicle may be called "heavy-duty" by one industry and "medium-duty" by another.

TRUCK AND BUS TYPES AND THEIR APPLICATIONS

The committee has sought to update and summarize key information for these vehicles. Table 2-1, "Comparison of Light-Duty Vehicles with Medium- and Heavy-Duty Vehicles," presents the committee's compilation of data for 2006 and 2007. It highlights weights, sales volumes and registrations, fuel economy, fuel consumption, mileage, and other information across the various vehicle classes. Even within a class, the range of applications signals the different uses or duty cycles experienced by medium- and heavy-duty vehicles across the transportation sector. These complexities within the industry indicate the difficulties of establishing effective policies to reduce fuel consumption.

Medium- and heavy-duty vehicles, defined as Classes 2b through 8, are the workhorses of industry. They are used in every sector of society and the economy, from carrying passengers to moving goods. This results in a broad range of duty cycles, from high-speed operation with few stops on highways to lower speed urban operation with dozens of stops per mile. The *Transportation Energy Data Book* (Davis and Diegel, 2007) reports (in Table 5-7) that the largest use of heavy-duty trucks is for moving goods and materials, noting that over 30 percent of Class 7 and 8 vehicles are used in for-hire transportation of freight. In addition, trucks carry 66 percent, by weight, of all goods shipped (in Table 5.4).

In the United States, for 2007, the largest company-owned fleet of heavy-duty vehicles had over 67,000 Class 8 vehicles (trucks), as shown in Figure 2-1. Bradley and Associates (2009) report that the 200 largest private and for-hire freight-hauling fleets controlled nearly 1 million Class 4 through 8 vehicles, representing 11 percent of heavy-duty vehicles. As shown in Figure 2-1, the Class 8 tractors are 86 percent company-owned and 14 percent owner-operator trucks. These larger fleets control more than 1.1 million trailers as well.

Small family-owned fleets are also important parts of the system. If the 200 largest fleets control 11 percent of the fleet, and owner-operators control 14 percent, then small fleets make up 75 percent of Class 4 through 8 trucks. In addition, small fleets may be the ones faced with the greatest potential burden of compliance in any regulation that the National Highway Traffic and Safety Administration (NHTSA) promulgates. Table 2-3 shows the top 10 for-profit fleets of heavy vehicles, as identified by the American Truckers Association. Table 2-4 identifies the 10 cities in North America with the largest transit bus fleets. Table 2-5 gives information on the top 10 U.S. and Canadian motor coach operators in 2008.

SALES OF VEHICLES BY CLASS AND MANUFACTURER

Medium- and heavy-duty vehicle sales have declined significantly across all classes of vehicles since 2004. As reported in the U.S. Department of Energy 2008 Vehicle Technologies Market Report (DOE/EERE, 2009, p. 20) *Ward's Motor Vehicle Facts and Figures* shows that over a

TABLE 2-1 Comparing Light-Duty Vehicles with Medium- and Heavy-Duty Vehicles

Class	Applications	Gross Weight Range (lb)	Empty Weight Range (lb)	Typical Payload Capacity Max (lb)	Payload Capacity Max (% of Empty)	2006 Unit Sales Volume	2006 Fleet Registrations (millions)	Typical mpg Range 2007	Typical Ton-mpg	Typical Fuel Consumed (1000 gals/ Ton-Mi) × 1000	Annual Fuel Consumption Range (gal)	Annual Fleet Fuel Consumption (Bgal)	Annual Mileage Range (1000 mi) est.	Annual Fleet Miles Traveled 2006 (B)
1c	Cars *only*	(3200)-6000	2400 to 5000	250-1,000	10-20	7,781,000	135	25-33	15	69.0	250-750	74.979	6-25	1,682
1t	Minivans, Small SUVs, Small Pick-Ups	(4000)-6000	3200 to 4500	250-1,500	8-33	6,148,000	70	20-25	17	58.8	300-1k	37.400	6-25	813
2a	Large SUVs, Standard Pick-Ups	6001-8500	4500 to 6000	250-2,500	6-40	2,030,000	23	20-21	26	38.5	500-1.2k	18.000	10-25	305
2b	Large Pick-Up, Utility Van, Multi-Purpose, Mini-Bus, Step Van	8501-10,000	5,000-6,300	3,700	60	545,000	6.2	10-15	26	38.5	1.5k-2.7k	5.500	15-40	93
3	Utility Van, Multi-Purpose, Mini-Bus, Step Van	10,001-14,000	7,650-8,750	5,250	60	137,000	0.69	8-13	30	33.3	2.5k-3.8k	1.462	20-50	12
4	City Delivery, Parcel Delivery, Large Walk-in, Bucket, Landscaping	14,001-16,000	7,650-8,750	7,250	80	48,000	0.29	7-12	42	23.8	2.9k-5k	0.533	20-60	4
5	City Delivery, Parcel Delivery, Large Walk-in, Bucket	16,001-19,500	9,500-10,800	8,700	80	41,000	0.17	6-12	39	25.6	3.3k-5k	0.258	20-60	2
6	City Delivery, School Bus, Large Walk-in, Bucket	19,501-26,000	11,500-14,500	11,500	80	65,000	1.71	5-12	49	20.4	5k-7k	6.020	25-75	41
7	City Bus, Furniture, Refrigerated, Refuse, Fuel Tanker, Dump,Tow, Concrete,Fire Engine,Tractor-Trailer	26,001-33,000	11,500-14,500	18,500	125	82,411	0.18	4-8	55	18.2	6k-8k	1.926	75-200	9
8a	Dump, Refuse, Concrete, Furniture, City Bus, Tow, Fire Engine (straight trucks)	33,001-80,000	20,000-34,000	20,000 to 50,000	100-150	45,600	0.43	2.5-6	115	8.7	10k-13k	3,509	25-75	12
8b	Tractor-Trailer: Van, Refrigerated, Bulk Tanker, Flat Bed (combination trucks)	33,001-80,000	23,500-34,000	40,000 to 54,000	125 to 200	182,395	1.72	4-7.5	155	6.5	19k-27k	28.075	75-200	142

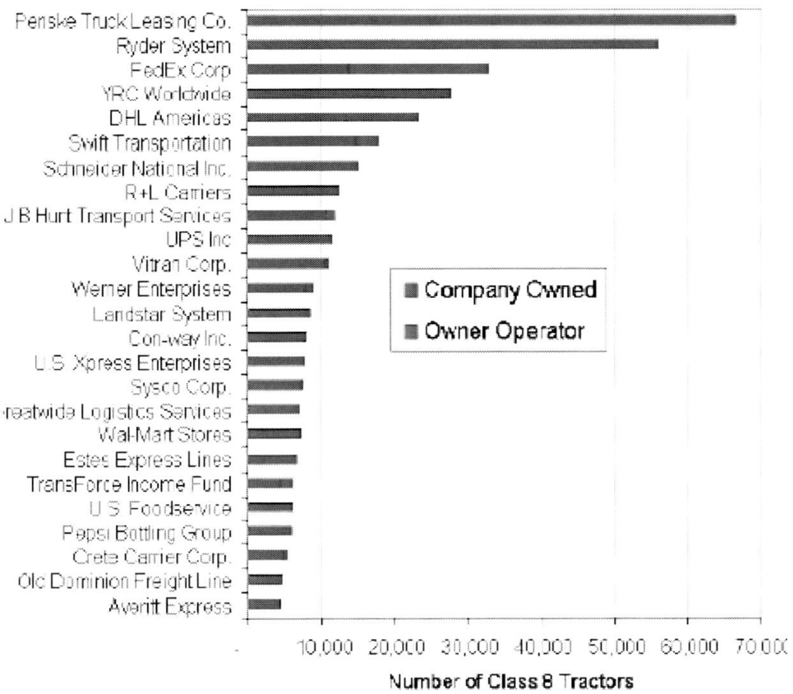

FIGURE 2-1 The 25 largest private and for-hire fleets. SOURCE: ATA (2007b). Used by permission of Transport Topics Publishing Group. Copyright 2009. American Trucking Associations, Inc.

5-year period (2004-2008), sales were down 30 percent for most classes, with only Class 5 showing a marginal increase of 6 percent (see Table 2-6). The *Ward's* data on vehicle classes and manufacturers (Table 2-7) show:

- Profound cycling of sales volumes, especially in higher weight classes.
- Though still down, sales between the dominant providers, Ford and General Motors, shifted significantly as the GM share went from 2 percent (2004) to 37 percent (2008). Sales were down 27 percent over the period.
- Classes 4 through 7 did not see significant shifts among the manufacturers—Ford, GM, International, Freightliner, Hino, and Sterling. Diesel emission requirements and general economic unknowns both contributed to a nearly 40 percent decline in sales over the 5-year period.
- Major manufacturers for Class 8 vehicles have not varied over the past 5 years with one exception—the case of Freightliner—whose market share has declined 5 percent since 2004.

As with vehicle sales, sales of engines manufactured for medium- and heavy-duty trucks declined from 764,000 units in 2004 to 557,000 in 2008 (Table 2-8).

The Class 8 vehicle and engine volumes illustrate profound fluctuations due to both a 2006 pre-buy to avoid cost increases and unknown reliability of 2007 emission controls, followed by the current U.S. recession.

INDUSTRY STRUCTURE

Chapter 1, among other things, reviews the economic power of the great industry built on heavy-duty vehicles and their users. Each year it accounts for billions of dollars in national income and millions of jobs: design engineers, drivers, manufacturing and maintenance technicians, materials handlers, and vehicle sales.

Unlike the makers of light-duty vehicles, which are dominated by a few very large companies (General Motors, Ford, and Toyota), manufacturers of trucks and buses are extremely varied in scale and depend on a web of suppliers, subcontractors, and service industries of all sizes and shapes. Even the largest builders of Class 8 trucks—Daimler, Navistar, PACCAR, and Volvo—each sell 18,000 to 80,000 units annually, and their relative market shares shift. For many medium-duty trucks, the manufacturer of record is essentially a body or equipment builder. The chassis and power train come from one of the major vehicle original equipment manufacturers (OEMs), but the body builder creates the final vehicle configuration. This approach is common for vehicles such as concrete mixers, school buses, utility trucks, and delivery trucks. In many cases the manufacturer of record has limited engineering resources and also limited influence over the fuel consumption of the vehicle. Even major vehicle OEMs sometimes buy components such as the engine, transmission, and axles, all of which have a significant impact on fuel consumption. Tractors and trailers are never built by the same company, and they are often not owned by the same company in actual operation. Even though the tractor-trailer truck's

TABLE 2-2 Product Ranges of U.S. Heavy-Duty Vehicle Manufacturers

STRAIGHT TRUCKS/CHASSIS		COMBINATION TRUCKS	
Class 6	**Class 8**	**Tractors**	**Trailers**
Daimler Trucks NA	General Motors	Daimler Trucks NA	Fontaine Trailer
Freightliner Custom Chassis	*Chevrolet*	*Freightliner*	Great Dane
Sterling Truck	Daimler Trucks NA	*Sterling Truck*	Hyundai
Navistar Intl.	*Freightliner*	*Western Star*	Stoughton Trailers
International Trucks	*Sterling Truck*	Navistar Intl.	Strick Corporation
Workhorse Custom Chassis	*Western Star*	*International Trucks*	Trailmobile Corporation
General Motors	Navistar Intl.	Paccar	Transcraft Corporation
Chevrolet	*International Trucks*	*Kenworth*	Utility Trailer Manufacturing
GMC	Paccar	*Peterbilt*	Vanguard National Trailer Corp.
Ford	*Kenworth*	Volvo	Wabash National Corporation
Isuzu	*Peterbilt*	*Mack*	
Hino Motors	Volvo		
Mitsubishi Fuso	*Mack*		
Paccar			
Kenworth			
UD Trucks			

VOCATIONAL TRUCK/BODY MANUFACTURERS[1]		
Refuse Trucks		*School Buses*
Amrep	Kann Manufacturing Corp.	Blue Bird Corporation
Autocar	Loadmaster	Collins Bus Corporation
Bridgeport Truck Manuf.	Mack Trucks	*Corbeil Bus Corporation*
Crane Carrier	McNeilus Companies, Inc.	*Mid Bus Corporation*
Dempster Equipment Co.	Oshkosh	Daimler Trucks NA
Leach	Pak-Mor Manufacturing	*Thomas Built Buses*
Haul-all Equipment Ltd.	Scranton Manufacturing Co.	Girardin Minibus
Heil Environmental Ltd.	Sterling Truck	Navistar International
Ingold's Hico, Inc.	Wayne Engineering Corp.	*IC Bus*[2]

Companies inset in *italics* are subsidiaries of the company listed above them.

[1] Example only. Other types of vocational trucks are built by different manufacturers. Many of these companies use chassis manufactured by the Truck/Chassis manufacturers.

[2] All Navistar school buses are now sold under the IC Bus brand. Navistar previously sold buses under the International and AmTran brands.

SOURCE: M.J. Bradley & Associates (2009).

fuel consumption is determined by features of both the tractor and the trailer, no single company is responsible for the development of the complete vehicle. This industry structure will complicate any effort to regulate fuel consumption.

Engine manufacturers are also quite numerous. At least a dozen are contenders, according to Table 2-8, and are highly competitive. The same highly competitive situation is true of the commercial users of vehicles. At one end the highway is home for the truly independent operator, the long-distance trucker. At the other end are large fleets with thousands of trucks supported by sophisticated logistics and maintenance systems.

METRICS TO DETERMINE THE FUEL EFFICIENCY OF VEHICLES

Fuel Economy versus Fuel Consumption

In the wake of the 1973 oil crisis and energy security issues, Congress passed the Energy Policy and Conservation Act (P.L. 94-163) in 1975 as a means of reducing the country's dependence on imported oil. The Act established the Corporate Average Fuel Economy (CAFE) program, which required automobile manufacturers to increase the average fuel economy of vehicles sold in the United States to a standard of 27.5 miles per gallon (mpg) for passenger cars. It also allowed the U.S. Department of Transportation (DOT) to set appropriate standards for light trucks. The standards are administered in DOT by the NHTSA on the basis of

TABLE 2-3 Top 10 Commercial Fleets in North America

Rank	Company Name and Location	Type of Business	Total Trucks, 2009	Fuel Types	Maintenance Services
1	UPS Inc. Atlanta	Package service	93,552	Gas, diesel, CNG, hybrid electric, LNG, electric	PM
2	FedEx Memphis, Tenn.	Package service	65,000	Gas, diesel, hybrid electric	PM, EO, HD, CM, EU
3	Quanta Services Houston	Utility construction	24,000	Diesel	PM
4	Waste Management Houston	Waste services	22,000	Diesel, natural gas, hybrid electric	PM, HD, EU
5	Republic Services Phoenix	Waste services	21,399	Diesel, gas, biodiesel, natural gas, hybrid electric	PM, EO, HD
6	PepsiCo/Frito-Lay Purchase, N.Y.	Food and beverage	19,424	Gas, diesel, hybrid electric	PM
7	ServiceMaster Co.	Home and business services	15,706	Gas	PM
8	Aramark Philadelphia	Uniform services and food and beverage	10,968	Gas, diesel	PM, EO, EU
9	Cintas Corp. Cincinnati	Uniform and business services	9,500	Gas, diesel	PM, EO
10	Coca-Cola Enterprises	Beverage bottler	9,500	Diesel, gasoline, biodiesel, hybrid electric, electric	PM, HD, CM

SOURCE: ATA (2009), p. 16.

TABLE 2-4 Top 10 Transit Bus Fleets in the United States and Canada

2008	2007	Agency	35 ft. and under	Over 35 ft.	Artic.	Trolley	2008 Total	2007 +/-
1	1	**MTA New York City Transit** New York City	0	3,919	629	0	4,548	-49
2	7	**Pace Suburban Bus** Chicago	2,163	477	0	7	2,647	1076
3	3	**Metro** Los Angeles	153	2,097	385	0	2,635	-51
4	2	**New Jersey Transit Corp.** Newark, N.J.	116	2,146	85	0	2,347	-728
5	4	**Chicago Transit Authority** Chicago	45	1,851	221	0	2,117	-83
6	5	**Toronto Transit Commission** Toronto	0	1,653	0	0	1,653	54
7	6	**Montreal Urban Transit** Montreal	4	1,591	0	0	1,595	20
8	8	**Washington Area Metropolitan Transit Authority** Washington, D.C.	125	1,320	56	0	1,501	2
9	11	**King County DOT/Metro Transit** Seattle	138	654	637	0	1,429	115
10	9	**Southeastern Pennsylvania Transportation Authority** Philadelphia	110	1,066	155	38	1,369	-2

SOURCE: Courtesy of *Metro Magazine* (2009), p. 14.

TABLE 2-5 Top 10 Motor Coach Operators, 2008, United States and Canada

Ranking 2008	Ranking 2007	Company Name Location	Total Buses and Coaches	Motorcoaches	Non-Motorcoach Buses	Total 40-Foot Buses and Coaches	Total Shorter Than 40 Ft.	Total Longer Than 40 Ft.	Plus/Minus 2007	Vans
1	2	**Coach America** Dallas	1,965	1,483	482	0	0	0	35	794
2	1	**Coach USA** Paramus, N.J.	1,650	1,070	580	0	0	0	-550	132
3	3	**Greyhound Lines** Dallas	1,266	1,266	0	317	0	949	2	0
4	4	**Pacific Western Transportation Ltd.** Toronto, Ontario	639	549	90	144	343	152	0	72
5	4	**Academy Bus LLC** Hoboken, N.J.	613	610	3	349	18	246	-26	8
6	7	**Greyhound Canada** Calgary, Alberta	436	436	0	154	0	282	0	0
7	8	**Coach Canada** Peterborough, Ontario	400	340	60	0	0	0	11	0
8	9	**Holland America Line Inc.** Seattle	344	329	15	226	19	99	-9	34
9	11	**Robert's Hawaii** Honolulu	289	195	94	102	94	93	12	74
10	10	**Peter Pan Bus Lines** Springfield, Mass.	274	240	34	41	8	225	-23	0

SOURCE: *Metro Magazine* (2009), p. 24.

TABLE 2-6 Medium- and Heavy-Duty-Vehicle Sales by Calendar Year

	Calendar Year					Percent Change,
Vehicle Class	2004	2005	2006	2007	2008	2004-2008
Class 3	136,229	146,809	115,140	156,610	99,692	–27
Class 4	36,203	36,812	31,471	35,293	21,420	–41
Class 5	26,058	37,359	33,757	34,478	27,558	6
Class 6	67,252	55,666	68,069	46,158	27,977	–58
Class 7	61,918	71,305	78,754	54,761	44,943	–27
Class 8	194,827	253,840	274,480	137,016	127,880	–34
TOTAL Sales	522,487	601,791	601,671	464,316	349,470	–33

SOURCE: DOE/EERE (2009), p. 20, based on *Ward's Motor Vehicle Facts and Figures*, available at http://www.wardsauto.com/about/factsfigures.

U.S. Environmental Protection Agency (EPA) city-highway dynamometer test procedures.[1]

The terms *fuel economy* and *fuel consumption* are both used to show the efficiency of how fuel is used in vehicles. These terms need to be defined.

- Fuel economy is a measure of how far a vehicle will go with a gallon of fuel and is expressed in miles per gallon (mpg). This is the term used by consumers, manufacturers, and regulators to communicate with the public in North America.
- Fuel consumption is the inverse measure—the amount of fuel consumed in driving a given distance—and is measured in units such as gallons per 100 miles or liters per kilometer. Fuel consumption is a fundamental engineering measure and is useful because it is related directly to the goal of decreasing the amount of fuel required to travel a given distance.

[1]A dynamometer is a machine used to simulate the forces on a drive train to test pollutant emissions, fuel consumption, and other operating characteristics of a vehicle or an engine under controlled and repeatable circumstances.

TABLE 2-7 Truck Sales, by Manufacturer, 2004-2008

	Calendar Year				
	2004	2005	2006	2007	2008
Class 3					
Chrysler	29,859	35,038	36,057	46,553	29,638
Ford	68,615	122,903	105,955	81,155	60,139
Freightliner[a]	270	14	0	0	0
General Motors	2,471	2,788	2,578	33,507	41,559
International	0	0	0	0	609
Isuzu	4,992	5,167	4,929	4,350	2568
Mitsubishi-Fuso	720	670	93	52	202
Nissan Diesel	352	276	232	279	112
Sterling	0	0	0	0	12
Total	107,279	166,856	149,844	165,896	134,839
Classes 4-7					
Chrysler	0	0	0	588	5,386
Ford	60,538	61,358	69,070	70,836	46,454
Freightliner[a]	51,814	51,639	51,357	42,061	30,809
General Motors	34,351	45,144	41,340	34,164	24,828
Hino	2,387	4,290	6,203	5,448	4,917
Navistar/ International	52,278	54,895	61,814	40,268	35,022
Isuzu	10,715	10,620	10,822	9,639	6,157
Kenworth	5,020	3,874	5040	4,239	3,710
Mack	21	0	0	0	0
Mitsubishi-Fuso	4,384	4,842	5,967	5,218	2,136
Nissan	0	0	0	0	0
Nissan Diesel	2,453	2,382	2,551	2,080	1,273
Peterbilt	4,495	4,739	6,307	5009	3,792
Sterling	0	0	102	578	467
Total	228,456	243,783	260,573	220,128	164,951
Class 8					
Freightliner[a]	73,731	94,900	98,603	51,706	42,639
Navistar/ International	38,242	46,093	53,373	29,675	32,399
Kenworth	23,294	27,153	33,091	19,299	15,855
Mack	20,670	27,303	29,524	13,438	11,794
Peterbilt	26,145	30,274	37,322	19,948	17,613
Volvo Truck	20,323	26,446	30,716	16,064	13,061
Other	792	623	1,379	835	112
Total	203,197	252,792	284,008	150,965	133,473
Grand Total	538,932	663,431	694,425	536,989	433,263

[a]Freightliner/Western Star/Sterling(domestic).
SOURCE: DOE/EERE (2009), pp. 21-22, based on *Ward's Motor Vehicle Facts and Figures*, available at http://www.wardsauto.com/about/facts figures.

TABLE 2-8 Engines Manufactured for Class 2b Through Class 8 Trucks, 2004-2008

	2004	2005	2006	2007	2008
Engines Manufactured for Heavy-Duty Trucks					
Cummins	64,630	79,100	91,317	65,228	75,307
Detroit Diesel	48,060	61,074	63,809	29,506	35,174
Caterpillar	74,224	86,806	97,544	33,232	20,099
Mack	25,158	36,211	36,198	18,544	16,794
Mercedes Benz	17,178	24,414	24,584	17,048	10,925
Volvo	12,567	19,298	23,455	9,850	8,822
Navistar	0	0	0	4	927
PACCAR	0	0	0	52	20
Total	241,817	306,913	336,907	173,464	168,068
Engines Manufactured for Medium-Duty Trucks					
Navistar	373,842	382,143	357,470	335,046	264,317
GM	74,328	77,056	83,355	87,749	72,729
Cummins	14,900	15,162	16,400	20,615	27,664
Mercedes Benz	16,075	20,038	27,155	19,330	9,066
Caterpillar	42,535	42,350	45,069	14,693	6,269
PACCAR	0	0	0	9,020	5,694
Hino	671	5,001	7,489	6,230	3,062
Detroit Diesel	0	958	8	0	0
Total	522,351	542,708	536,946	492,683	388,801
Engines Manufactured for Medium- and Heavy-Duty Trucks					
Navistar	373,842	382,143	357,470	335,050	265,244
Cummins	79,530	94,262	107,717	85,843	102,971
GM	74,328	77,056	83,355	87,749	72,729
Detroit Diesel	48,060	62,032	63,817	29,506	35,174
Caterpillar	116,759	129,156	142,613	47,295	26,368
Mercedes Benz	33,253	44,452	51,739	36,378	19,991
Mack	25,158	36,221	36,198	18,544	16,794
Volvo	12,567	19,298	23,455	9,850	8,822
PACCAR	0	0	0	9,072	5,714
Hino	671	5,001	7,489	6,230	3,062
Total	764,168	849,621	873,853	666,147	556,869

The CAFE for light-duty vehicles is calculated from fuel consumption data using a "harmonic average."[2] The harmonic average in the CAFE standards is determined as the sales weighted average of the fuel consumption for the Urban and Highway schedules, converted into fuel economy. The average is calculated using the fuel consumption of individual vehicles times the number of vehicles sold of each model, summed over the whole fleet and divided by the total fleet.

Because fuel economy and fuel consumption are reciprocal, each of the two metrics can be computed in a straightforward manner if the other is known. In mathematical terms, if fuel economy is X and fuel consumption is Y, their relationship is expressed by $XY = 1$. This relationship is not linear, as illustrated by Figure 2-2. In this figure, fuel consumption is shown in units of gallons/100 miles, and fuel economy is shown in units of miles/gallon. The figure also shows that a given percentage improvement in fuel economy saves less and less fuel as the baseline fuel economy increases. Each bar represents an increase in fuel economy by 100 percent, which corresponds to a decrease in fuel consumption by 50 percent. The data on the graph show the resulting decrease in fuel consumption per 100 miles and the total fuel saved in driving 10,000 miles. The dramatic decrease in the impact of increasing fuel economy by 100 percent for a high fuel economy vehicle is most visible in the case of increasing the fuel economy from 40 to 80 mpg, where the total fuel saved in driving 10,000 miles is only 125 gallons, compared to

[2]Harmonic average weighted CAFE = $\dfrac{\sum_1^n N_n}{\sum_1^n N_n \dfrac{1}{FE_1} + \ldots + N_n \dfrac{1}{FE_n}}$

where N_n = number of vehicles in class n, FE_n = fuel economy of class n vehicles and n = number of separate classes of vehicles.

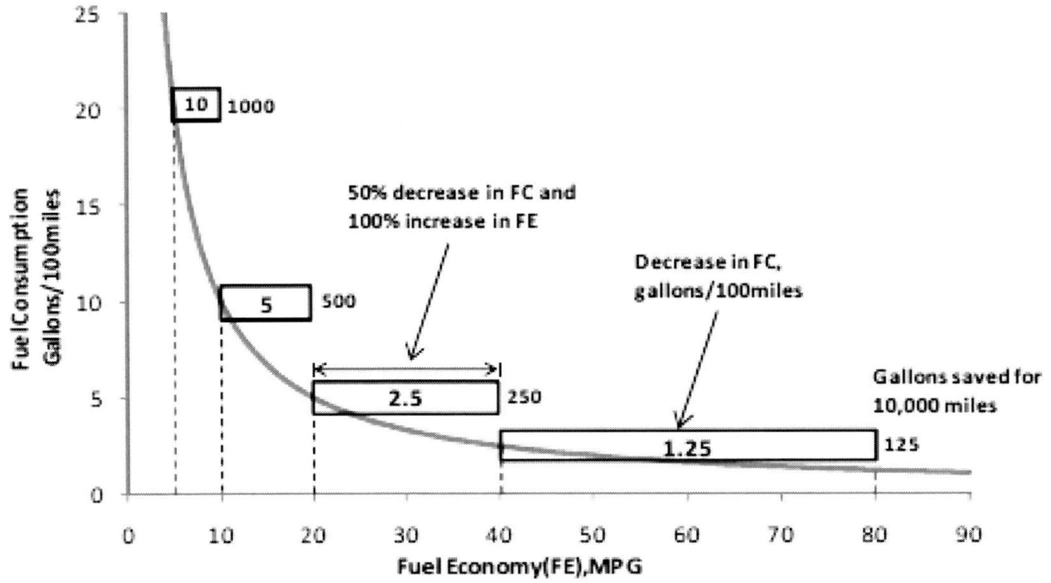

FIGURE 2-2 Fuel consumption (FC) versus fuel economy (FE), showing the effect of a 50 percent decrease in FC and a 100 percent increase in FE for various values of FE, including fuel saved over 10,000 miles. Results are based on Eq 2-1.

1,000 gallons for a change from 5 to 10 mpg. Appendix E discusses further implications of the relationship between fuel consumption and fuel economy for various fuel economy values.

Fuel consumption difference is also the metric that determines the yearly fuel savings in going from a given fuel economy vehicle to a higher fuel economy vehicle:

$$\text{Yearly Fuel Savings} = \text{Yearly Miles Driven} \times \frac{(FC_1 - FC_2)}{100} \qquad \text{(Eq. 2.1)}$$

where FC_1 = fuel consumption of existing vehicle, gallons/100 miles, and FC_2 = fuel consumption of new vehicle, gallons/100 miles.

The amount of fuel saved for a light-duty vehicle in going from 14 to 16 mpg for 12,000 miles per year is 107 gallons. This savings is the same as a change in fuel economy for another vehicle in going from 35 to 50.8 mpg. The amount of fuel saved for a heavy-duty truck in going from 6 to 7 mpg for 12,000 miles per year is 286 gallons, which is more than double the fuel savings of the light-duty vehicle examples. Once the average long-haul tractor vehicle miles traveled of 120,000 miles per year is considered, the fuel savings for an increase from 6 to 7 mpg is 2,857 gallons. This is 26.7 times more fuel savings than for the two car examples. The fuel savings achieved by a heavy truck going from 6 to 7 mpg is also the same as a change in fuel economy for a medium-duty vehicle in going from 10 to 13.1 miles per gallon, assuming identical driving distance. In practice, medium-duty trucks tend to drive fewer miles, so a higher fuel economy improvement would be required to save an equal amount of fuel. Equation 2.1 and these examples again show how important the use of fuel consumption metric is to judge yearly fuel savings.

Because of the nonlinear relationship in Figure 2-2, consumers of light-duty vehicles have been shown to have difficulty using fuel economy as a measure of fuel efficiency in judging the benefits of replacing the most inefficient vehicles. Larrick and Soll (2008) conducted three experiments to test whether people reason in a linear but incorrect manner about fuel economy. These experimental studies demonstrated a systemic misunderstanding of fuel economy as a measure of fuel efficiency. Using linear reasoning about fuel economy leads people to undervalue small improvements (1 to 4 mpg) in lower-fuel-economy (15 to 30 mpg range) light-duty vehicles, despite the fact that there are large decreases in fuel consumption in this range, as shown in Figure 2-2. This problem worsens when fuel economy numbers typical of trucks and busses are considered (3 to 12 mpg).

Clearly, fuel economy is not a good metric for judging the fuel efficiency of a vehicle. The CAFE standards for light-duty vehicles are expressed in terms of fuel economy, although fuel consumption of individual vehicles is used in the calculation of the sales weighted harmonic average fuel economy. To be consistent throughout this report, fuel consumption is used as the metric. It is the fundamental measure of fuel efficiency both in the regulations and for judging fuel savings by consumers and truck operators. Figure 2-3 was derived from Figure 2-2 to show how percent of fuel

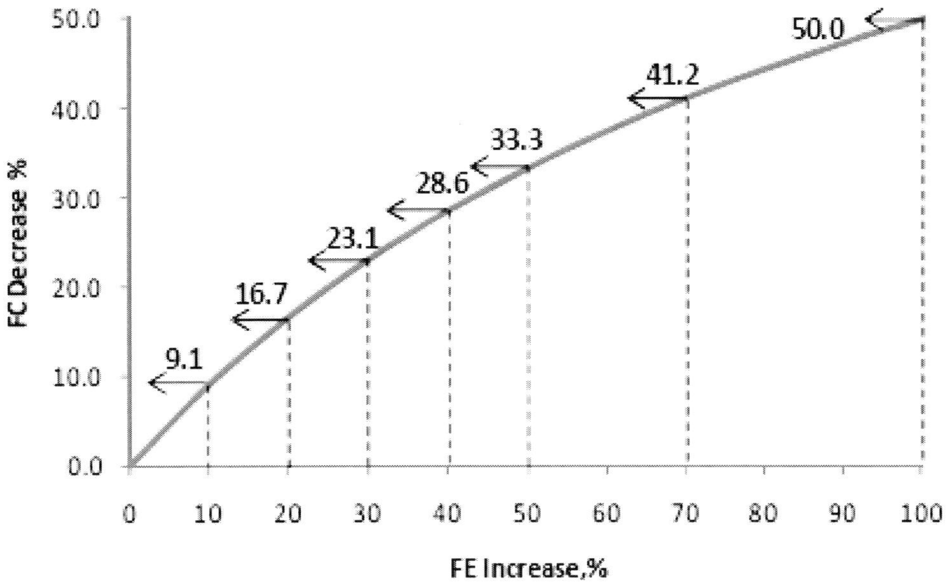

FIGURE 2-3 Percentage fuel consumption (FC) decrease versus percentage fuel economy (FE) increase.

consumption decrease is related to percent increase of fuel economy. The curve in Figure 2-3 is independent of the value of fuel economy. Where fuel economy increase data have been used from the literature, or from fleets, manufacturers of vehicles, and component suppliers, this figure or an equation[3] has been used to convert the data to a fuel consumption decrease in percent.

Load-Specific Fuel Consumption

Medium- and heavy-duty vehicles are unlike light-duty vehicles in that they are clearly designed to carry loads in an efficient and timely manner. In the EPA light-duty vehicle fuel economy tests, the only load in the vehicle during the test is one 150-lb person as the driver. This is the typical way these vehicles operate, although different light-duty vehicles have the capacity to carry additional passengers and cargo, depending on their size. Delivering the driver and passengers to a destination can be considered the primary purpose of light-duty vehicles. On the other hand, the primary purpose of most medium- and heavy-duty vehicles is to deliver freight or passengers (the payload). A simple way to reduce the fuel consumption of a truck is to leave the cargo on the loading dock. This approach, however, ignores the purpose of these vehicles. In view of these facts, the way to represent an appropriate attribute-based fuel consumption metric is to normalize the fuel consumption to the payload that the vehicle hauls. This is represented by the following equation:

$$\text{Load-Specific Fuel Consumption (LSFC)} = \frac{\text{FC}}{\text{payload in tons}} \qquad \text{(Eq. 2.2)}$$

where FC = fuel consumption on a given cycle, gallons/100 miles. The literature also shows data represented by the following equation:

$$\text{Load-Specific Fuel Economy (LSFE)} = \frac{1}{\text{LSFC}} \qquad \text{(Eq. 2.3)}$$

$$\text{FC} = \frac{100}{\text{FE}} \qquad \text{(Eq. 2.4)}$$

where FE = fuel economy on a given cycle, miles/gallons.

It is important to note that the payload of a vehicle significantly affects the fuel economy (FE), fuel consumption (FC), and LSFC as shown in Figures 2-4, 2-5, and 2-6. These results are from simulations for a line-haul vehicle and an urban delivery vehicle in operations based on real-world routes recorded by Cummins. Table 2-9 shows a few of the variables used for the simulations in Figures 2-4 through 2-6. Note that adding payload to a vehicle increases fuel consumption, but the higher payload actually improves

[3]If $FE_f = (FE_2 - FE_1)/FE_1$ and $FC_f = (FC_1 - FC_2)/FC_2$ where FE_1 and FC_1 = FE and FC for vehicle baseline and FE_2 and FC_2 = FE and FC for vehicles with advanced technology, then, $FC_f = FE_f/(FE_f + 1)$ where FE_f = fractional change in fuel economy and FC_f = fractional change in fuel consumption. This equation can be used for any change in FE or FC to calculate the values shown in Figure 2-2. Also, $FE_f = FC_f/(1 - FC_f)$ and % FC = 100 FC_f, % FE = 100 FE_f.

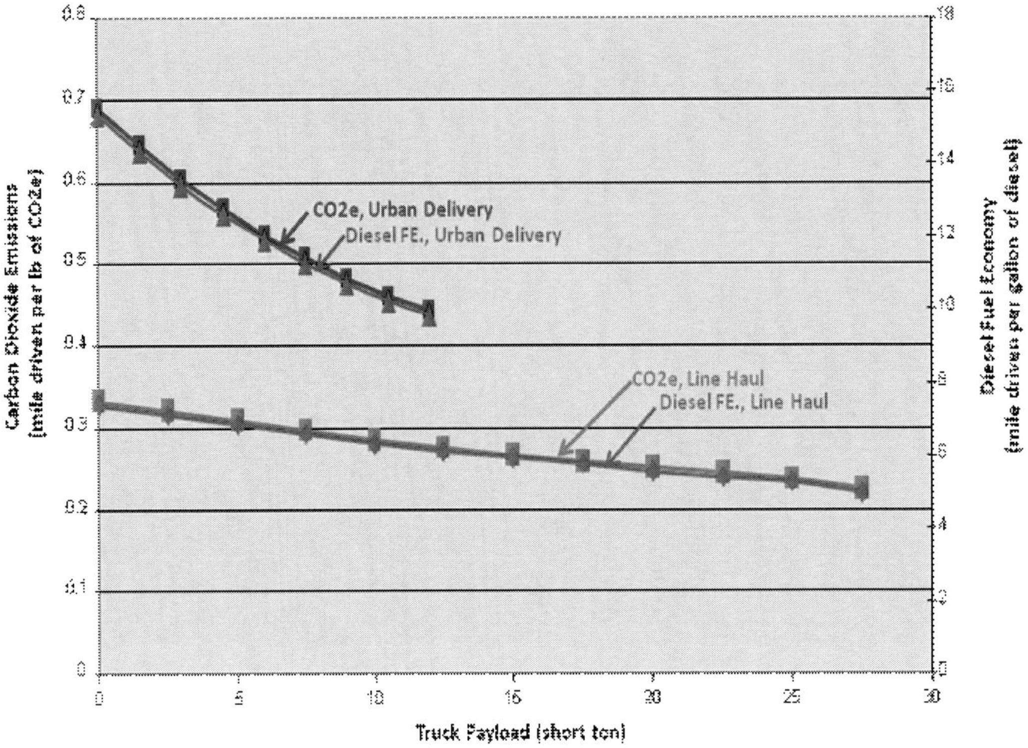

FIGURE 2-4 Fuel economy versus payload. SOURCE: Jeffrey Seger, Cummins, Inc., personal communication, June 6, 2009.

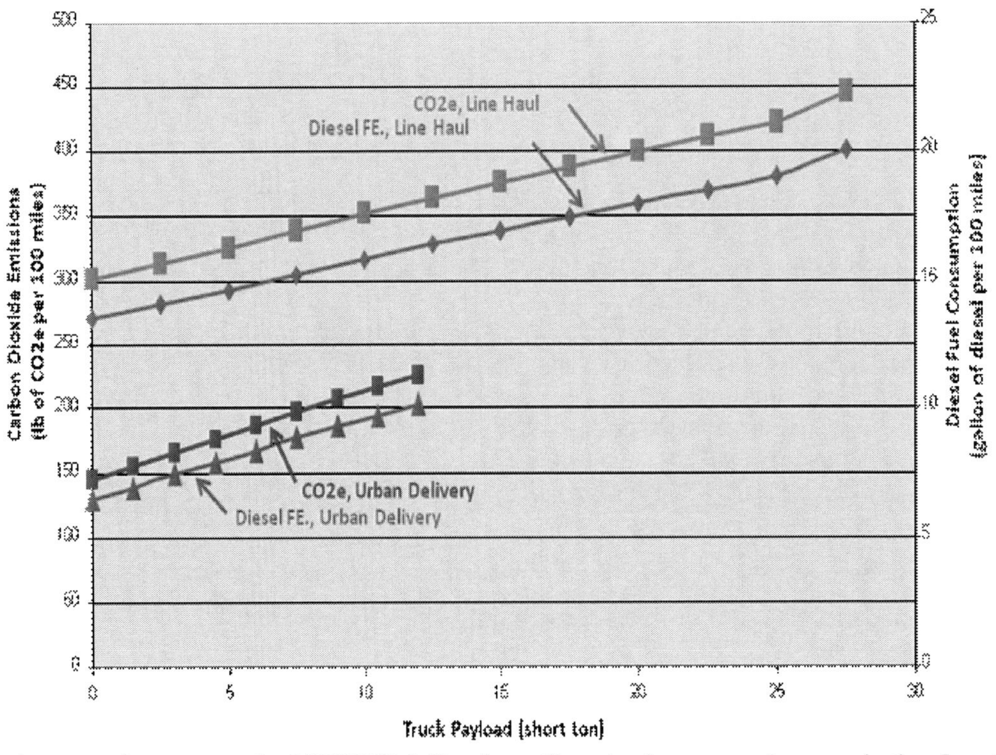

FIGURE 2-5 Fuel consumption versus payload SOURCE: Jeffrey Seger, Cummins, Inc., personal communication, June 6, 2009.

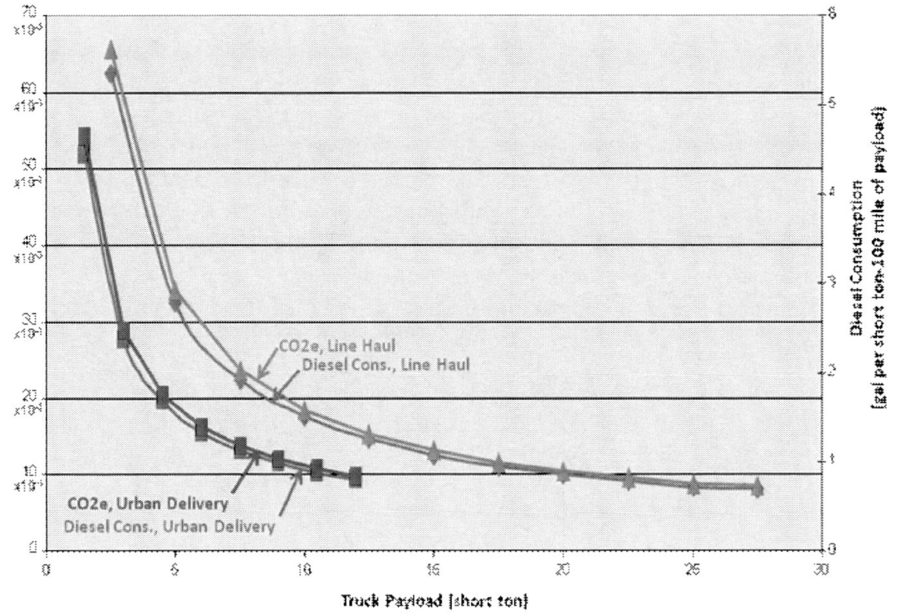

FIGURE 2-6 Load-specific fuel consumption versus payload. SOURCE: Jeffrey Seger, Cummins, Inc., personal communication, June 6, 2009.

the efficiency of the vehicle (in terms of LSFC). Failure to understand this counterintuitive fact can lead to regulations with severe unintended consequences.

Payload is an important variable to input for either a vehicle computer simulation or an experimental test for determining a vehicle's fuel consumption. The duty cycle that the vehicle operates on is also important. Another important variable is average vehicle speed. It is important that any regulation use an average payload based on national data representative of the class and duty cycle of the vehicle. Appendix E gives national data for the average payload of various classes of vehicles. Buses could use the average number of typical passengers times an average weight (150 lb as used in light-duty standards) plus some average baggage weight for each passenger (perhaps 25 to 35 lb).

NHTSA would use the data in Appendix E or other payload data to arrive at a simple specific average or typical payload for each class and for each separate vehicle application within a class e.g. tractor trailer, box truck, bucket truck, refuse truck, transit bus, motor coach, etc, for carrying out vehicle certification testing/simulation. For example, this payload would be at a given point in Figures 2-4 to 2-6.

If the payload for a line-haul truck was 20 tons, LSFC would be about 0.9 gallons/ton-100 miles. This might be the example for a typical grossed-out (at maximum cargo weight) vehicle. If the payload was 6 tons for the example of an urban delivery cubed-out (at maximum cargo volume) vehicle, the LSFC would be about 1.3 gallons/ton-100 miles. Now, how can the LSFC be reduced? Since the payload for the test/simulation for a given vehicle is fixed, the engine and vehicle technology discussed in Chapters 4 and 5 can be used to reduce FC, increase FE and reduce LSFC. Weight reduction of the vehicle can also be used to reduce LSFC at the specified payload which would allow full-load payload to be increased for the grossed-out vehicle. In the cubed-out vehicle example, the payload volume can be increased, new technology added, and weight reduced to reduce FC, increase FE, and reduce LSFC. This would allow the cubed-out vehicle to carry more low-density cargo.

Using LSFC in these two examples provides an incentive for industry to reduce FC and LSFC. The key to this approach is a specified typical payload: payload cannot be changed to improve LSFC. The other important point is that this approach is not a full-payload test/simulation unless the vehicle always operates at this load. Clearly, because the levels of FC and LSFC from Figures 2-5 and 2-6 vary widely depending on the type of vehicle and payload, there will be a need for different standards for different vehicle classes and corporate fleet averaging.

Further, it is important that any standard for fuel efficiency be based on LSFC, since it focuses on reducing the fuel consumed by medium-and heavy-duty vehicles sold in the United States, when operating on cycles representative of their work-duty cycles. LSFC can be used directly times the number of vehicles and averaged over the fleet if NHTSA desires to use a fleet average standard for vehicles of a given class that operate in a similar manner. Payload is

TABLE 2-9 Vehicle, Engine, and Cycle Variables

	Line Haul	Urban Delivery
Vehicle weight empty (lb)	33,500	7,500
Engine power (hp)	450	245
Length of route (miles)	65.66	100
Average vehicle speed (mph)	60.5	19.2
Payload (lb)	0-55,000	0-24,000

SOURCE: Jeffrey Seger, Cummins, Inc., personal communication, June 6, 2009.

an important variable that affects FC and LSFC; therefore, any reported values or labels should state FC = gallon/100 miles and LSFC = gallons/ton-100 miles at specific tons of payload.

TRUCK TRACTIVE FORCES AND ENERGY INVENTORY

It is instructive to review the fundamental vehicle attributes that account for fuel consumption before examining the technologies that could reduce fuel consumption.

Road Load

The force or power required to propel a vehicle at any moment in time is customarily presented as a "road load equation." For the case of force, the equation has four terms to describe tire rolling resistance, aerodynamic drag, acceleration, and grade effects:

$$F_{RL} = mgC_{rr} + 0.5C_D A \rho_a V^2 + m(dV/dt) + mg\sin(\theta)$$

where mg is vehicle weight, C_{rr} is tire rolling resistance, A is the frontal area, C_d is a drag coefficient based on the frontal area, ρ_a is the air density, V is the vehicle velocity, m is vehicle mass, t is time, and $\sin(\theta)$ is the road gradient (uphill positive). Neither C_D nor C_{rr} need be constant with respect to speed, and the term $C_D A$ should not be split without careful thought.

For road load power, the force equation is merely multiplied by the velocity:

$$P_{RL} = mgC_{rr}V + 0.5C_D A \rho_a V^3 + mV(dV/dt) + mg\sin(\theta)V.$$

In conventional vehicles the road load power is supplied by an engine, via a transmission and one or more drive axles characterized by an efficiency (η).The engine may also supply power for auxiliary loads (P_{aux}), including cooling fan loads, so that a simple engine power demand (P_E) model is given by:

$$P_E = \frac{P_{RL}}{\eta} + P_{aux}.$$

The force F_{RL} may become negative while the vehicle is decelerating or traveling on a sufficiently steep downgrade, with "negative" power being absorbed through engine braking or friction brakes. For hybrid-drive vehicles, some of the "negative" power may be absorbed and stored for use in future propulsion of the vehicle. Since hybrid vehicles have at least two sources of power during part of their duty cycle, the engine power demand model must be adjusted to account for the flow of power to or from other sources during operation.

A specific engine type may be used in a variety of vehicle applications and may be coupled to the wheels via a variety of drivetrains, so that the in-use engηpe the average power demand, fuel consumption, and energy required to travel a specific distance vary substantially with the vehicle activity, or duty cycle. The average engine efficiency will also be impacted by the duty cycle, as is the contribution of each major element of the road load equation (aerodynamics, weight, tires) to the overall vehicle fuel consumption. Figure 2-7 illustrates how the extremes of duty cycles can create a wide range of impacts of the specific vehicle attributes to the overall vehicle fuel consumption.

When either engines or vehicles are to be certified for efficiency or emissions standards, it is necessary to establish test cycles to challenge the vehicle or engine, but it has historically been accepted in regulations that these tests cannot hope to represent every in-use behavior. This is discussed further later in this chapter and in Chapter 3.

TEST PROTOCOLS

Fuel consumption may be measured directly from a vehicle on the road, a test track, or a chassis dynamometer. It is important to distinguish between comparative testing, where fuel consumption values used by two trucks of different technology are compared, and absolute testing, where fuel consumption is measured using a standardized procedure so that the results may be compared with results from tests conducted at different times or in different locations. If on-road measurement is conducted over a long distance or long period of time, the resulting average fuel consumption values may be compared fairly with those from another vehicle operated over a sufficiently similar route with sufficiently similar operating conditions. The purpose of a test track is to provide sufficiently repeatable conditions and vehicle activity that a comparison between the performances of two vehicles is possible with a reduced distance or time of operation relative to less controlled on-road tests.

A chassis dynamometer simulates road load on a vehicle while the vehicle drive wheels operate on rollers rather than a road surface. This provides a high degree of repeatability in testing but requires that the effective vehicle mass is known and that road load constants are available. These constants are associated with the rolling resistance C_{rr} and $C_d A$ but cannot be computed directly from them because there is an offset associated with drive train losses. It is customary, especially for passenger cars, to perform an on-road coast-down test of the vehicle to obtain the road load constants, discussed in subsequent sections.[4]

Both on-road and chassis dynamometer measurement methods are described in EPA SmartWay documents.

The Recommended Practices of the Society of Automotive Engineers (SAE) present details of road testing and of chassis dynamometer methods to determine hybrid and conventional vehicle fuel economy.[5]

Fuel-use data from on-road tests or chassis dynamom-

[4]SAE Recommended Practice J1263.

[5]SAE J1082, SAE J1711, SAE J2711, SAE J1321, and SAE J1264.

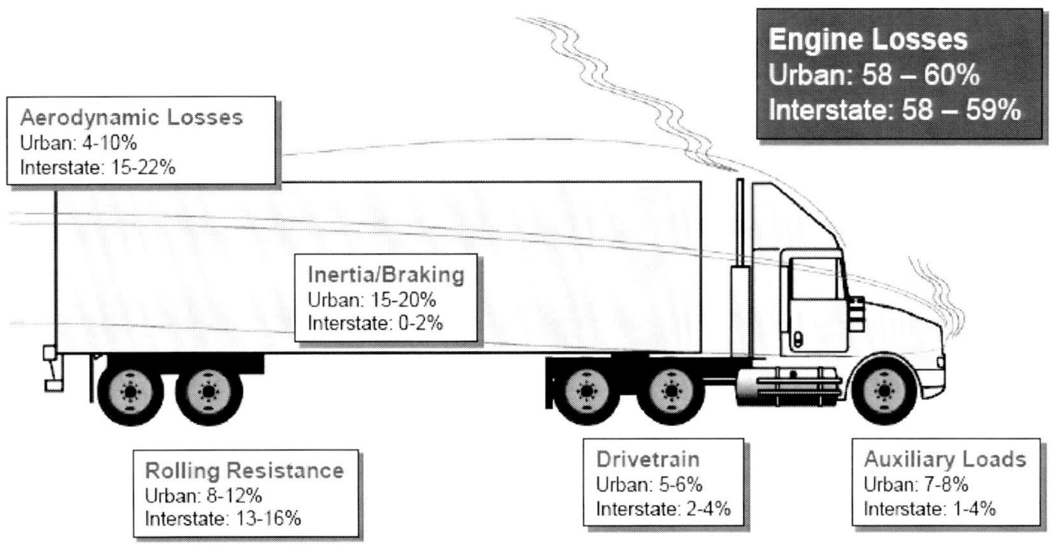

FIGURE 2-7 Energy "loss" range of vehicle attributes as impacted by duty cycle, on a level road.

eter tests may be used indirectly to calibrate whole-vehicle models where C_{rr} and C_DA are not known directly or independently. The model may be used, in turn, to predict fuel consumption on unseen cycles.

On-Road Testing

Physical Testing—Powered: SAE J1321 Fuel Consumption Test Procedure, Type II

This procedure measures on-road fuel consumption utilizing a similarly equipped, unchanging control vehicle operated in tandem with a test vehicle to provide reference fuel consumption data. This procedure has become the de facto test for both carrier and manufacturer fuel economy evaluations, largely due to its ability to use real-world vehicles and routes.

The specification requires both careful control of potential operational variables and numerous replications to validate the difference statistics. The procedure is claimed to provide precision within ±1 percent. The committee's analysis suggests that the current precision, acquired from three T/C[6] ratios within a 2 percent range, results in a standard deviation in the range of 0.9 to 1.1 percent. So, the precision of the SAE J1321 result is more nearly ±2 percent for 95 percent confidence and ±3 percent for 99 percent confidence.

This full-truck system validation includes aerodynamic losses and produces results in percent of reduction in fuel consumption over whatever road type is selected, e.g., from a track or to a specific carrier route. When conducted by expert third-party labs, evaluation of a base case plus three variables can cost $33,000.

For evaluation of aerodynamic systems alone, it may be helpful to use unladed trucks. This process decreases the total fuel consumed so that the incremental consumption of the test truck is larger than in the laded condition. Unfortunately, the procedure has no systematic process for accounting for side winds (yaw conditions), which is a clear and significant aerodynamic shortcoming.

EPA modified the SAE J1321 test procedure (TP) to require use of a test track environment, and each test segment incurs only one acceleration and deceleration. It measures fuel consumption and requires that average speed be controlled to 55 to 62 mph preferred, 65 mph maximum (EPA, 2009).

Coast Down: SAE J1263 Test Procedure

Coast-down testing, as mentioned earlier, is performed to define the rolling resistance and characteristic aerodynamic drag of a vehicle as inputs for a chassis dynamometer load setting. The coast-down process[7] must be well regulated and avoid uncharacteristic wind drag or gradients and will be dependent on vehicle mass and the nature of the road surface.[8]

This procedure is now used infrequently as the other test procedures have gained increased use due to their more acceptable precisions. Coast-down tests are complicated by prevailing winds that reduce the overall precision of the procedure.

[6]T/C is the ratio of test track data to the data of a control.

[7]SAE J2263, SAE J2452, and SAE J2264.

[8]SAE J1263.

Physical Testing—Wind Tunnel: SAE J1252 Test Procedure

The SAE J1252 test procedure measures aerodynamic drag force directly, from which the C_d is calculated. A wind tunnel is the only accurate method to measure the yaw force and thereby the C_d in yaw. This TP also provides for the calculation of a wind average drag coefficient. The drag curve for a tractor with a 45-ft trailer in Figure 5-7 would have a wind average C_d about 15 percent higher than the 0° C_d. That fact begs for a wind average measurement, particularly since certain devices are better at reducing drag in yaw than at 0°. The gap region and trailer (rear) base are particularly sensitive to oblique wind conditions.

After construction of a base tractor and trailer models, evaluation of three variables can cost $7,000, in addition to the base models' fabrication.

The National Aeronautics and Space Administration has developed a correlation between complete truck C_d and fuel consumption.

Computational Fluid Dynamics

Over the past 6 years, computational fluid dynamics (CFD) codes have found increased application to the flow and drag conditions in truck aerodynamics management, encouraged by the DOE. CFD uses numerical methods and algorithms to analyze and solve problems that involve fluid flows. Computers are used to perform the millions of calculations required to simulate the interaction of fluids and gases with the complex surfaces used in engineering. The computer codes/procedures often embody unique individualities of their various developers, and no single practice has emerged as a standard.

Manufacturers are increasingly using this tool to provide details of aero effects helpful to differentiate multiple design features even before building models for wind tunnel evaluation. They have found CFD complements wind tunnel results, which can directly provide C_d results (TMA, 2007, pp. 7, 20). Another recent study concluded that through the example of the Jaguar XF program a combination of (CFD) simulation and relatively simple full-scale wind tunnel testing can deliver competitive aerodynamic performance (Gaylard, 2009).

Chassis Dynamometers

Chassis dynamometers must mimic vehicle inertia and road load for transient cycle evaluations. Simpler dynamometers developed to measure only vehicle power output are unsuited for general fuel consumption measurement. In most cases both inertia and road load forces are applied between the wheel and the roller, but in other cases the drive hubs themselves may be connected mechanically to a dynamometer system. The inertia effect may be applied by either using flywheels or applying torque generated by a substantial electric motor/generator and controlled to apply the torque in proportion to vehicle acceleration and deceleration. Road load may be applied by the same substantial electric machine, or by a smaller electric motor/generator, eddy current power absorber or hydraulic power absorber used in conjunction with flywheels. Flywheels offer the advantage of mimicking inertia faithfully at very low speeds, while systems with a large electric motor/generator may also be used to mimic gradients.

Light-duty vehicle dynamometers for U.S. emissions certification use are well described and employ a single 4-ft-diameter roll under the drive axle. Use of these dynamometers is closely prescribed in the *Code of Federal Regulations*. Other common light-duty designs use four rolls for inspection and maintenance implementation and for garage-grade testing. Heavy-duty units are few in number and vary in design.

A dynamometer test sequence consists of a coast-down (or equivalent, explained in previous section) method to set road load for a given inertial weight, followed by exercising the vehicle through a cycle by a human driver instructed by a video screen speed-time graph. Fuel used may be measured using emissions measurement equipment to determine carbon dioxide and fuel analysis to determine carbon content. Alternatively, fuel mass used may be determined directly by a scale or measured volumetrically. Fuel flow rate is also broadcast by most modern engines but is insufficiently accurate for fuel consumption determination.

Validation of Test Results

The SAE has tasked its Truck and Bus Aerodynamic and Fuel Economy Committee to bring the various current SAE procedures and practices into the needs of the 21st century, reflective of prevailing engineering and scientific data analysis to facilitate robust validation. An early assessment of the SAE committee is that "uncertainty analysis" must play a key role in achieving the overarching goal of providing unified industry standards for validating fuel consumption of heavy trucks and buses, including their aerodynamic properties. Indeed, this study will also assess if new procedures are required. This SAE committee is represented by wide participation across industry and academia.

The committee believes that this SAE committee should be specifically requested to provide a summary and rationale for the completion of Table 2-10. This table considers the

TABLE 2-10 Validation, Accuracy, and Precision

Parameter	SAE J1321	EPA-Mod J1321	Coast Down	Wind Tunnel	CFD	Full-Truck Computer Simulation
Accuracy	%	%	%	%	%	%
Precision	%	%	%	%	%	%

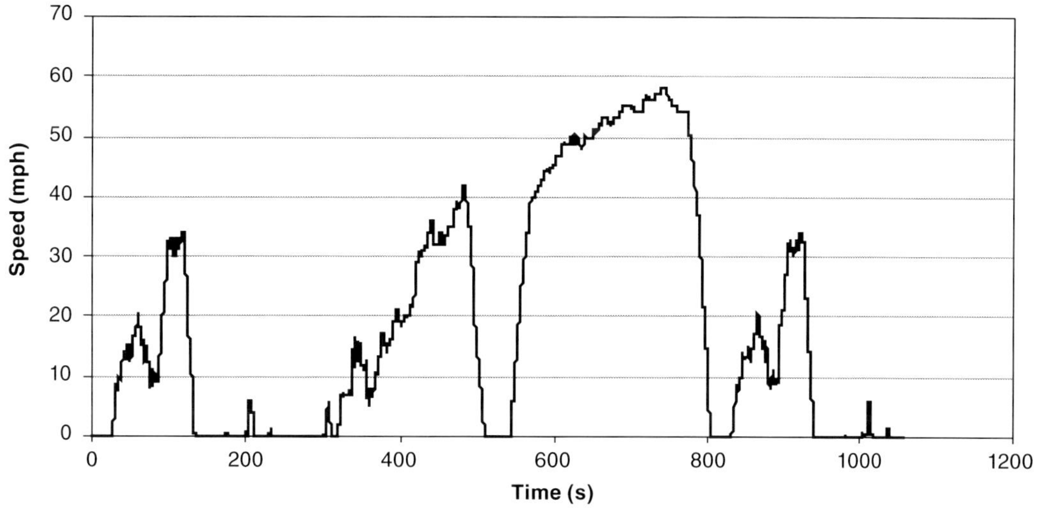

FIGURE 2-8 The Heavy-Duty Urban Dynamometer Driving Schedule.
SOURCE: Clark (2003). Reprinted with permission from SAE. © 2003 SAE International.

adequacy of influencing parameter control pertinent to each validation process. Variables of concern include vehicle speed, wind speed and direction (yaw), temperature, humidity, wind tunnel variables, geometry modeling, flow modeling, fuel, lubricants, and driver.

TEST-CYCLE DEVELOPMENT AND CHARACTERISTICS

Development of Test Cycles

In characterizing the fuel efficiency of a whole vehicle (or of a chassis or mule created to mimic a whole vehicle) against a standard, it is essential to exercise the vehicle through a prescribed speed-time sequence that reasonably reflects actual use. Such has been the case for passenger vehicles. For emissions regulations for heavy-duty vehicles, the representative test cycles are applied to only the engine on an engine dynamometer. However, many nonregulatory test cycles have been developed and documented for heavy vehicles for a variety of purposes. The EPA's Heavy-Duty Urban Dynamometer Driving Schedule (UDDS) is set by regulation (40 CFR 86, App. I) as a vehicle conditioning cycle. The UDDS (Figure 2-8) was created using Monte Carlo simulation with a statistical speed-acceleration basis, and it has origins similar to those of the heavy-duty engine certification test used for implementation of emissions standards for diesel engines. The UDDS includes "freeway" and "nonfreeway" activity.

Engineers typically assemble cycles in this way, by combining real-world truck activity data. An activity database may be created by logging speed from one or many trucks over a representative period of time. The log is then divided into "trips" or "microtrips," either with idle activity separated or included with microtrips. A number of microtrips are then connected to form a cycle of desired length. Many such cycles are created from the database, and the cycle that is statistically most representative of the whole database, using metrics such as average speed and standard deviation of speed, is chosen as a representative cycle. Examples include the suite of "modes" of the Heavy Heavy-Duty Diesel Truck (HHDDT) schedule used in the E-55/59 California truck emissions inventory program. The idle, creep, transient, cruise, and high-speed cruise modes represent progressively higher average speeds of operation (Gautam et al., 2002; Clark et al., 2004). The creep and cruisecreep and cruise modes are shown in Figure 2-9. In a similar fashion, a Medium Heavy-Duty Schedule was also created (Clark et al., 2003).

Cycles have also been created to represent vocational truck and bus behavior. The National Renewable Energy Laboratory has proposed a refuse truck cycle for use in the EPA SmartWay program (EPA, 2009). The Hybrid Truck Users Forum Class 4 and Class 6 Parcel Delivery Cycles are also reported here. The "William H. Martin" cycle has been developed for refuse truck operation, which is acknowledged to vary widely in characteristics.

Transit bus fuel consumption has traditionally been established on test tracks.[9] The SAE, in Recommended Practice J1376, provides a test procedure with three segments (Central Business District, Arterial, and Commuter) that mimic stop-and-go track testing for transit buses. These have been applied to bus testing on chassis dynamometers (Wang et al., 1994, 1995) and are "geometric" in nature. Figure 2-10 shows the Central Business District, which consists of idle, acceleration, cruise, and deceleration periods, with the acceleration and deceleration portions reflecting the abilities of a particular bus at the time of the cycle's creation.

[9]See "Bus Research and Testing Facility (Test Track)" at http://www.vss.psu.edu/BTRC/btrc_test_track.htm (accessed September 22, 2009).

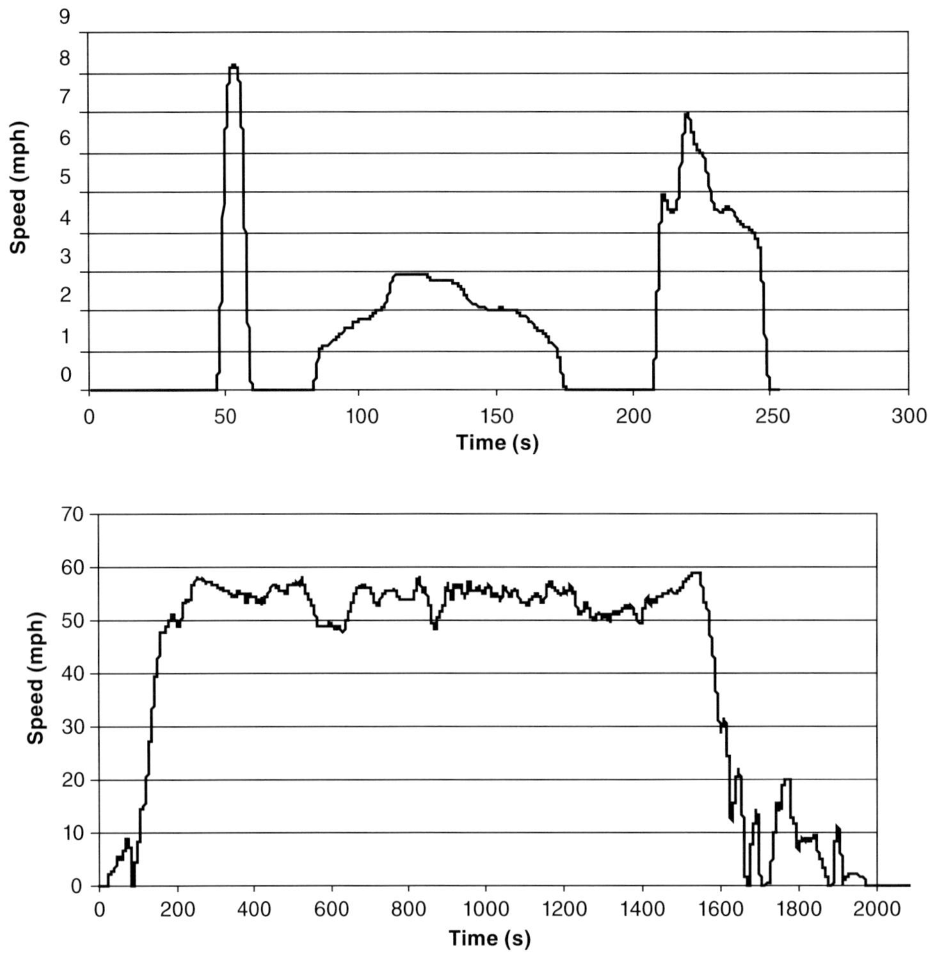

FIGURE 2-9 The creep (top) and cruise (bottom) modes of the HHDDT schedule.
SOURCE: Clark (2003). Reprinted with permission from SAE. © 2003 SAE International.

Bus cycles developed from microtrips include the Manhattan and Orange County Transit Authority (OCTA) cycles[10] (see Figure 2-11) and the Washington Metropolitan Area Transit Authority cycle (Wayne et al., 2008). Numerous additional bus and truck cycles receive attention on the website dieselnet.com and by Wayne et al. (2008) and Davies et al. (2005).

Application of a Cycle

On a chassis dynamometer, the vehicle speed provides for unambiguous wind drag and rolling resistance terms provided that the frontal area, drag coefficient, air density, vehicle mass, gravitational acceleration, and tire rolling resistance coefficient are known. The vehicle mass, acceleration, and deceleration derived from the speed plot provide the inertial term. Usually, no grade term is assumed, although limited research has been conducted on cycles incorporating grades (Walkowicz, 2006; Thompson et al., 2004). The dynamometer may be configured to mimic loads directly, or the dynamometer may be set to match a speed-time coast-down curve obtained from the vehicle during an on-road test.[11]

Cycle Characteristics

The average speed of a real-world cycle implies the level to which the cycle includes transient speed behavior. Very low speed cycles have high idle content, and idle content diminishes. In the same way, values such as "stops per unit distance," average instantaneous acceleration or deceleration, and coefficient of variance of speed become smaller as average speed rises. Table 2-11 shows selected parameters from four truck cycles.

Consider a specific truck being operated at a defined weight. The fuel efficiency of that truck, in units of fuel consumed per unit distance, will vary substantially with respect

[10]SAE J2711.

[11]SAE J2264 and SAE J2263.

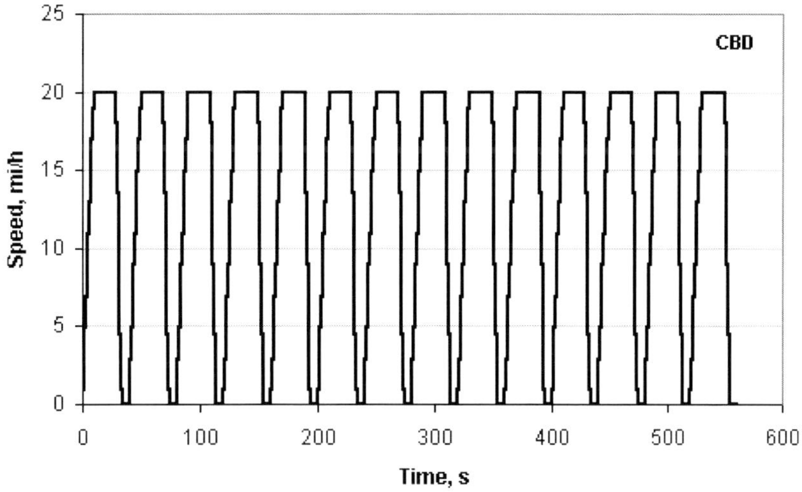

FIGURE 2-10 Central Business District segment of SAE Recommended Practice J1376.

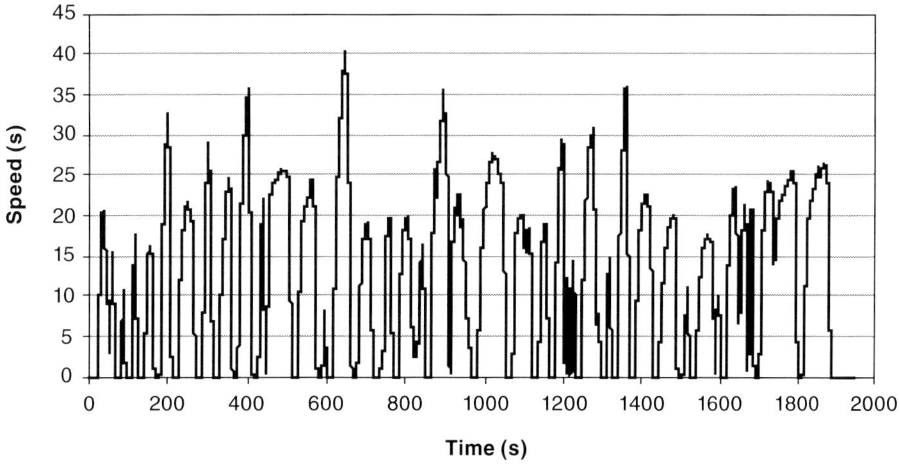

FIGURE 2-11 Orange County Transit Authority cycle derived from transit bus activity data. SOURCE: SAE.

TABLE 2-11 Characteristics of Selected Cycles

Parameter	Filtered Creep Mode of HHDDT	Filtered Transient Mode of HHDDT	Filtered Cruise of HHDDT	Test-D (UDDS)
Duration (sec)	253	668	2083	1063
Distance (miles)	0.124	2.85	23.1	5.55
Average speed (mph)	1.77	15.4	39.9	18.8
Stops/mile	24.17	1.8	0.26	2.52
Maximum speed (mph)	8.24	47.5	59.3	58
Maximum acceleration (mph/s)	2.3	3	2.3	4.4
Maximum deceleration (mph/s)	–2.53	–2.8	–2.5	–4.6
Total KE (mph-squared)	3.66	207.6	1036	373.4
Percentage idle	42.29	16.3	8	33.4

SOURCE: Data from CRC (2002).

to the vehicle activity or duty cycle (Graboski et al., 1998; Nine et al., 2000).

The effect of drive cycle is also well documented for light-duty vehicles and is known to affect emissions in addition to fuel economy (Nam, 2009; Wayne et al., 2008). It is essential to define the activity or cycle that the truck will follow before stating the associated fuel efficiency. The road load equation may be used to compute the power needed to propel a defined vehicle at steady speed over level terrain. The fuel consumed by the vehicle reflects this power requirement, but disproportionately more fuel is consumed at light loads for most conventional vehicles due to the inefficiency of an engine at light load conditions. The plot of fuel consumed (as l/100 km) against the steady speed is a curve that is concave upward. The fuel consumption tends to infinity at idle (zero speed) because fuel is consumed with no distance gained. The curve has a minimum at some midspeed where aerodynamic drag forces are not yet excessive and the engine is at high efficiency, and the curve turns upward at high speed where aerodynamic forces start to dominate the energy required for propulsion. The minimum occurs at low speeds for vehicles with a high ratio of drag to rolling resistance. In this way the minimum occurs at low speeds for automobiles and at high speed for heavily loaded large trucks. Figure 2-12 shows the results of Argonne National Laboratory's PSAT (Power Train Systems Analysis Toolkit) simulations for steady-state operation of two classes of heavy-duty vehicles, with a clear minimum in fuel consumption.

Vehicles in the real world do not operate at steady speed. For a given segment of activity, or for a cycle, it is therefore important to use the metric of average speed in discussing fuel use. Trucks operating at high average speed on freeways tend to be driven at a sustained, fairly steady speed, but trucks operating at lower speed in suburban or urban environments tend to vary their speed substantially, and urban activity is associated with frequent stops. A measure of speed variability is the standard deviation of speed (taken at one-second intervals) over a cycle. The standard deviation of speed does not vary linearly with the average speed. Figure 2-13 shows data for a number of cycles used in transit bus testing and shows a correlation between the standard deviation of speed and the average speed. This suggests that the average speed of a cycle conveys more information than the value of average speed itself: it also conveys the inherent transient nature of lower average speed operation. Figure 2-14 shows that the average speed also offers correlation with the percentage of time that a vehicle idles in the cycle and the number of times that the vehicle stops per mile of travel. Both idle operation and stop-start behavior are more common at low average

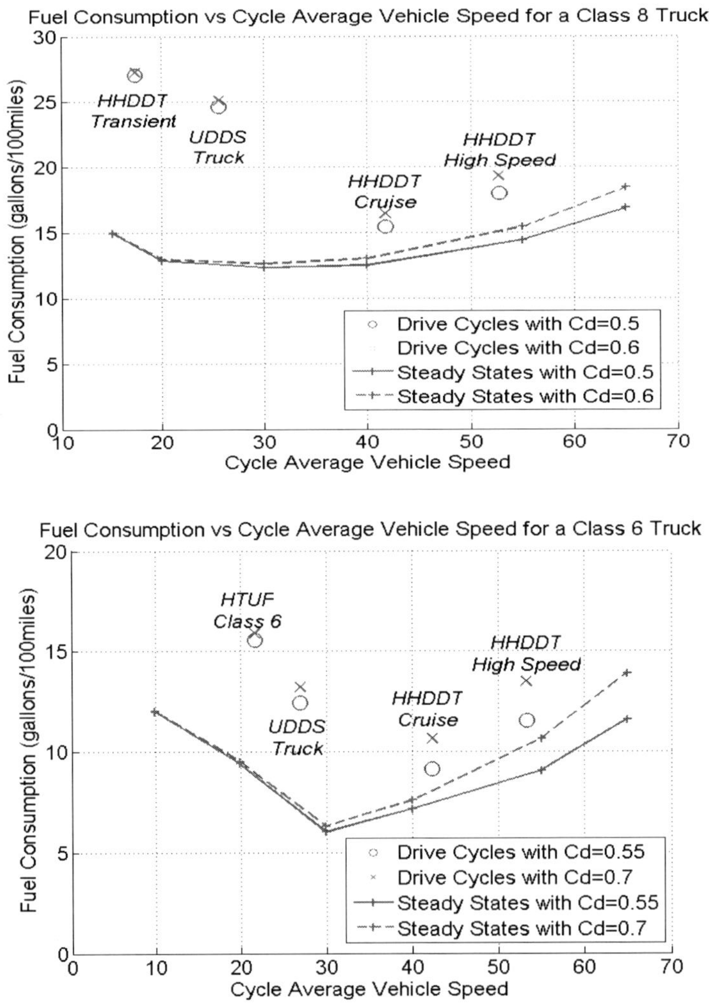

FIGURE 2-12 PSAT simulation results for steady-state operation and for selected transient test cycles for a Class 8 truck (top) and a Class 6 truck (bottom). The Class 6 truck modeled at 9,070 kg was based on a GMC C Series, and the Class 8 truck modeled at 29,931 kg was based on a Kenworth T660 with Cummins 14.9 L ISX.
SOURCE: ANL (2009), Figures 26 and 28.

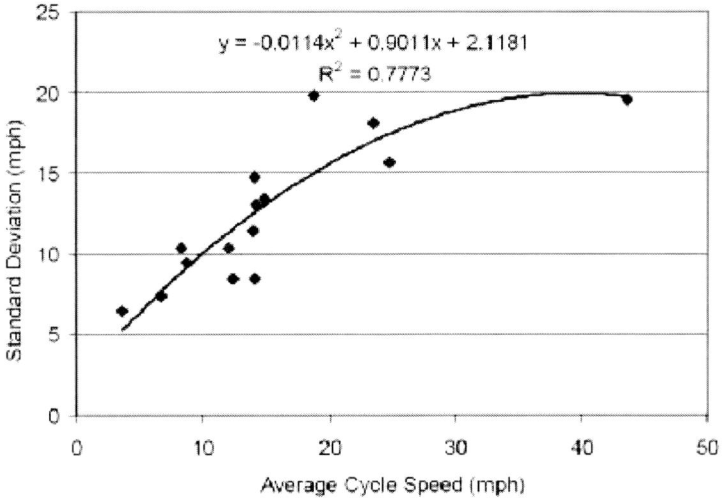

FIGURE 2-13 Standard deviation of speed changes (coefficient of variance rises) as the average speed drops for typical bus activity. SOURCE: Wayne et al. (2008). Reprinted with permission from the Transportation Research Forum.

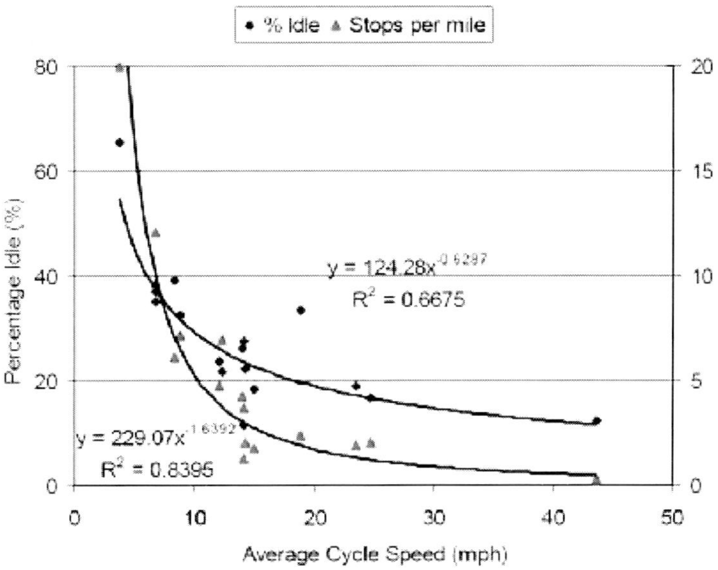

FIGURE 2-14 Percentage of time spent idling rises and there are more stops per unit distance as the average speed drops for typical bus activity. SOURCE: Wayne et al. (2008). Reprinted with permission from the Transportation Research Forum.

speed operation than on freeways. Freeways operating in choked condition will imply low average truck speeds, and the truck activity will more closely resemble urban activity than open freeway activity. Further evidence supporting the correlation between the nature of activity and the average speed of activity is provided elsewhere in a plot for data from automobiles.[12]

The effect of the increased transient behavior at low speeds is to raise the quantity of fuel consumed at low speeds. This is mainly due to the wasting of energy with service brakes and the associated need for propulsion energy during the next acceleration event. In addition, some power trains are less efficient under transient operation than under steady operation. If distance-specific fuel consumption is plotted against average speed, a curve is produced that is concave upward, with high values near zero speed, a minimum at midspeed, and rising values at very high speeds when aerodynamic forces start to dominate. The four cycles in Figure 2-12 also show the role that aerodynamic forces play in determining the speed at which the curve turns upward for typical Class

[12]Available from California Air Resources Board, http://www.arb.ca.gov/msei/onroad/downloads/tsd/Speed_Correction_Factors.pdf.

6 and Class 8 trucks. Curves of this kind have long been used in normalized form for emissions inventory models as "speed correction factors" to adjust distance-specific emissions when average speed deviates from the average speed of a reference cycle used to measure emissions (Frey and Zheng, 2002; Nam, 2009).

Real-world bus data to support the concept further are shown in Figure 2-15. Hybrid vehicles, which store braking energy for reuse during acceleration, and which may increase transient and light load power train efficiency, will primarily produce benefits at low speed. Figure 2-15 shows two best-fit curves for a 40-ft conventional (automatic transmission, diesel) transit bus and a hybrid (diesel) transit bus of similar size and weight. The curves are fitted to chassis dynamometer data taken using numerous transient cycles, each with a representative average speed. The fuel efficiency advantage of the hybrid bus at low operating speeds is evident.

Reporting Fuel Consumption from Different Cycles

The fuel efficiency of a truck is not readily characterized by a single number, but rather by a curve against average speed. Figures 2-14 and 2-15 suggest an approach that may be used to represent the fuel efficiency of a truck to an interested party. If varying operating weight is also considered a factor, fuel efficiency information forms a surface of values against the axes of average speed and operating weight. Creating curves or surfaces of this kind would require exhaustive chassis dynamometer measurements, but they may also be created using models that are calibrated with more limited chassis dynamometer data. Curves or surfaces would show that some technology has low-speed benefits and some has high-speed benefits and that some technology is more sensitive to payload than other technology.

Vehicle Simulation

As new power train and vehicle technologies appear, there will be an on-going challenge to make sure that the simulation tools provide an adequate representation of actual vehicle performance and fuel consumption. In this report, vehicle modeling and simulation will be used to assess the impact of current and future technologies on fuel consumption (see Appendixes G and H). While numerous modeling studies are available in the literature, the assumptions associated with the results are not always available. The committee decided to perform simulation studies using PSAT to analyze the impact of metric selection and assess the impact of current and future technologies. In addition, vehicle modeling will be assessed as part of the regulatory process.

In a world of growing competitiveness, the role of simulation in vehicle development is constantly increasing to allow engineers to bring new technologies to the market faster by reducing the need for hardware testing. Because of the number of possible advanced power train architectures and component technologies that can be employed, the development of the next generation of vehicles requires accurate, flexible simulation tools. Such tools are necessary to quickly narrow the technology focus to those configurations and components that are best able to reduce fuel consumption and performance.

Because models are a mathematical representation of physical components, different levels of fidelity will be used to represent different phenomena. As such, different approaches will be used to answer specific questions. At a high

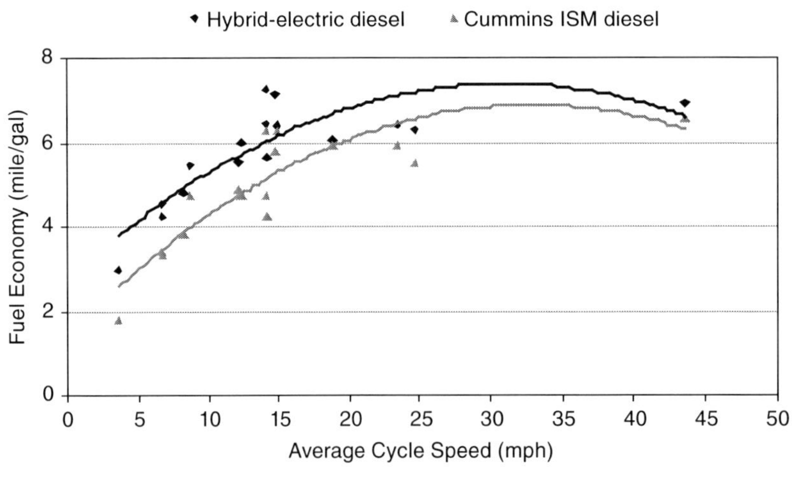

FIGURE 2-15 Curves based on chassis dynamometer for fuel economy versus average speed for conventional and hybrid buses. SOURCE: Wayne et al. (2008). Reprinted with permission from the Transportation Research Forum.

level, a model required to analyze the effects of technologies on fleets (e.g., VOLPE[13] and MOBIL6[14]) will be radically different from ones developed to focus on specific vehicles (e.g., PSAT,[15] CRUISE,[16] RAPTOR,[17] ADVISOR,[18] and PERE[19]).

For fleet analysis, average efficiency or fuel consumption gains are usually considered (e.g., VOLPE). In other instances, vehicle fuel consumptions are assumed for specific operating conditions through the use of Bins (e.g., MOBIL6). In all cases, however, the values implemented to assess fleet impacts are generated from more detailed models developed to analyze specific vehicles.

Two main philosophies are used to model specific vehicles: backward-looking model (or vehicle-driven) and forward-looking model (or driver-driven). In a forward-looking model, the driver model will send an accelerator or a brake pedal to the different power train and component controllers (e.g., throttle for engine, displacement for clutch, gear number for transmission, or mechanical braking for wheels) in order to follow the desired vehicle speed trace. The driver model will then modify its command depending on how close the trace is followed. As components react as in reality to the commands, advanced component models can be implemented, transient effects (such as engine starting, clutch engagement/disengagement, or shifting) can be taken into account, or realistic control strategies can be developed that would later be implemented in real-time applications. By contrast, in a backward-looking model, the desired vehicle speed goes from the vehicle model back to the engine to finally find out how each component should be used to follow the speed cycle. Because of this model organization, quasi-steady models can only be used and realistic control cannot be developed. Consequently, transient effects cannot be taken into account. Backward-models are usually used to define trends, while forward-looking models allow selection of power train configurations, technologies as well as development of controls that will later be implemented in the vehicles.

Simulation tools, more specifically forward-looking models that target specific vehicles, are widely used in the industry to properly address the component interactions that affect fuel consumption and performance. With systems becoming increasingly complex, predicting the effect of combining several systems (whether between components or subsystems) is becoming a difficult task due to the nonlinearity of some phenomena.

The models and controls required to accurately model fuel consumption are well defined. For hot conditions and with accurate plant[20] data, conventional vehicles can achieve fuel consumptions within 1 to 2 percent compared to dynamometer testing. Advanced vehicles, such as hybrid electric vehicles, are more difficult to validate because the power management system selected by the power train manufacturer has a higher impact on fuel consumption and is subject to many variations as discussed in Chapter 6. The plant models used for fuel consumption are usually based on steady-state look-up tables representing the component losses for different operating conditions. The main datasets are captured from dynamometer testing (e.g., fuel rate for different engine torque/speed points).

Lately, simulation tools have been used to further minimize the time required for the vehicle development process using advanced techniques such as model-based design. Advanced techniques are used to develop/test new control algorithms or plant design, including hardware-in-the-loop (HIL), rapid control prototyping or component-in-the-loop. For example, the component control algorithms are currently developed in simulation using detailed plant models (e.g., GTPower for engine or AMESIM for transmission) and can later be tested using the plant hardware.

To represent any technology properly, such models must be established using the appropriate datasets. One of the critical elements in generating accurate results relies on both selection of the proper level of modeling and collection of the data that will populate the model.

While some phenomena are currently well understood and can be properly modeled (e.g., fuel consumption, performance within 1 or 2 percent), others remain difficult to address properly (e.g., emissions or extreme thermal conditions).

Because criteria emissions cannot be simulated with the fidelity available to simulate fuel consumption and vehicle performance, there can be inherent disconnects and inaccuracies in modeling fuel consumption in an emission-constrained vehicle, meaning all vehicles. For example, engine-off modes that would be used with hybrids might result in lower aftertreatment temperatures and thus lower aftertreatment performance. Without aftertreatment constraints in the simulation, the model might allow engine system operation outside the emission-constrained envelope. At the same time, a hybrid might allow the engine to operate in modes where emissions are lower than they would be in a conventional drive train. More investigation needs to be conducted regarding the influence of fuel consumption reduction technology on actual in-service emissions.

[13]DOT/NHTSA, "Corporate Average Fuel Economy Compliance and Effects Modeling System Documentation," DOT HS 811 112, April 2009.

[14]EPA, "The MOVES Approach to Model Emission Model," CRC On-Road Vehicle Emission Workshop, March 2004.

[15]See www.transportation.anl.gov.

[16]See www.avl.com.

[17]SwRI, "RAPTOR Vehicle Modeling and Simulation," November 2004.

[18]See www.avl.com.

[19]EPA, "Fuel Consumption Modeling of Conventional and Advanced Technology Vehicles in the Physical Emission Rate Estimator (PERE)," EPA420-P-05-001, February 2005.

[20]A "plant" is defined as a system that can be controlled.

Model-Based Design

Model-based design (MBD) is a mathematical and visual method of addressing the problems of designing complex control systems and is being used successfully in many motion control, industrial equipment, aerospace, and automotive applications. It provides an efficient approach for the four key elements of the development process cycle: modeling a plant (system identification), analyzing and synthesizing a controller for the plant, simulating the plant and controller, and deploying the controller, thus integrating all these multiple phases and providing a common framework for communication throughout the entire design process.

This MBD paradigm is significantly different from the traditional design methodology. Rather than using complex structures and extensive software code, designers can now define advanced functional characteristics using continuous-time and discrete-time building blocks. These built models along with some simulation tools can lead to rapid prototyping, virtual functional verification, software testing, and validation. MBD is a process that enables faster, more cost-effective development of dynamic systems, including control systems, signal processing, and communications systems. In MBD a system model is at the center of the development process, from requirements development, through design, implementation, and testing. The control algorithm model is an executable specification that is continually refined throughout the development process.

MBD allows efficiency to be improved by:

- Using a common design environment across project teams
- Linking designs directly to requirements
- Integrating testing with design to continuously identify and correct errors
- Refining algorithms through multidomain simulation
- Automatically generating embedded software code
- Developing and reusing test suites
- Automatically generating documentation
- Reusing designs to deploy systems across multiple processors and hardware targets.

The different phases of MBD are shown in Figure 2-16 (see also Appendix G). The methodology is increasingly being implemented by vehicle manufacturers as part of their vehicle development process. As such, one can envision that some of the same techniques used to accelerate the introduction of new technologies on the market could also be part of the portfolio of options available for regulation. One example is the use of HIL for medium- and heavy-duty-vehicle regulation in Japan. However, one can envision that any step of the MBD approach, from pure simulation to a combination of hardware and software to complete vehicle testing, can be part of the process.

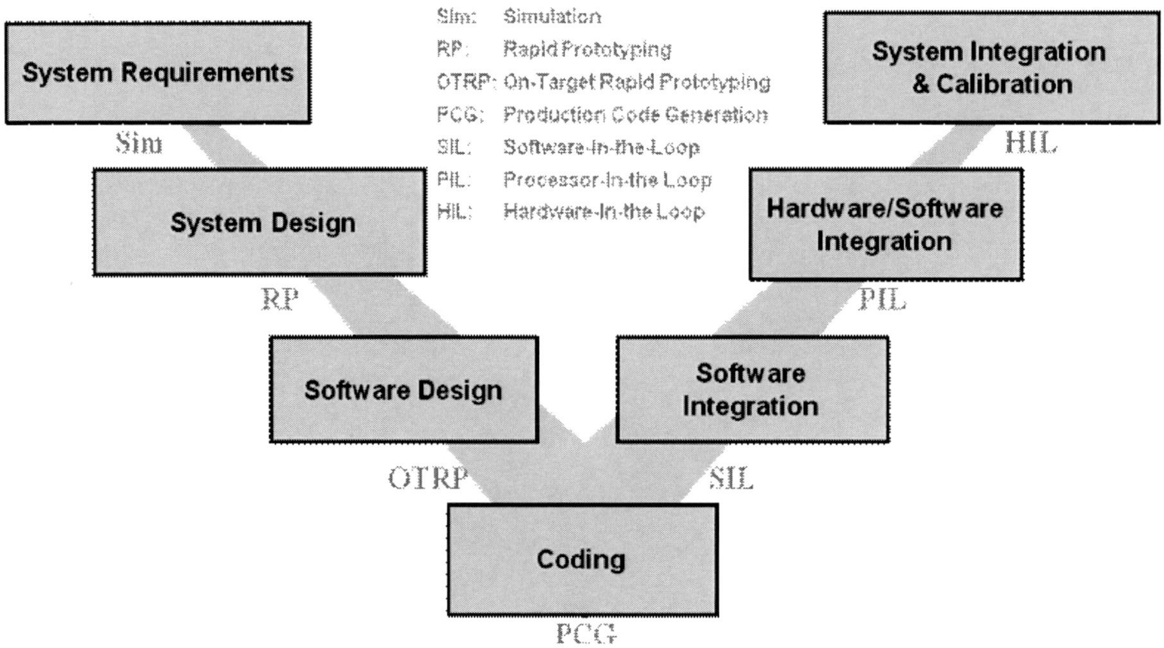

FIGURE 2-16 "V" diagram for software development.

FINDINGS AND RECOMMENDATIONS

Finding 2-1. Fuel consumption (fuel used per distance traveled; e.g., gallons per mile) has been shown to be the fundamental metric to properly judge fuel efficiency improvements from both engineering and regulatory viewpoints, including yearly fuel savings for different technology vehicles. The often-used reciprocal, miles per gallon, called fuel economy, was shown in studies to mislead light-duty vehicle consumers to undervalue small increases (1 to 4 mpg) in fuel economy in lower-fuel-economy vehicles, even though there are large decreases in fuel consumption for small increases in fuel economy. This is because the relationship between fuel economy and fuel consumption is nonlinear. Truck and bus buyers could also likely be misled by using fuel economy data since their fuel economy values are in the lower range (3 to 15 mpg).

Finding 2-2. The relationship between the percent improvement in fuel economy (FE) and the percent reduction in fuel consumption (FC) is nonlinear, and the relationship between change in FE and FC is as follows:

% Increase in Fuel Economy	% Decrease in Fuel Consumption
10	9.1
50	33.3
100	50

Finding 2-3. Medium- and heavy-duty vehicles are designed as load-carrying vehicles, and consequently their most meaningful metric of fuel efficiency will be in relation to work performed, such as fuel consumption per unit payload carried, which is load-specific fuel consumption (LSFC). Because the main social benefit of trucks and buses is the efficient and reliable movement of goods or passengers, establishing a metric that includes a factor for the work performed will most closely match regulatory with societal goals. Methods to increase payload may be combined with technology to reduce fuel consumption to improve LSFC. Future standards might require different values to accurately reflect the applications of the various vehicle classes (e.g., buses, utility, line haul, pickup, and delivery).

Finding 2-4. Yaw-induced drag can be accurately measured only in a wind tunnel. Standard practice in wind tunnel testing reports a wind average drag (coefficient) that can be 15 percent higher than the drag neglecting yaw.

Finding 2-5.* The large per-vehicle annual miles traveled and fuel use by many heavy-duty vehicles magnify the importance, especially to the user, of technologies or design alternatives that can reduce fuel consumption by as little as 1 percent. As a result, accurate test procedures are required to reliably determine the potential benefit of technologies that reduce fuel consumption. Unfortunately, it is very difficult to achieve, at the 90 or 95 percent confidence interval, a precision of less than ±2 percent for vehicle fuel consumption measurements with the current SAE test procedures. The recently convened SAE Truck and Bus Aerodynamic and Fuel Economy Committee effort is a good start toward developing high-quality industry standards.

Recommendation 2-1. Any regulation of medium- and heavy-duty-vehicle fuel consumption should use load-specific fuel consumption (LSFC) as the metric and be based on using an average (or typical) payload based on national data representative of the classes and duty cycle of the vehicle. Standards might require different values of LSFC due to the various functions of the vehicle classes, e.g., buses, utility, line haul, pickup, and delivery. Regulators need to use a common procedure to develop baseline LSFC data for various applications, to determine if separate standards are required for different vehicles that have a common function. Any data reporting or labeling should state an LSFC value at specified tons of payload.

Recommendation 2-2.* Uniform testing and analysis standards need to be created and validated to achieve a high degree of accuracy in determining the fuel consumption of medium- and heavy-duty vehicles. NHTSA should work with industry to develop robust test and analysis procedures and standards for fuel consumption measurement.

BIBLIOGRAPHY

ANL (Argonne National Laboratory). 2009. *Evaluation of Fuel Consumption Potential of Medium and Heavy Duty Vehicles Through Modeling and Simulation*

ATA (American Trucking Associations, Inc.). 2007a. Top 100 private carriers 2007. *Transport Topics.* Available at http://www.ttnews.com/tt100.archive.

ATA. 2007b. Top 100 for-hire fleets 2007. *Transport Topics.* Available at http://www.ttnews.com/tt100.archive.

ATA. 2009. Top 100 commercial fleets 2009. *Light and Medium Truck.* July. Available at http://www.lmtruck.com/lmt100/index.asp.

Bradley, M.J., and Associates LLC. 2009. Setting the Stage for Regulation of Heavy-Duty Vehicle Fuel Economy and GHG Emissions: Issues and Opportunities. Washington, D.C.: International Council on Clean Transportation. February.

Chapin, C.E. 1981. Road load measurement and dynamometer simulation using coastdown techniques. SAE Paper 810828. Warrendale, Pa.: SAE International.

Clark, N.N., M. Gautam, W.S. Wayne, R.D. Nine, G.J. Thompson, D.W. Lyons, H. Maldonado, M. Carlock, and A. Agrawal. 2003. Creation and evaluation of a medium heavy-duty truck test cycle. *SAE Transactions: Journal of Fuels & Lubricants*, Vol. 112, Part 4, pp. 2654-2667.

**Note added in proof:* Recommendation 2-2 in the prepublication version of this report implied that a 1 percent level of accuracy is achievable, which may not be possible. The committee thus corrected and refined Recommendation 2-2 to make it a more general and actionable statement and added Finding 2-5 to summarize the motivation for the recommendation.

Clark, N.N., M. Gautam, W. Riddle, R.D. Nine, and W.S. Wayne. 2004. Examination of a heavy-duty diesel truck chassis dynamometer schedule. SAE Paper 2004-01-2904. SAE Powertrain Conference, Tampa, Fla., Oct.

Clark, N.N., M. Gautam, W.S. Wayne, D.W. Lyons, and G.J. Thompson. 2007. Heavy-duty vehicle chassis dynamometer testing for emissions inventory, air quality modeling, source apportionment and air toxics emissions inventory. Coordinating Research Council, Inc. Report No. E-55/59, Available at http://www.crcao.com/reports/recentstudies2007/E-55-59/E-55_59_Final_Report_23AUG2007.pdf. Accessed July 8, 2009.

CRC (Coordinating Research Council), March 2002. Qualification of the Heavy Heavy-Duty Diesel Truck Schedule and Development of Test Procedures. Final Report, CRC Project E-55-2.

Davies, C., J. Findsen, and L. Pedraza 2005. Assessment of the Greenhouse Gas Emission Benefits of Heavy Duty Natural Gas Vehicles in the United States. Final Report, September 22. Prepared for the U.S. Department of Transportation, Center for Climate Change and Environmental Forecasting, Washington, D.C.

Davis, S., and S. Diegel. 2007. Transportation Energy Data Book Edition 26. Oak Ridge, Tenn.: Oak Ridge National Laboratory.

DOE/EERE (U.S. Department of Energy/Office of Energy Efficiency and Renewable Energy). 2009. 2008 Vehicle Technologies Market Report. Golden, Colo.: National Renewable Energy Laboratory. July. Available at http://www.nrel.gov/docs/fy09osti/46018.pdf.

EPA (U.S. Environmental Protection Agency). 2004. The MOVES Approach to Model Emission Model. Presentation to the Coordinating Research Council's 14th On-Road Vehicle Emission Workshop, San Diego, Calif., March 29-31. Available at http://www.epa.gov/otaq/models/ngm/may04/crc0304a.pdf.

EPA (U.S. Environmental Protection Agency). 2009. SmartWay Truck Emissions Test Protocol Workshop. Drive Cycle Development. Available at http://www.epa.gov/SmartWayshipper/transport/documents/tech/drive-cycle-development.pdf. Accessed September 22, 2009.

Frey, C., and J. Zheng. 2002. Probabilistic analysis of driving cycle-based highway vehicle emission factors. *Environmental Science and Technology*, Vol. 35 (23), October 30, pp. 5184-5191.

Gautam, M., N.N. Clark, W. Riddle, R. Nine, W.S. Wayne, H. Maldonado, A. Agrawal, and M. Carlock. 2002. Development and initial use of a heavy duty diesel truck test schedule for emissions characterization. *SAE Transactions: Journal of Fuels & Lubricants*, Vol. 111, pp. 812-825.

Gaylard, A.P. 2009. The appropriate use of CF in the automotive design process. SAE Paper 2009-01-1162. Presented at the SAE World Congress and Exhibition, Detroit, Mich. April.

Gaylard, Adrian Philip. The appropriate use of CFD in the Automotive Design Process, SAE 2009-01-1162. April 2009.

Graboski, M.S., R.L. McCormick, J. Yanowitz, and L. Ryan. 1998. Heavy-duty diesel vehicle testing for the northern front range air quality study. Fort Collins, Colo.: February.

Larrick, R.P., and J.B. Soll. 2008. The MPG illusion. *Science*, Vol. 320, June 20.

Metro Magazine. 2009. 2009 Fact Book.

Nam, E. 2009. Drive cycle development and real-world data in the United States: Why drive cycles are important. Presentation by the U.S. Environmental Protection Agency at WLTP meeting Geneva, January 15.

Nine, R.D., N.N. Clark, and P. Norton. 2000. Effect of emissions on multiple driving test schedules performed on two heavy duty vehicles. SAE Paper 2000-01-2818. Fall Fuels and Lubricants Meeting and Exposition, Baltimore, Md. October.

SAE (Society of Automotive Engineers) Standards, Procedures, and Recommended Practices. Warrendale, Pa.: SAE International.Various years:

SAE J1082. Fuel Economy Measurement Road Test Procedure. February 2008.

SAE 1252. SAE Wind Tunnel Text Procedure for Trucks and Buses. Warrendale, Penn.: SAE International. July 1981; update in progress.

SAE J1263. Road Load Measurement and Dynamometer Simulation Using Coastdown Techniques. January 2009.

SAE J1264. Joint Rccc/SAE Fuel Consumption Test Procedure (Short Term In-Service Vehicle), Type I. October 1986; update in progress.

SAE J1321. Joint Tmc/SAE Fuel Consumption Test Procedure—Type II. October 1986; update in progress

SAE J1376. Fuel Economy Measurement Test (Engineering Type) for Trucks and Busses. July 1982.

SAE J1711. Recommended Practice for Measuring the Exhaust Emissions and Fuel Economy of Hybrid-Electric Vehicles. March 1999; update in progress.

SAE J2263. Road Load Measurement Using Onboard Anemometry and Coastdown Techniques. December 2008.

SAE J2264. Chassis Dynamometer Simulation of Road Load Using Coastdown Techniques. April 1995; update in progress.

SAE J2452. Stepwise Coastdown Methodology for Measuring Tire Rolling Resistance. June 1999; update in progress.

SAE J2711 Recommended Practice for Measuring Fuel Economy and Emissions of Hybrid-Electric and Conventional Heavy-Duty Vehicles. September 2002.

Thompson, G.J., N.N. Clark, R.J. Atkinson, Z. Luzader, F.L. Vanscoy, V. Baker, and J. Chandler. 2004 Development of an interface method for implementing road grade in chassis dynamometer testing. American Society of Mechanical Engineers (ASME) Paper ICEF2004-896. Internal Combustion Engine Division 2004, ASME Fall Technical Conference, October 24-27, Long Beach, Calif.

TMA (Truck Manufacturers Association). 2007. Test, Evaluation, and Demonstration of Practical Devices/Systems to Reduce Aerodynamic Drag of Tractor/Semitrailer Combination Unit Trucks. Contract Number DE-FC26-04NT42117. National Energy Technology Laboratory, Morgantown, W.Va. April.

Walkowicz, K. 2006. Testing and evaluation of the GM/Allison Hybrid System in the King County Metro Transit Fleet. Presentation to the APTA Bus and Paratransit Conference, Seattle, Wash., May.

Wang, W., M. Gautam, X. Sun, R. Bata, N.N. Clark, G.M. Palmer, and D.W. Lyons. 1994. Emission comparisons of twenty-six heavy duty vehicles operated on conventional and alternative fuels. *SAE Transactions, J. Commercial Vehicles*, Vol. 102, Section 2, pp. 31-40.

Wang, W., D.W. Lyons, R. Bata, N.N. Clark, and M. Gautam. 1995. In-use emissions tests of alternatively fueled heavy-duty vehicles with a transportable chassis dynamometer. *Proc. Inst. Mech. Eng., Part D., J. Automobile Engineering*, Vol. 209.

Wayne, W.S., N.N. Clark, A.B.M.S. Khan, M. Gautam, G.J. Thompson, and D.W. Lyons. 2008. Regulated and non-regulated emissions and fuel economy from conventional diesel, hybrid-electric diesel and natural gas transit buses. *Journal of the Transportation Research Forum*, Vol. 47, No. 3, October, pp. 105-126.

3

Review of Current Regulatory Approaches for Trucks and Cars

The industry built around the use, servicing, and manufacture of medium- and heavy-duty vehicles is large and diverse. Manufacturers, for example, produce many customer-specific configurations assembled with components from multiple suppliers. To aid consideration of what part of the industry should be regulated and how regulations could be implemented, this chapter reviews current regulations and certification protocols as they might apply to medium- and heavy-duty truck fuel consumption. Regulations for medium- and heavy-duty truck fuel economy already exist in Japan and are under development in the European Union (EU). In the United States, California has promulgated rules for heavy-duty truck fuel economy based on the SmartWay voluntary program of the U.S. Environmental Protection Agency (EPA). Regulation of the fuel economy of passenger vehicles began in the United States in 1975, building on the emissions certification procedures already in place since the 1960s. Emissions regulations for heavy-duty trucks have been in force since the early 1970s. Because the test procedures and standards have been revised several times, regulators have repeatedly been faced with the challenge of regulating an extremely diverse industry. In addition, regulation of safety cuts across component suppliers and truck assemblers. Other regulations of interest govern truck size and weight, which are covered in Chapter 7.

EUROPEAN APPROACH

In June 2007 the European Commission (EC) began a study to explore test procedures and metrics for measuring fuel consumption and carbon dioxide (CO_2) emissions. The study initially focused only on the engine and on efficiencies that could be gained through technologies related to the engine. Through active collaboration with heavy-duty truck manufacturers, the EC began to define a variety of duty cycles for the various vocational uses of such trucks. As a result of the collaborative work, in June 2008 the scope of the study and test procedure was expanded to include the whole vehicle and was planned to include alternative driveline concepts such as hybrids, as well as the aerodynamics of both tractor and trailer, gross combination vehicle weight, loading capacity, rolling resistance, and all other technologies that could be applied to reduce fuel consumption.

The absence of a uniform vehicle size and configuration and the existence of myriad possible duty cycles, however, contribute to an enormous degree of complexity in defining a regulation that can be applied. The result is that a simple "one size fits all" approach to measuring CO_2 emissions or fuel efficiency is not feasible. Further, full-vehicle testing of even a fraction of the possible combinations of vehicles and duty cycles would be prohibitively costly.

The absence of uniform vehicles and duty cycles also led the EC to conclude that any metric used should include some indication of the work done as well as the fuel used (e.g., liters of fuel per ton-kilometer).

As a result of the study, the European Automobile Manufacturers Association and the European Council for Automotive R&D have proposed a project to develop a methodology to evaluate the fuel efficiency of heavy-duty vehicles using computer simulation. This will provide a common tool for determining the fuel efficiency and CO_2 generation of heavy-duty vehicles, buses, and coaches over a wide range of duty cycles, taking into account the many possible configurations and mission profiles. Figure 3-1 provides an overview of the proposed approach and simulation tool. The program began in 2009 with a projected 4-year timeline, resulting in a fuel economy regulation for the European Union in 2013-2014. It is expected that the truck manufacturer will be the regulated entity, given the widespread integration of truck and engine manufacturers. As is evident in Figure 3-1, the EC is considering a work-based metric of fuel consumed per payload mass, payload volume, or number of passengers carried per distance traveled.

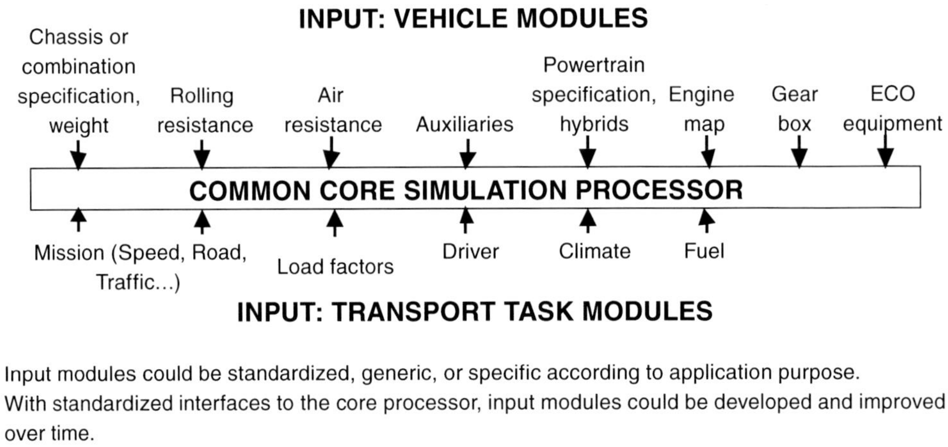

FIGURE 3-1 Overview of simulation tool and methodology proposed for use in the European Union. Both inputs and results are declared transparently. SOURCE: Stefan Larsson, European Auto Manufacturers, presentation to the committee.

JAPANESE APPROACH

Regulation of fuel consumption in Japan is directly linked to Japan's commitment under the Kyoto Protocol to reduce greenhouse gases.[1] In Japan, heavy-duty trucks account for 25 percent of the greenhouse gases generated by automotive sources. The "Top-Runner Standard" for measuring fuel consumption was started for heavy-duty trucks in 2006, with a target implementation date of 2015. The vehicle manufacturer is the regulated entity. In Japan the engine and heavy-duty vehicle manufacturers are integrated and few in number, so the point of regulation is more obvious than in the United States.

As in Europe, the process began with collaborative meetings with the heavy-duty truck manufacturers to collect data on vehicles and technologies that could improve fuel consumption. However, unlike in Europe, the primary focus in Japan is on improvements due to changes in engine technology only, rather than to the whole vehicle, and the metric used is "kilometers/liter," with differing standards for different weight classes (Figure 3-2).

Fuel consumption is evaluated through computer simulation based on a combination of an urban duty cycle defined in JE005, used for emissions testing, and an interurban cycle developed for fuel economy testing. The simulation tool, which is available online for manufacturers to use, requires vehicle specifications and engine fuel maps as input data. An overview of the simulation methodology is given in Figures 3-3 and 3-4.

The vehicle simulation tools used by Japan's Ministry of Land, Infrastructure, Transport, and Tourism evaluate the fuel consumption and performance of conventional vehicles. The software, available in both FORTRAN and C++, allows users to modify the transmission ratio, the final drive ratio, the wheel radius, and the main engine characteristics (including wide-open-throttle and closed-throttle torque curves as well as fuel rate map). However, it forces most of the remaining parameters to remain constant. As such the vehicle characteristics (weight, frontal area, drag coefficient) or component losses (efficiencies) cannot be modified. Moreover, advanced shifting control algorithms that might be available cannot be implemented. The impact of active regeneration of diesel particle filters is handled by calculating the ratio of vehicles with this feature to those without it (Sato, 2007). Overall, the tool allows evaluation of new engine technologies while keeping the rest of the power train and vehicle unchanged. Finally, only two drive cycles can currently be selected.

Because the Japanese program focuses on engines, new methodologies for measurement must be developed as new technologies are introduced to account for their contribution to improving fuel consumption. There is currently no provision in the simulations to take these contributions into account.

Because of the large reductions in fuel consumption achievable with hybrid electric trucks, a measurement method for this technology was included in the Japanese regulation. To measure the contribution of hybrid technology, the Japanese developed hardware-in-the-loop simulation (HILS) testing (Figure 3-5; see also Appendix H) and used it for measuring emissions, as well as calculating fuel consumption. HILS substitutes for the conversion program (see Figure 3-4) used in the process for nonhybrid vehicles (Morita et al., 2008). Details of the method and validation are available in Morita et al. (2008). The HILS approach was recently recommended for further study and potentially wider implementation (in Europe and beyond) for hybrid vehicles by an international committee of engine and vehicle manufacturers.

[1]The Kyoto Protocol, an international agreement linked to the United Nations Framework Convention on Climate Change, sets binding targets for 37 industrialized countries and the European community for reducing greenhouse gas emissions by an average of 5 percent against 1990 levels over the 5-year period 2008-2012.

Target Standard Values for Trucks

Class by GVW (t)	3.5 -7.5				7.5 -8	8 -10	10 -12	12 -14	14 -16	16 -20	20-
Payload (t)	-1.5	1.5-2	2-3	3-							
Target (km/L)	10.83	10.35	9.51	8.12	7.24	6.52	6.00	5.69	4.97	4.15	4.04

Target Standard Values for City Buses

Class by GVW (t)	-8	8-10	10-12	12-14	14-
Target (km/L)	6.97	6.30	5.77	5.14	4.23

FIGURE 3-2 Japanese fuel economy targets for heavy-duty vehicles by weight class. The target is for fiscal year 2015. SOURCE: Presentation to the committee by Akihiko Hoshi, Ministry of Land, Infrastructure, Transport, and Tourism, Japan.

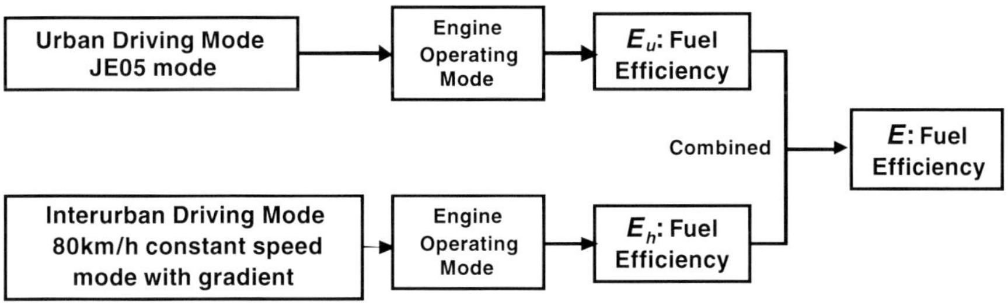

$$E=1/(\circ_u/E_u+\circ_h/E_h)$$

E: Heavy vehicle mode fuel efficiency (km/L)
E_u: Urban driving mode fuel efficiency (km/L)
E_h: Interurban driving mode fuel efficiency (km/L)
$\circ_u$: Proportion of urban driving mode
$\circ_h$: Proportion of interurban driving mode

FIGURE 3-3 Japanese simulation method incorporating urban and interurban driving modes. SOURCE: Presentation to the committee by Akihiko Hoshi, Ministry of Land, Infrastructure, Transport, and Tourism, Japan.

U.S. APPROACH: EPA SMARTWAY VOLUNTARY CERTIFICATION PROGRAM

In 2004 the EPA began development and implementation of SmartWay, an organized effort to specify a collection of current and emerging technologies for creating efficient tractor-trailer combinations with the best environmental performance in terms of both air pollution and emission of greenhouse gases. The certification program uses existing (2007) EPA test methods, supplemented by additional testing, available industry test data, and ongoing research. SmartWay designations are limited to new passenger vehicles (cars, light trucks, sport utility vehicles, vans), new Class 8 sleeper trucks, new 53-ft dry van trailers, and retrofit 53-ft dry van trailers but will include other truck types in the future. The partnership program under SmartWay includes other types of trucks above 8,500 gross vehicle weight rating, in that some participants include their medium-duty trucks in the partnership. Certification allows carriers, manufacturers, and shippers to apply the SmartWay logo (Figure 3-6) to their products as a signal to consumers and the community that they are taking actions to limit the negative environmental impacts of their business operations.

To attain SmartWay certification, tractors must have an aerodynamic profile that includes a high roof sleeper, integrated roof fairings, cab side extenders, fuel tank side fairings, and aerodynamic bumpers and mirrors (Figure 3-7). They must be powered by a 2007 or newer engine, with a SmartWay-approved option for idle reduction. The tires must be SmartWay-approved, low-rolling-resistance tires; the use of aluminum wheels for weight reduction is an option. The tractor specification is a design attribute only. EPA sets no vehicle-level performance targets in SmartWay and requires

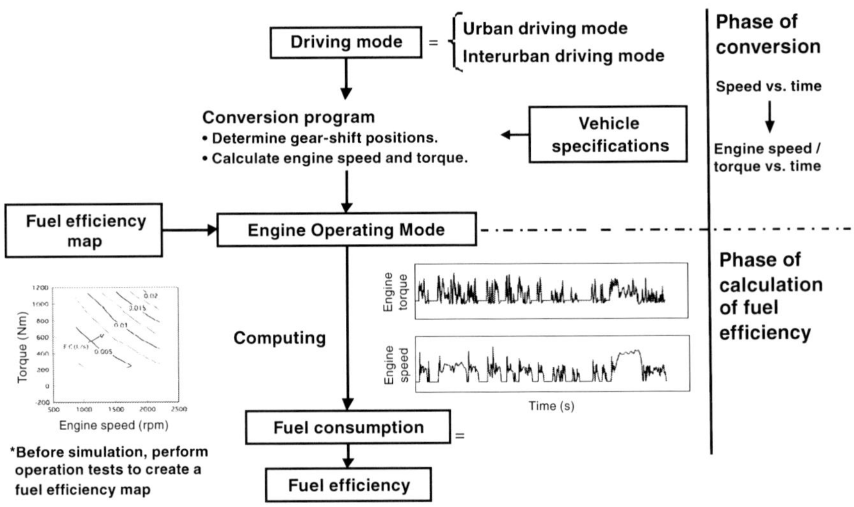

FIGURE 3-4 Japanese simulation method overview. SOURCE: Presentation to the committee by Akihiko Hoshi, Ministry of Land, Infrastructure, Transport, and Tourism, Japan.

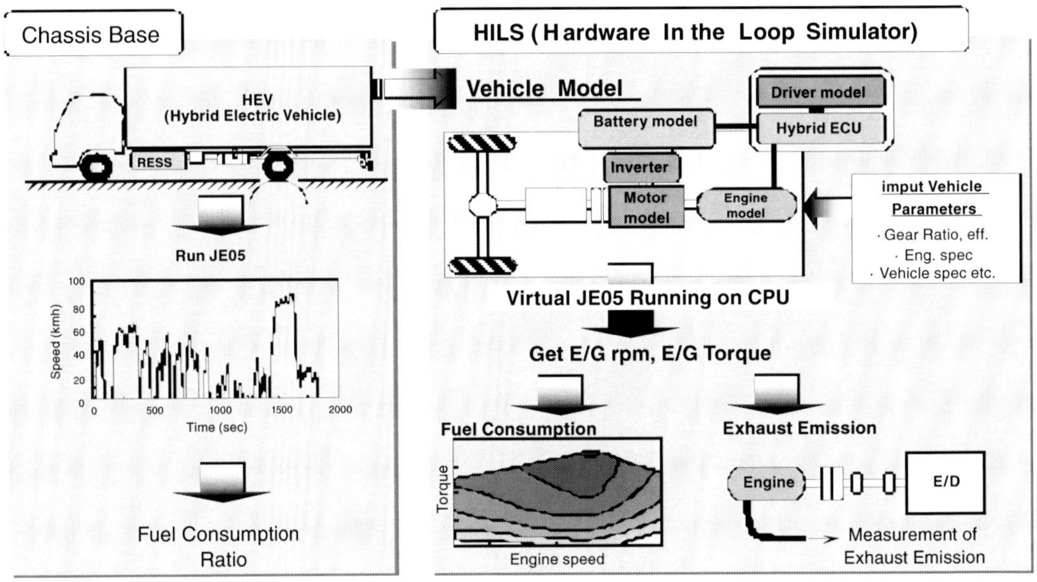

FIGURE 3-5 Japanese hardware-in-the-loop simulation (HILS) testing of hybrid vehicles.
SOURCE: Presentation to the committee by Akihiko Hoshi, Ministry of Land, Infrastructure, Transport, and Tourism, Japan.

no technical validation (although in the case of tires, for example, data proving low rolling resistance must be supplied for the designation). All six major U.S. truck manufacturers have one or more complying tractors. Manufacturer compliance with the SmartWay specification is completely voluntary.

SmartWay certification applies to 53-ft or longer dry box van trailers as well. To attain certification, these trailers must have side skirt fairings, a front gap or rear fairing, and SmartWay-approved low-rolling-resistance tires and can have aluminum wheels. The fairings, tires, and wheels can be either provided by the original equipment manufacturer (OEM) on new trailers or retrofitted to older trailers. Several OEMs currently offer at least one SmartWay trailer model, and several manufacturers have developed aerodynamic components that can be retrofitted. EPA has validated trailer side skirts, trailer boat tails, and trailer gap reducers.

As an alternative to the aerodynamic specification for the trailer, the EPA will grant certification upon review of demonstrated equivalent environmental performance for the aerodynamic specification, defined as 5 percent or greater fuel savings using an SAE J1321[2] test track procedure as modified by EPA.

[2]SAE (Society of Automotive Engineers) J1321: Joint TMC/SAE Fuel Consumption Test Procedure–Type II. October 1986; update in progress.

FIGURE 3-6 EPA's SmartWay logos. SOURCE: "External SmartWay Marks: SmartWay Tractors and Trailers," available at http://www.epa.gov/otaq/smartway/transport/what-smartway/tractor-trailer-markuse.htm.

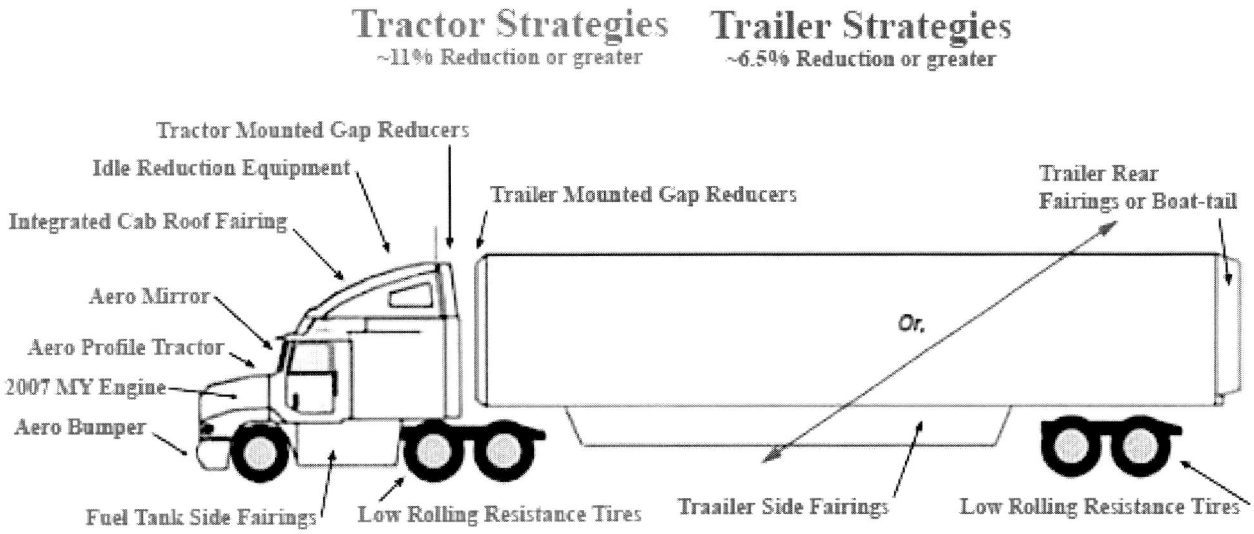

FIGURE 3-7 Some of the aerodynamic technologies included in the SmartWay certification program. SOURCE: Mitch Greenberg, EPA, "SmartWay Voluntary Certification Program," presentation to the committee.

CALIFORNIA REGULATION BASED ON EPA SMARTWAY PROGRAM

One of the most significant consequences of the SmartWay certification program is action taken by the California Air Resources Board. With the intention to reduce the state's greenhouse gas emissions to 80 percent of 1990 levels by 2020, in 2006 the California legislature approved the Global Warming Solutions Act. A resulting action in December 2008 was the adoption of a measure that defines a schedule by which all tractor-trailer combinations that operate in California will be required to implement SmartWay technologies.

Beginning in January 2010, with the 2011 model year, all sleeper cab tractors that pull 53-ft or longer box van trailers must be SmartWay certified. Day cab tractors must have SmartWay-approved low-rolling-resistance tires. At the same time, in model year 2011 and beyond, all 53-ft or longer van trailers must also be SmartWay certified, based on either OEM equipment or retrofits. The legislation also calls for retrofitting older trailers with SmartWay-approved technologies, providing a phase-in period for larger fleets from 2010 to 2015 and for smaller fleets from 2013 to 2016.

LIGHT-DUTY-VEHICLE FUEL ECONOMY STANDARDS

The development of fuel consumption standards for passenger vehicles provides useful lessons for consideration in trucks and buses. In 1975, Congress enacted the Energy Policy and Conservation Act (EPCA, P.L. 94-163), with the goal of reducing the nation's dependence on foreign oil by a number of measures, including doubling the fuel economy of passenger vehicles. The statute established a Corporate Average Fuel Economy (CAFE) program, which requires automobile manufacturers to increase the sales-weighted average fuel economy of their passenger car and light-truck fleets sold in the United States. The metric chosen by Congress for expressing the standard was fuel economy expressed in miles per gallon (mpg).

Measurement

Passenger vehicle fuel economy is determined by testing a vehicle on a chassis dynamometer. In the laboratory the vehicle's drive wheels are placed on two 48-in.-diameter metal rollers—the dynamometer—that simulate the loads experienced by the vehicle in the real world. The energy required to move the rollers is adjusted to account for the vehicle's weight and wind resistance and for the tire rolling resistance of the two undriven wheels. The laboratory testing is very repeatable and precise, allowing consistent results across the vehicle fleet.

The levels of hydrocarbons (HCs), carbon monoxide (CO), carbon dioxide (CO_2), and nitrous oxides (NO_x) in the exhaust are measured using a constant-volume sampling system. Fuel consumption is determined by summing all the carbon in the exhaust and converting this to gallons using the amount of carbon per gallon, although direct measurement of fuel consumed is greatly improved and is now routinely used in parallel.

Manufacturers test their own vehicles and report the results to EPA. EPA reviews the results and confirms them by testing 10 to 15 percent of the certified models at its own laboratory. About 1,250 vehicle models are certified annually (EPA, 2006a). EPA is empowered to audit production-line vehicles for compliance. For post-production vehicles in use, EPA conducts tests of vehicles from customers in real service and gathers data from onboard diagnostics records and manufacturer in-use verification testing. EPA is empowered to require a recall of vehicles to correct defects in emission control systems.

Test Cycles

A test cycle is a series of driving routines that specify the vehicle speed for each second during a particular test. Two test cycles were originally used for determining compliance with CAFE standards: (1) a city cycle originally developed in the mid-1960s to represent home-to-work, urban commuting, commonly referred to as the FTP (federal test procedure), and (2) a highway test cycle that represents a mix of rural and interstate highway driving.

The FTP was developed for emission control purposes, not fuel economy, and the highway test reflects the 55-mph speed limit that was in effect when the cycle was created. The EPA quickly realized that these factors caused the tests to overstate the average in-use fuel economy. EPA spent several years analyzing the average offset and in 1984 issued adjustments for the fuel economy label values that had been required on window stickers for all new cars since 1978. For the labels the city test results were discounted by 10 percent and the highway test results were discounted by 22 percent. However, the unadjusted values continued to be used for CAFE purposes, as the EPCA specified that the test procedures in place in 1975 must continue to be used for calculating CAFE values.

In 2006, EPA reevaluated the difference between the test results and the fuel economy experienced by the average consumer. It found that the gap had widened since the early 1980s, in part because the underlying test procedures did not fully represent real-world driving conditions. Thus, three additional tests have been added to the original city and highway estimates to adjust for higher speeds, air-conditioning use, and colder temperatures. These tests, applied beginning in 2008, were developed in two previous rulemakings for the purpose of controlling emissions under conditions not included in the original FTP test. For the revised fuel economy labels, rather than basing the city mpg estimate solely on the adjusted FTP test result, and the highway mpg estimate solely on the adjusted highway fuel economy test (HFET) result, each estimate will be based on a multiequation "composite" calculation of all five tests, weighting each appropriately to arrive at new city and highway mpg estimates. The new city and highway estimates will each be calculated according to separate city and highway "five-cycle" formulas based on fuel economy results over these five tests. A simplified approach, called the mpg-based method, will be an interim option in the first 3 years of the program and an available option under certain circumstances in subsequent years. It relies on data from the FTP and HFET cycles (EPA, 2006b).

Table 3-1 lists the key parameters of the five tests used to determine the fuel economy (of passenger vehicles) that is listed on the familiar window stickers.

In summary, light-duty-vehicle emissions and fuel consumption regulations exemplify the compromises necessary due to the diversity of vehicle types and uses. One set of drive cycles is applied to about 1,200 vehicle models, over a wide range of weights, with the heaviest about twice the weight of the lightest. Similarly, drive cycles cover a wide range of powers, with the highest about five times as powerful as the lowest. Light-duty fuel economy regulations adopted the same testing protocols used for emissions certification, and numerous adjustments and correction factors have been needed to bring reported fuel economy values closer to consumer (real-world) experience. Concepts and recommendations for CAFE predated the regulation by 1 to 2 years, and the regulation was not implemented until 1978. Hence, even though there was the foundation of the emissions regulations, the time, and presumably effort, needed to implement CAFE and its test protocols were very substantial (Greene and DeCicco, 2000).

HEAVY-DUTY-ENGINE EMISSIONS REGULATIONS

Background, Test Methods, and Cycles

Pollutant emissions from heavy-duty vehicles are regulated in terms of emissions from their engines. Faced with how to regulate emissions given the great diversity of heavy-

TABLE 3-1 Fuel Economy Vehicle Testing

	Original Test Cycles		New Test Cycles		
Driving Schedule Attributes	City	Highway	High Speed (US06)	Air-Conditioning (AC) (SC03)	Cold Temperature
Trip type	Low and moderate speeds in urban traffic	Free-flow traffic at highway speeds	Higher speeds; harder acceleration and braking	AC use under hot ambient conditions and full sun load	City test w/cold outside temperature
Top speed	56 mph	60 mph	80 mph	54.8 mph	56 mph
Average speed	21.2 mph	48.3 mph	48.4 mph	21.2 mph	21.2 mph
Maximum acceleration	3.3 mph/sec	3.2 mph/sec	8.46 mph/sec	5.1 mph/sec	3.3 mph/sec
Simulated distance	11 miles	10.3 miles	8 miles	3.6 miles	11 miles
Time	31.2 min.	12.75 min.	9.9 min.	9.9 min.	31.2 min.
Stops	23	None	4	5	23
Idling time	18% of time	None	7% of time	19% of time	18% of time
Engine startup[a]	Cold	Warm	Warm	Warm	Cold
Lab temperature	68°-86°F			95°F	20°F
Vehicle air conditioning	Off	Off	Off	On	Off

[a]A vehicle's engine does not reach maximum fuel efficiency until it is warm.
SOURCE: Department of Energy.

duty vehicles, EPA chose many years ago to regulate the engine manufacturers. A number of reasons can be cited: in many cases the engine and chassis are produced by different manufacturers; it is more efficient to hold a single entity responsible; and testing an engine cell is more accurate and repeatable than testing a whole vehicle. About 275 engine models are certified for heavy trucks each year, considerably fewer than the number of passenger vehicles that are certified.

Emissions standards for heavy-duty engines are expressed in terms of mass (grams) per unit of energy output. Energy output is expressed in horsepower-hours (hp-hr) or kilowatt-hours (kWh), so the standards are expressed as g/hp-hr or g/kWh. The *Code of Federal Regulations* (40 CFR Part 86) requires that the engine be connected to a dynamometer in a test cell and be exercised through a transient test procedure (the FTP). The FTP was created from data logged from trucks and buses operating in Los Angeles and New York as part of the CAPE-21 Coordinating Research Council (CRC) program of 1973 to 1977 (CRC, 1973, 1974, 1977). Speed data were converted to percentage values with idle speed set at zero percent and rated speed set at 100 percent; torque data were converted to a zero to 100 percent torque referenced to zero torque and maximum engine torque at each speed. The FTP was created using a Monte Carlo simulation based on percent speed and percent torque probabilities. The diesel engine procedure consists of a time sequence of rotational speed and torque set points determined from a table of dimensionless speeds and torques and a full-power map of the engine (Figure 3-8). Torque and speed regression criteria are available to determine whether the engine has been properly loaded during the test. A separate FTP schedule exists for heavy-duty gasoline engines.

The whole engine exhaust is fed to a full-flow dilution tunnel and is mixed with dilution air to produce a constant-volume flow. Gaseous concentrations are measured at a sampling point downstream in the dilution tunnel and are integrated with respect to the flow to yield a total mass for the cycle. A background correction is applied to remove the mass of pollutant that may be in the dilution air. The net mass is divided by the energy out of the engine over the cycle to yield brake-specific emissions. Particulate matter (PM) is measured by filtering a slipstream of dilute exhaust, weighing the filter, and projecting the total PM mass from the ratio of total flow to slipstream flow. Although NO_x, CO, and HCs are the regulated emissions, CO_2 is also measured in most test cells, and this measurement may be used to project the engine brake-specific fuel consumption. Fuel mass or volume flow is also measured in many test cells, providing a separate check. During the testing, certain loads, such as water pump and oil pump demands, are met by the engine, but engine coolant heat exchange and intercooler heat removal are supplied by the test cell.

The practice of testing and certifying only the engine (across selected cycles), and not the entire vehicle, was an accepted compromise for pollutant emissions. Even with this simplified engine-only approach, the FTP transient test procedures were under development at EPA for 5 years, and several years more were needed for the industry to employ electric dynamometers, constant-volume samplers, dilution systems, and new measurement systems for PM (Merrion, 2002).

How representative of on-road emissions are the current engine dynamometer procedures? John Wall presented an analysis to the committee comparing CO_2 emissions from the certification test process to over-the-road vehicle fuel consumption data.[3] For line-haul truck duty cycles, CO_2 emissions from vehicle field data were within 3.7 percent of the CO_2 emissions predicted from the Supplementary

[3]John Wall, Cummins, Inc., presentation to the committee, August 6, 2009.

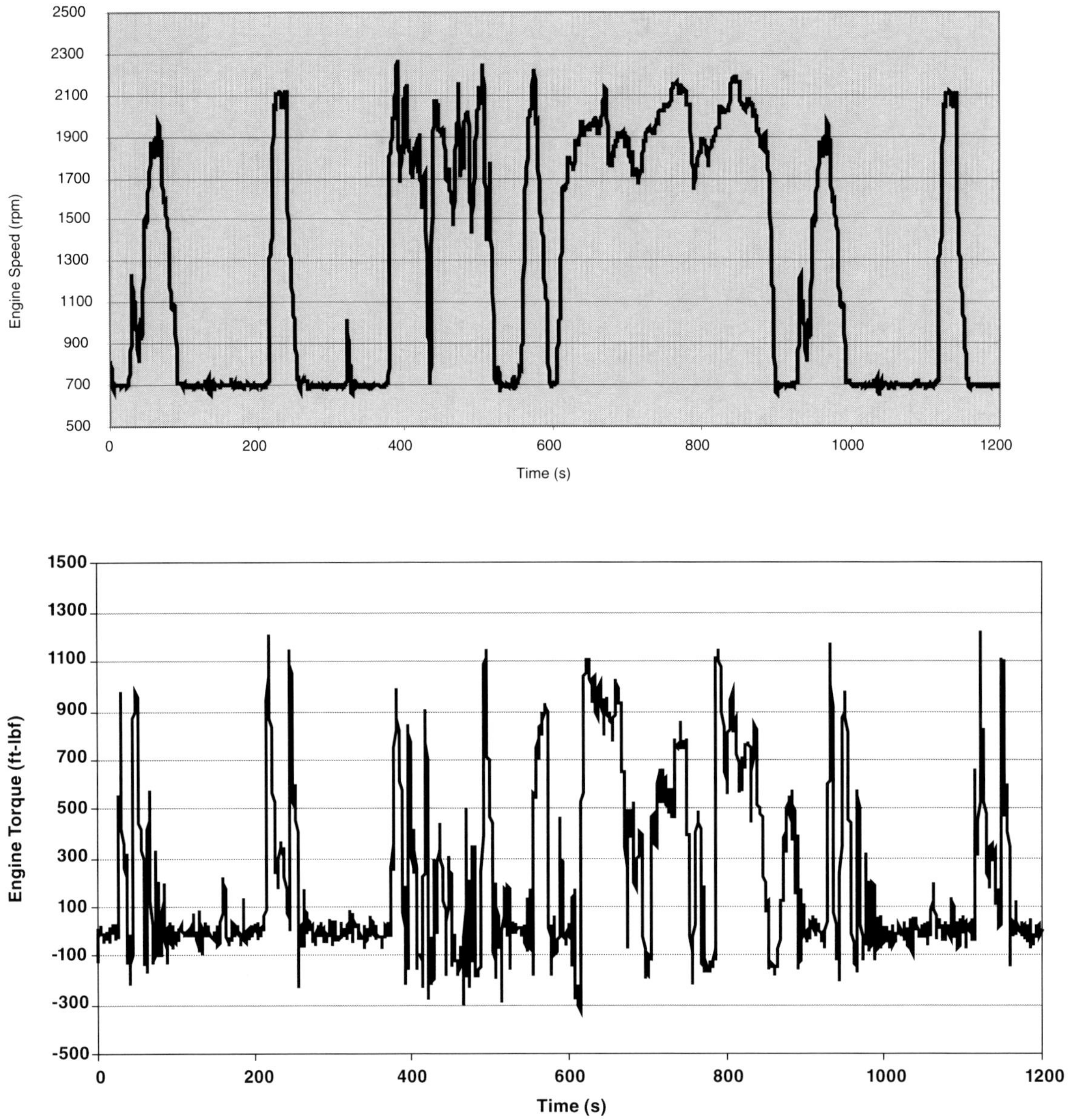

FIGURE 3-8 FTP speed (top) and torque (bottom) from a specific engine following the transient FTP on a dynamometer. SOURCE: Based on data from West Virginia University.

Emission Test (SET) 13 mode engine data. For a set of field-monitored vocational trucks, the FTP transient engine test data predicted CO_2 emissions to within 4.0 percent of the on-road results. Further analysis would be needed to determine the breadth of vehicles and duty cycles that an engine-only test could accurately cover, and this correlation would not be expected to hold true for hybrid vehicles.

Certification and Enforcement

The emissions certification process begins with a meeting with EPA to review the manufacturer's product line and to describe the emission control systems. Agreement is reached on engine families that have similar displacements, number of cylinders, fuel injection systems, turbocharging systems, and after-treatment systems. Several ratings of horsepower, torque, and rpm can be included in one family. Emissions testing then follows, conducted by the engine manufacturer. This testing involves one "data engine" and a durability/deterioration factor (DF) engine for each family. The data engine is run for 125 hours and emissions measured on the FTP and on the ramp modal cycle (RMC). The RMC replaced the SET but is still a 13-mode steady-state test with emissions also measured during the ramp between modes. The DF engine is run for a period of time that is representative of 35 percent of the engine's useful life, such as 1,000 hours at full load, to demonstrate the durability of the emissions control system and to establish the emissions deterioration. Emissions are measured at three points during the DF test (125 hours, midpoint, and end of test) on the FTP and RMC. Following testing, all data are submitted to EPA as part of the application for certification.

Following certification and production, an enforcement process is in effect. EPA can come into the production facility and perform a Selective Enforcement Audit. It can sequester 25 to 30 engines and start an audit process under which five of these engines must be under the emission standards as measured on the FTP and RMC after a 125-hour conditioning period. If failures occur, a statistical process is used that may require all (25 to 30) engines to be tested. EPA can also do in-use emissions testing on vehicles in regular service. This testing is done using a Portable Emission Measurement System either over-the-road or on a chassis dynamometer. Emissions are measured over a 30-minute period (constantly updating) and a determination is made as to whether emissions are exceeding the not-to-exceed values (usually either 125 or 150 percent of the regulated values depending on the engine family).

Compliance flexibility is provided with averaging, banking, and trading, which can take place among engine families or with other manufacturers. Also, a provision to pay noncompliance penalties is available if the manufacturer wants to certify to an emissions level higher than the standard.

Chassis Testing for Certain Heavy Vehicle Classes

A chassis cycle or schedule may be used for the emissions certification of certain Class 2B vehicles as an alternative to the heavy-duty engine dynamometer procedure. Historically, the Class 2B vehicle manufacturer was also the manufacturer of the engine, because Class 2B vehicles were gasoline powered and usually produced by a light-duty-vehicle manufacturer. EPA Tier 2 light-duty emissions requirements were extended to include medium-duty passenger vehicles, which are between 8,501 and 10,000 lb gross vehicle weight (GVW) and may include diesel-powered vehicles. The EPA's 2006 regulatory announcement (EPA, 2006b) explains that larger sport utility vehicles and vans (8,501- to 10,000-lb GVW) will require light-duty-style fuel economy measurements from 2011. California's Low Emission Vehicle (LEV II) regulations cover emissions from vehicles in the 8,501- to 14,000-lb range through chassis testing. A federal option also exists for chassis testing of vehicles up to 14,000 lb GVW.

Nonroad Engines

EPA began regulation of nonroad engines in the mid-1990s, with regulations now covering the immense range from handheld spark ignition engines (such as leaf blowers) to locomotive engines. Stated by EPA as being one of the most complex sets of emissions regulations undertaken, the certification process focuses again on the engine only, tested over prescribed cycles or steady-state modes. The engines are divided into many classes depending on the size and type of use, and test cycles are prescribed for each class.

REGULATORY EXAMPLE FROM TRUCK SAFETY BRAKE TEST AND EQUIPMENT

Heavy-duty vehicle regulations can be complex; however, there are examples of compliance mechanisms that provide flexibility and minimize the burden on industry. An illustrative example is performance-based Federal Motor Vehicle Safety Standard (FMVSS) 121, which requires that a vehicle stop within a certain distance from an initial speed when loaded to the GVW rating (Table 3-2). The stopping distance is dependent on vehicle type. For example, from an initial speed of 60 mph, truck tractors must stop within 355 ft, single-unit trucks within 310 ft, and buses within 280 ft (NHTSA, 2004). This example illustrates clearly that regulatory requirements can differ for the various vehicle types within the general class of heavy-duty vehicles.

The brake performance evaluation of trailers is carried out differently than for trucks and buses in that, because they are not self-powered, trailers have a dynamometer requirement rather than a test track requirement (NHTSA, 1990). This illustrates that the test methods for a given heavy-duty vehicle regulation can vary significantly depending on the vehicle unit (e.g., truck-tractors or trailers).

Because the heavy-duty truck market is so complex, and many of the vehicles are custom-built (as discussed in Chapter 2), regulations applied at the final stage of manufacture can pose a heavy burden on the manufacturer. However, this burden can be lessened by "type approval" at the component level. For example, in the case of brake regulations, truck axle manufactures supply brakes with the axle assemblies properly sized and rated for the load that the axle is designed to carry. By using these axles the final-stage manufacturer can assure compliance with the brake regulations without

TABLE 3-2 Stopping Distances Required by FMCSS 121 Regulation

	Service Brake Stopping Distance (ft)			
Vehicle Speed (mph)	Loaded and Unloaded Buses	Loaded Single-Unit Trucks	Unloaded Truck Tractors and Single-Unit Trucks	Loaded Truck Tractors
20	32	35	38	40
25	49	54	59	62
30	70	78	84	89
35	96	106	114	121
40	125	138	149	158
45	158	175	189	200
50	195	216	233	247
55	236	261	281	299
60	280	310	335	355

SOURCE: Adapted from NHTSA (2004).

having to test each vehicle. In effect the regulatory effort has cascaded down to the component manufacturer, providing much-needed design flexibility to the final-stage manufacturer (which would not have to test every vehicle design variation). Aspects of this regulatory model may prove useful in the development of regulatory instruments for governing heavy-duty truck fuel consumption.

FINDINGS

Finding 3-1. Regulators have dealt effectively with the diversity and complexity of the vehicle industry for current laws on fuel consumption and emissions for light-duty vehicles. Engine-based certification procedures have been applied to address emissions from heavy-duty vehicles and the myriad of nontransportation engines. Regulators and industry have reached consensus in these cases, but years of development of procedures and equipment for certification, compliance, and the defining the standards themselves have been required. Standardized drive or operating cycles are utilized in all emissions and fuel consumption regulations to represent actual use of the vehicle or engine.

Finding 3-2. The heavy-duty-truck fuel consumption regulations in Japan, and those under consideration and study by the European Commission (EC), provide valuable input and experience to the U.S. plans. In Japan the complexity of medium- and heavy-duty vehicle configurations and duty cycles was determined to lend itself to the use of computer simulation as a cost-effective means to calculate fuel efficiency. The EC studies thus far indicate plans to develop and use simulations in the expected European regulatory system. Japan is not using extensive full-vehicle testing in the certification process, despite the fact that its heavy-duty-vehicle manufacturing diversity is less than in the United States, with relatively few heavy-duty-vehicle manufacturers and no independent engine companies.

Finding 3-3. The existing regulations pertaining to medium- and heavy-duty vehicle safety and emissions provide examples indicating that the industry's diversity is addressed by requiring compliance, or at least conformity, at the component level, reducing the regulatory burden on the final-stage manufacturer and thus preserving the flexibility of assembly to meet customer demands.

Finding 3-4. The legislation passed by California requiring tractor-trailer combinations to be SmartWay certified will have a significant impact on the number of vehicles in the United States that are specified with fuel-efficient technologies beginning in 2010.

REFERENCES

CRC (Coordinating Research Council). 1973. Heavy-Duty Vehicle Driving Pattern and Use Survey. Columbia, S.C.: Wilbur Smith and Associates. May.

CRC. 1974. Heavy-Duty Vehicle Driving Pattern and Use Survey: Part II—Los Angeles Basin. Columbia, S.C.: Wilbur Smith and Associates. February.

CRC. 1977. Truck Driving Pattern and Use Survey—Phase II. Columbia, S.C.: Wilbur Smith and Associates. June.

EPA (U.S. Environmental Protection Agency). 2006a. Fuel Economy Labeling of Motor Vehicles: Revisions to Improve Calculation of Fuel Economy Estimates, Final technical support document. 40 CFR Parts 86 and 600 [EPA-HQ-OAR-2005-0169; FRL-8257-5]. Federal Register: Dec. 27, 2006 (Vol. 71, No. 248): Rules and Regulations: Pp. 77871-77969.

EPA. 2006b. Regulatory Announcement: EPA Issues New Test Methods for Fuel Economy Window Stickers. EPA fact sheet, document number EPA420-F-06-069. Available at http://www.epa.gov/fueleconomy/420f06069.htm.

Greene, D., and J. DeCicco. 2000. Engineering and economic analyses of automotive fuel economy potential in the United States. *Annual Review of Energy and the Environment*. Vol. 25, pp. 477-535. November.

Merrion, D. 2002. Heavy Duty Diesel Emissions, Fifty Years, 1960-2010. Proceedings of the 2002 Fall Technical Conference of the ASME Internal Combustion Engine Division, New Orleans, La., Sept. 8-11. New York: American Society of Mechanical Engineers.

Morita, K., K. Shimamura, S. Yamaguchi, K. Furumachi, N. Osaki, S. Nakamura, K. Narusawa, K-J. Myong, and T. Kawai. 2008. Development of a fuel economy and exhaust emissions test method with HILS for heavy-duty HEVs. SAE Paper 2008-01-1318. Also in SAE *International Journal of Engines*.

NHTSA (National Highway Traffic Safety Administration, U.S. Department of Transportation). 1990. Laboratory Test Procedure for Federal Motor Vehicle Safety Standards and Regulations No. 121D, Air Brake Systems—Dynamometer, TP-121D-01. May. Available at http://www.nhtsa.dot.gov/cars/rules/import/fmvss/index.html#SN121.

NHTSA. 2004. Laboratory Test Procedure for FMVSS, No. 121, Air Brake Systems, TP-121V-05. March 12.

Sato, S. 2007. Fuel Economy Test Procedure for Heavy Duty Vehicles, Japanese Test Procedures, Presented at the IEA/International Transport Forum Workshop on Standards and Other Policy Instruments on Fuel Efficiency for HDVs, Paris France, June 21-22.

4

Power Train Technologies for Reducing Load-Specific Fuel Consumption

Technologies for reducing fuel consumption of medium- and heavy-duty vehicles depend on the power train type. For instance, replacing gasoline engines with diesel engines in medium-duty trucks is a very effective technology, but heavy-duty trucks are already more than 99 percent dieselized. This chapter discusses the energy balance for a typical diesel engine that leads to a resulting brake power or brake thermal efficiency. It presents technologies for improving the efficiency of diesel and gasoline engines (including fuels and emission systems) as well as technologies for transmissions and drive axles. It also discusses the role of hybrid power trains (electric and hydraulic) in reducing fuel consumption.

DIESEL ENGINE TECHNOLOGIES

Diesel engines use the high gas temperatures generated by compression as the ignition source. The timing of ignition is determined mainly by when the fuel is injected. These engines operate on the four-stroke-cycle principle and are arranged either in-line or in a "vee" configuration. Displacements range from 3.0 to 16.0 liters. These engines typically burn diesel fuel, and also some kerosene and some biodiesel blends. Some engines that were originally designed as diesel engines are converted to use spark ignition to take advantage of alternative fuels. These engines typically burn gaseous fuels such as compressed natural gas (CNG), liquefied natural gas (LNG), or propane, but other spark ignition fuels can also be used. Essentially all of the diesel engines used today in medium- and heavy-duty vehicles are turbocharged, direct fuel injected, and electronically controlled; most are intercooled or after cooled. In addition, they use exhaust gas recirculation (EGR) to limit in-cylinder formation of nitrogen oxides (NO_x) and some form of exhaust aftertreatment (diesel oxidation catalyst [DOC] diesel particulate filter [DPF], or other system) to control particulate matter (PM) emissions. Starting in 2010, most diesel engines will add selective catalytic reduction systems (SCR) as a form of NO_x aftertreatment to meet 2010 requirements. A typical diesel engine energy audit is shown in Figure 4-1, where the fuel energy is converted to brake power and the efficiency associated with the output power will be referred to as brake thermal efficiency. The accessory losses are for engine-driven pumps that are necessary to run the engine on a dynamometer or on the road (fuel, lubricating oil, cooling water). Auxiliary loads such as alternator, air compressor, and power steering pump will use a portion of the brake power.

The following material summarizes various technologies for reducing fuel consumption from diesel engines. Some of the engine technologies listed here are the products of participants in the multiagency, multicompany 21st Century Truck Partnership. The partnership's goals for engines are to achieve 50 percent thermal efficiency, while meeting 2010 emissions standards, by 2010 and to develop technologies to achieve 55 percent thermal efficiency by 2013 (NRC, 2008).

Turbochargers

In a turbocharger the radial exhaust-driven turbine drives the radial compressor to increase the air density going into the engine. The turbochargers can have a fixed geometry or more commonly a variable geometry turbine, or they can have a "wastegated" turbine (a bypass). Improved efficiency of the compressor or turbine will improve fuel consumption. Higher pressure ratio radial compressors or axial compressors are emerging technologies. Improvements in compressor efficiency and/or turbine efficiency can contribute to improved fuel consumption. A presentation[1] to the committee on Japan's Top Runner fuel efficiency regulation estimated 0.3 to 0.5 percent improvement from increased supercharging efficiency. The TIAX investigation, by contract to the committee, put the improvement at up to 2 percent (TIAX,

[1]Akihiko Hoshi, Ministry of Land, Infrastructure, Transport, and Tourism, "Japanese Fuel Efficiency Regulation," presentation to the committee by teleconference, Washington, D.C., February 4, 2009.

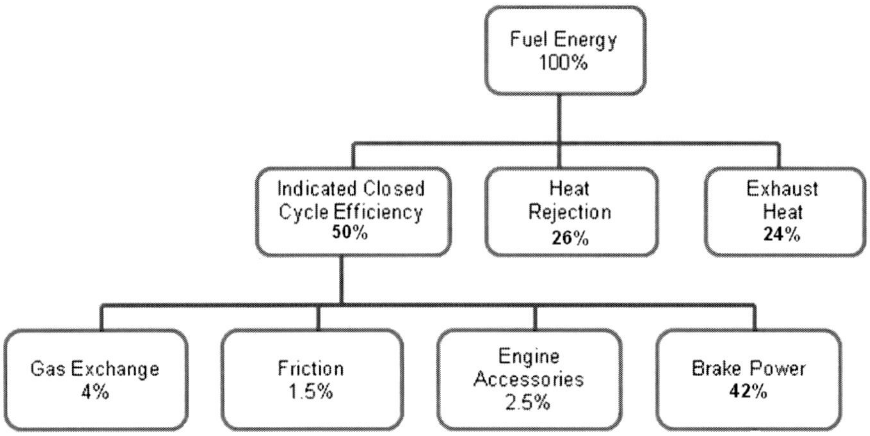

FIGURE 4-1 Energy audit for a typical diesel engine.
SOURCE: Adapted from Vinod Duggal, Cummins, Inc., "Industrial Perspectives of the 21st Century Truck Partnership," presentation to the committee, Dearborn, Mich., April 6, 2009, Slide 14 (and TIAX (2009), p. 4-3, Table 4-1).

2009, pp. 3-5 and 4-14). NESCCAF/ICCT (2009, p. 83) estimates the fuel savings of an improved-efficiency single-stage turbocharger at 1 percent. Another source projects that a high-pressure-ratio axial compressor will reduce fuel consumption by 1.1 to 3.6 percent.[2]

Almost all heavy-duty diesel engines sold in North America today use high-pressure loop EGR for control of engine-out NO_x levels. To get EGR to flow from the exhaust manifold to the intake manifold, the pressure in the exhaust manifold must be higher than the pressure in the intake manifold. When the exhaust manifold pressure is higher than the intake manifold pressure, this is called having a negative Δp, where Δp refers to the difference in pressure between the intake and exhaust manifolds. High-efficiency turbochargers naturally produce a positive Δp over much of their operating range, so turbocharger efficiency must be intentionally compromised in order to facilitate EGR flow. If it is possible to produce adequate EGR flow without reducing turbo efficiency, the overall engine efficiency will increase.

Dual-Stage Turbocharging with Intercooling

Modern engines use high-pressure ratios, which limit the efficiency of turbochargers. Using two turbochargers in series with intercooling would allow higher turbocharger efficiency, but this adds cost and packaging complexity and requires an EGR pump or other device such as a turbocompound system to facilitate EGR flow. Air-to-water intercooling is used after the first-stage compressor, in some applications, and air-to-air aftercooling is used after the second-stage compressor.

Conventional two-stage turbocharging employs two turbochargers working in series at all times. True sequential turbocharging switches turbochargers in and out of use as required, but they are normally connected in parallel. A modulated two-stage system brings some of the benefits of each of these two approaches. At low engine speeds it works as a two-stage system, delivering high-boost pressure despite the low engine speed. At high engine speeds it bypasses the small high-pressure turbocharger, allowing the bigger, low pressure turbocharger to work on its own and produce the higher flows at high engine speeds. Modulated two-stage systems offer the benefits of both high-boost pressure and wide-flow range, mainly due to the fact that two different-sized compressors are used. Using two compressors replicates the effect of a variable compressor without the need for a complex housing. The modulated two-stage system can have a high-pressure turbocharger far smaller than that of a conventional two-stage system, improving transient performance by reducing the turbo lag that affects both drivability and emissions.[3]

Dual-stage turbocharging is used in production by Navistar, Daimler Trucks, and Caterpillar in the United States and by MAN and Mercedes in Europe. Ford has announced that the 2011 diesel engine used in its Class 2b to 7 trucks will use a twin-compressor turbocharger (back-to-back compressors on the same shaft). Another source estimates a 2 to 5 percent reduction in fuel consumption.[4] These benefits are only available if a way to provide the required EGR flow is available.

[2]Personal communication between Steve Edmonds and David F. Merrion, committee member, September 2008.

[3]See www.cummins.com/turbos.

[4]Private communication from S.M. Shahed to David F. Merrion, October 1, 2009.

Mechanical Turbocompound

The base turbocharged engine remains unchanged and a power turbine is added to the exhaust stream to extract additional energy from the exhaust. The power turbine is connected to the crankshaft to supply additional power (NESCCAF/ICCT, 2009, p. 81). Typically, the attachment includes a fluid coupling (to allow for speed variation and to protect the power turbine from engine torsional vibration) and a gear set to match power turbine speed to crankshaft speed. Published information on the fuel consumption reduction from mechanical turbocompounding varies, as evidenced by the following: 3 percent, according to the Detroit Diesel Corporation,[5] which has a turbocompound engine in production; 2.5 to 3 percent (NESCCAF/ICCT, 2009, p. 54); 3 percent (K.G. Duleep, Energy and Environmental Analysis)[6] and 4 to 5 percent (Kruiswyk, 2008, pp. 212-214); TIAX (2009, pp. 4-17) used 2.5 to 3 percent. Some of these differences may depend on the operating condition or duty cycle that was considered by the different researchers. The performance of a turbocompound system tends to be highest at full load and much less or even zero at light load.

Electric Turbocompound

This approach is similar in concept to mechanical turbocompound, except that the power turbine drives an electrical generator (NESCCAF/ICCT, 2009, p. 29). The electricity produced can be used to power an electrical motor supplementing the engine output, to power electrified accessories, or to charge a hybrid system battery. Electric turbocompound is a technology that fits particularly well with a hybrid electric power train for long-haul applications where regenerative braking opportunities are limited. The benefits of electric turbocompound and an electric hybrid power train can be additive. Energy and Environmenal Analyis[7] has said that "electric turbo-compound is more efficient and possible as part of hybrid packages." Fuel consumption reduction benefits as large as 10 percent are claimed. The NESCCAF/ICCT study (p. 54) modeled an electric turbocompound system and estimated benefits at 4.2 percent, including electrification of accessories. Caterpillar, Inc., as part of Department of Energy (DOE) funded work, modeled a system that showed 3 to 5 percent improvement, while John Deere investigated a system (off-highway) that offered 10 percent improvement (Vuk, 2006; TIAX, 2009, p. A-10). None of these systems have been demonstrated commercially. TIAX (2009, pp. 3-5) used a range of 4 to 5 percent for its estimates, which included the benefits of electric accessories. Achieving the full benefit of electric turbocompound requires the electrification of vehicle accessories, the addition of an electric motor to apply turbocompound energy to supplement engine output, and an electric storage system (battery) to store any energy from the power turbine that is not immediately required. Making all of these changes to the vehicle will pose significant development and cost challenges.

Variable Valve Actuation

Variable valve actuation (VVA), also called variable valve timing or discrete variable valve lift, allows the valve actuation to be adjusted independently from the crankshaft angle. There are many implementations of VVA. Some are hydromechanical, such as the system used on some BMW passenger car engines. Other designs use electromagnets or high-pressure hydraulic systems. Some versions offer "full authority," or unlimited, control of valve timing and lift, while other implementations offer limited control, such as variable duration only, variable lift only, or even more limited control, such as with the system used on some Caterpillar engines to permit a Miller cycle to be used under some operating conditions. VVA technology can also be used for cylinder deactivation. One of the primary drivers for introducing VVA in diesel engines is to facilitate the use of nonconventional combustion modes. According to several sources, variable valve timing can improve fuel consumption by about 1 percent when standard diesel combustion is used (NESCCAF/ICCT, 2009, p. 55).

Low-Temperature Exhaust Gas Recirculation (Also Called Advanced EGR Cooling)

Most medium- and heavy-duty vehicle diesel engines sold in the U.S. market today use cooled EGR, in which part of the exhaust gas is routed through a cooler (rejecting energy to the engine coolant) before being returned to the engine intake manifold. EGR is a technology employed to reduce peak combustion temperatures and thus NO_x. Low-temperature EGR uses a larger or secondary EGR cooler to achieve lower intake charge temperatures, which tend to further reduce NO_x formation. If the NO_x requirement is unchanged, low-temperature EGR can allow changes such as more advanced injection timing that will increase engine efficiency slightly more than 1 percent (NESCCAF/ICCT, 2009, p. 62). Because low-temperature EGR reduces the engine's exhaust temperature, it may not be compatible with exhaust energy recovery systems such as turbocompound or a bottoming cycle.

Electrification of Engine-Driven Accessories

Accessories that are traditionally gear or belt driven by a vehicle's engine can be converted to electric power. Ex-

[5]Detroit Diesel Corporation, DD15 Brochure, DDC-EMC-BRO-0003-0408, April 2008.

[6]K.G. Duleep, Energy and Environmental Analysis, "Heavy Duty Trucks Fuel Economy Technology," presentation to the committee, Washington, D.C., December 5, 2008, slide 17.

[7]K.G. Duleep, Energy and Environmental Analysis, "Heavy Duty Trucks Fuel Economy Technology," presentation to the committee, Washington, D.C., December 5, 2008, slide 17.

amples include the engine water pump, the air compressor, the power-steering pump, cooling fans, and the vehicle's air-conditioning system. In many cases this can result in a reduction in power demand, because electrically powered accessories (such as the air compressor or power steering) operate only when needed if they are electrically powered, but they impose a parasitic demand all the time if they are engine driven. In other cases, such as cooling fans or an engine's water pump, electric power allows the accessory to run at speeds independent of engine speed, which can reduce power consumption. Electrification of accessories can individually improve fuel consumption, but as a package on a hybrid vehicle it is estimated that 3 to 5 percent fuel consumption reduction is possible.[8] The TIAX (2009, pp. 3-5) study used 2 to 4 percent fuel consumption improvement for accessory electrification, with the understanding that electrification of accessories will have more effect in short-haul/urban applications and less benefit in line-haul applications.

Engine Friction Reduction

Reduced friction in bearings, valve trains, and the piston-to-liner interface will improve efficiency. Any friction reduction must be carefully developed to avoid issues with durability or performance capability. An example would be to develop heavy-duty diesel engines to run on 10W-30 oil instead of the current standard of 15W-40. The lower viscosity oil would reduce friction, at the expense of bearing capability. Fuel consumption improvement from one source[9] was 2 percent, whereas another source[10] claims 1 to 1.5 percent. The use of a thermatic oil cooler (thermostatically controlled oil cooler) in conjunction with lower viscosity lubricating oils could yield 1.5 percent improvement.[11] The effect of friction reduction and oil temperature control will be greatest during cold starts and under light load operation, where friction accounts for a larger portion of total energy consumption.

Alternative Combustion Cycles

Alternatives to the standard diesel combustion cycle are available, such as low-temperature combustion (LTC), homogeneous charge compression ignition (HCCI), and premix charge compression ignition (PCCI). These combustion modes can be more efficient than standard diesel combustion under some conditions, particularly when very low NO_x is a requirement. There are significant control requirements to make these alternative combustion modes work, and these modes cannot generally be used over the whole operating range of the engine, nor have they demonstrated inherent fuel consumption advantages (NRC, 2008, Finding 3-8, p. 42). The primary purpose of alternative combustion cycles is to lower engine-out emissions. This can lead to either lower overall emissions or lower cost for exhaust aftertreatment.

Effects of DPF and SCR on Engine Efficiency

The use of emissions control devices has an influence on engine efficiency. This is true whether the emissions are controlled on an in-cylinder basis or via the aftertreatment. In most cases, the effect of adding an emissions control device increases fuel consumption, either directly by reducing the efficiency of energy extraction from the combustion process or indirectly by requiring the use of additional fuel to maintain the performance of an aftertreatment system.

Improved SCR Conversion Efficiency

NO_x is formed in a reaction that occurs naturally whenever nitrogen and oxygen are heated above a certain temperature. The higher the temperature, the more rapid the NO_x-forming reaction occurs. In-cylinder technologies to control NO_x formation in diesel engines are aimed at reducing the maximum temperature reached by the gases in the combustion chamber. The approaches used include retarded injection timing, multiple injection events and injection rate shaping, EGR, charge air cooling, and alternative combustion modes (such as HCCI, PCCI, LTC). Some of these approaches leads to a decrease in work output of the engine due to exhaust emissions control (NRC, 2008), except charge air cooling.

The DPF is used to eliminate PM on an aftertreatment basis. A DPF requires energy for regenerating the filter on a periodic basis. This energy most commonly comes from injecting diesel fuel into the exhaust stream. By definition, fuel injected into the exhaust stream will not contribute to crankshaft power and thus represents a decrease in efficiency. A DOC or other device oxidizes the fuel in the exhaust stream, providing the heat required for DPF regeneration and increasing the fuel consumption of the vehicle. Another method to provide the heat required for DPF regeneration is to revise the air/fuel ratio of the engine to produce exhaust constituents and heat that are used to regenerate the DPF. This approach also increases the fuel consumption of the vehicle.

The SCR aftertreatment system for reducing NO_x also requires a fluid, which is urea mixed with water (called Adblue in Europe and DEF [Diesel Exhaust Fluid] in the United States), to supply the reducing agent. The urea is made from natural gas. The energy use of this fluid and/or its cost must be accounted for in the calculation of energy consumption. The use of SCR can allow a higher engine-out NO_x level, which in turn can be used to reduce fuel consumption, but

[8] Anthony Greszler, Volvo Powertrain, "Reducing Emissions in Heavy Vehicles," presentation to the committee. Washington, D.C., December 5, 2008, slide 23.

[9] Anthony Greszler, Volvo Powertrain, "Reducing Emissions in Heavy Vehicles," presentation to the committee. Washington, D.C., December 4, 2008, page 14.

[10] Site visit to Daimler/Detroit Diesel, April 7, 2009.

[11] Site visit to Daimler/Detroit Diesel, April 7, 2009.

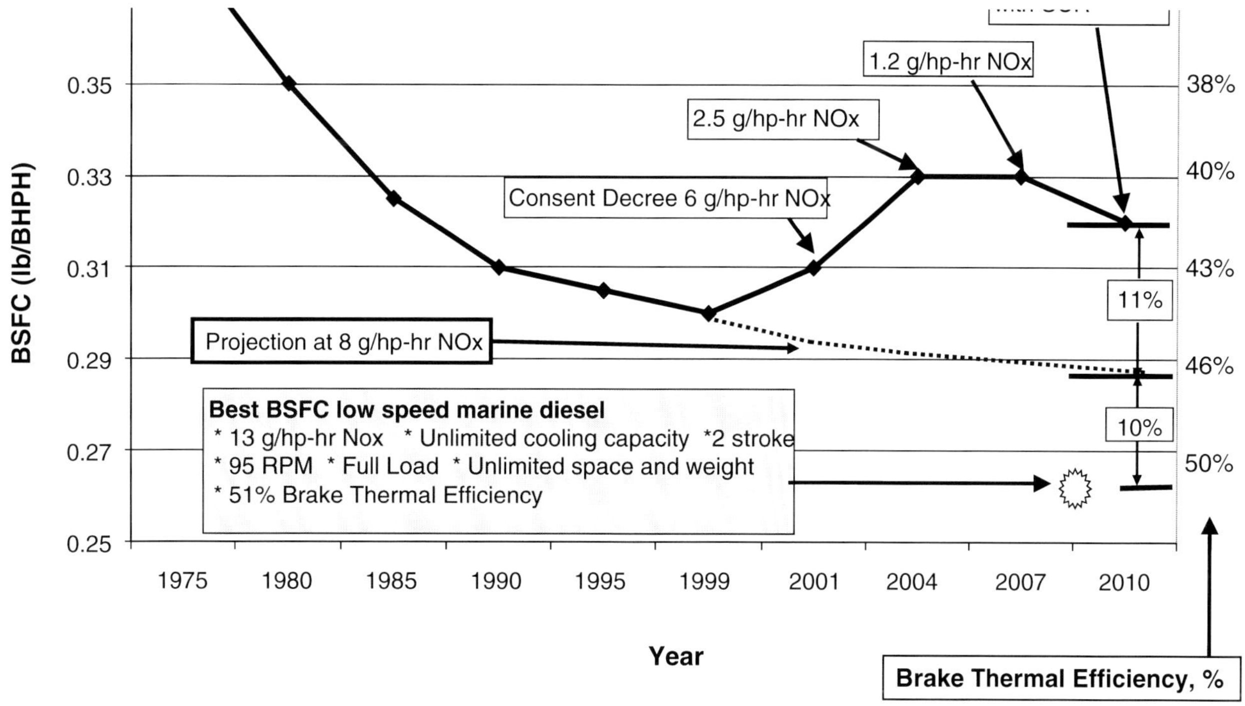

FIGURE 4-2 Historical trend of heavy-duty truck engine fuel consumption as a function of NO_x requirement. SOURCE: Tony Greszler, Volvo, October 2009.

this improvement must be weighed against the urea consumption of the SCR aftertreatment system. The upcoming 2010 heavy-duty emissions standards reduce the allowable NO_x level by a factor of 6 from 1.2 g/hp-hr to 0.2 g/hp-hr, which limits the ability of some manufacturers to use SCR to increase engine-out emissions.

There is a close relationship between emissions control requirements and fuel consumption. In particular, certain technologies that are used to control NO_x emissions have the effect of increasing fuel consumption. See Figure 4-2 for an example of this trend. Figure 4-2 also compares truck engines with the most efficient large marine engines, which so far do not face any emissions constraints. The efficiency of large marine engines is due to several factors that cannot be reproduced in vehicle applications. Marine engines are very large and heavy, they run at very low speeds, they have a source of unlimited cooling capacity (seawater), and they face (for the time being) no emissions constraints. All four of these factors contribute to the high efficiency of marine engines. According to information provided by Volvo,[12] the fuel consumption of truck diesel engines is about 10 percent higher than for marine engines due to the size, weight, and cooling factors that are limits faced in vehicle applications. The 2010 model truck engines will suffer an 11 percent fuel consumption penalty for NO_x control (compared with an 8 g/hp-hr NO_x engine, see Figure 4.2), which is less than the penalty of 2002 and 2007 engines. This penalty does not include the energy content of DEF (urea) for the SCR system.

Most 2010 engines will use SCR aftertreatment to reduce NO_x emissions. Daimler's Detroit Diesel estimates that, with the installation of SCR in 2010, fuel consumption will be reduced 3 to 4 percent by a combination of higher engine-out NO_x, controls and fuel system improvements, reduced DPF regeneration frequency, and other efficiency gains, while Volvo estimates the potential improvement of its 2010 engines at 2 percent. As the SCR system conversion efficiency improves, it allows higher engine-out NO_x emissions. Engine manufacturers can take advantage of this by making changes in fuel injection timing or by using less EGR, in order to reduce fuel consumption.

Thermal Insulation of Ports and Manifolds

Thermal insulation would reduce heat rejection to the engine coolant (from exhaust ports) or to the ambient air (from manifolds). The energy retained in the exhaust can

[12]Tony Greszler, Volvo, private communication with Thomas Reinhart, October 2009.

be used by downstream devices such as a turbocompound system or bottoming cycle. Caterpillar Inc. made components, such as air gap pistons and exhaust port liners as part of the 21st Century Truck Partnership, but reported results are not available (NRC, 2008, p. 30). The anticipated benefit is small.

Improved Work Extraction from Combustion Process

The compression ratio, expansion ratio, combustion chamber shape, injection spray pattern, injection pressure, injection timing, injection rate shaping, air/fuel mixing, peak cylinder pressure limit, air/fuel ratio, and EGR rate are all parameters that can be modified in an effort to reduce fuel consumption. Improved combustion chamber design allows for improved air management and mixing. Improved materials and structural design enable higher cylinder pressures. These enhancements allow more precise control of the timing and rate of heat release (combustion) as well as higher combustion temperatures, both of which can improve thermal efficiency. Unfortunately, higher combustion temperatures also lead to higher NO_x. Combustion chamber design enhancements may require more advanced materials and a more complex machining process. More complex and expensive fuel systems allow greater control of injection pressure, timing, and rate shaping. In addition, because higher cylinder pressures must not result in higher NO_x, measures must be taken to allow the improved fuel consumption without creating an increase in emissions. These measures may include improved NO_x conversion efficiency by the aftertreatment, advanced fuel injection techniques (which enable more detailed control of combustion), and improved engine controls. The efficiency benefit of these improvements is estimated at 1 to 3 percent.[13]

Finely controlled, high-pressure fuel injection is a key enabler for more fuel efficient combustion and a cleaner, more consistent fuel burn. Current state-of-the-art systems planned for deployment in 2010 engines include very high pressure (2,000 to 2,400 bar) common rail injection systems with advanced nozzle designs that are capable of finely shaped and controlled spray, along with multiple injection events per cycle.

Potential future improvements will continue to improve control, allow more accurate timing and metering of injection with combustion events, and further increase fuel injection pressure. Improved material properties and controls could enable pressures of up to 3,000 bar in the 2015 time frame and perhaps 4,000 bar by 2020. Future systems will also utilize increasingly sophisticated injection techniques such as variable-spray nozzles, piezo-electric nozzles, or supercritical fuel injection (fuel changing instantaneously from liquid state to supercritical gaseous state at injection based on site visits). These advances may be possible in the 2013 to 2015 time frame (TIAX, 2009). Fuel injection systems were estimated on site visits to offer between 1 and 4 percent improvement in fuel consumption; Vyas et al. (2002) estimates fuel injection systems have the potential to improve fuel consumption by 6 percent, although this estimate is now several years old, and considerable improvement in fuel systems has already been made since 2002. Real-time combustion control with start of combustion sensors can also yield a fuel consumption reduction of 1 percent to 4 percent.[14]

Engine Electronic Controller Calibration Management

Advanced engine controls will be enabled in part by the onboard diagnostic systems that are mandated for medium and heavy trucks beginning in 2010 on one family and across the board in 2013. Increasingly sophisticated engine controls, particularly a transition to closed-loop control approaches, will enable engine efficiency improvements. Closed-loop controls will feed information about the engine's operating regime and emissions back to the system controls. This improved feedback will allow manufacturers to optimize emissions and fuel consumption within the constraints of emissions requirements across a variety of operating conditions.

Better use of calibration tools to improve control of EGR, injection rate shapes, multiple injection events, and increased injection pressure can yield 1 to 4 percent fuel consumption reduction. These benefits are redundant with those described above for improved work extraction from the combustion process. Another feature already in use on some long-haul trucks is adding 200 lb-ft of torque in the top two transmission gears, which manufacturers claim can give 2 percent reduction in fuel consumption by reducing the need for downshifting on modest grades. With the next generation electronic controller, using model-based controls, it is predicted that another 1 to 4 percent fuel consumption reduction will be achieved.[15] Note that the reductions listed in here may be repeats of reductions from previous sections, and there may be some redundancy in the percentages quoted, but the concepts and percentages presented here came from committee site visits where engineers talked about these reductions. The overall approach of more sophisticated control of the combustion process tends to include several building blocks, such as upgraded fuel system capabilities, sophisticated control algorithms, additional sensor inputs for feedback control, and technologies such as model-based controls. The benefits of this approach are often claimed by each of the individual building blocks, leading to redundant claims on the same fuel consumption benefit.

[13]Presentations by and discussions with Cummins, Detroit Diesel, and Volvo.

[14]Committee site visit, Daimler/Detroit Diesel, April 8, 2009.

[15]Committee site visit, Daimler/Detroit Diesel, April 8, 2009.

Bottoming Cycle

A bottoming cycle is basically a secondary engine that uses exhaust energy or other heat sources from the primary engine to develop additional power without using additional fuel. The energy sources used by the bottoming cycle are sources that normally go to waste in a conventional engine. A typical bottoming cycle includes the following components: a feed pump to drive the working fluid from the condenser to the evaporator (or boiler); the evaporator, which transfers waste heat energy from the primary engine to the working fluid; an expander, which takes energy from the working fluid to make mechanical power; and a condenser that rejects unused heat energy from the bottoming cycle working fluid before starting a new cycle. The power generated by the expander can be used to make electricity, which in turn can power an electric motor supplementing the engine output, power electrified accessories, or charge a hybrid system battery. Sources of energy to power a bottoming cycle can include the EGR stream, exhaust stream, charge air stream, and engine coolant circuit (NESCCAF/ICCT, 2009, pp. 85-88). Cummins, Inc. has shown a projected increase of thermal efficiency from 49.1 to 52.9 percent (7.2 percent decrease in fuel consumption) using an organic Rankine cycle. Cummins also lists turbocompounding and a Brayton cycle as alternative methods of extracting work from unused energy in the exhaust stream. Cummins reports recovering 2.5 thermal efficiency points from the exhaust and 1.3 thermal efficiency points from the coolant and EGR stream.[16] The NESCCAF/ICCT report (2009, pp. 55-56) showed the effect of a steam bottoming cycle to reduce fuel consumption by up to 10 percent.

Other Technologies

Other technologies for reducing the fuel consumption of diesel engines are discussed in the press almost every day (e.g., *Automotive News*, *Transport Topics*, *Diesel Fuel News*, *DieselNet.com*). Some are emerging technologies and may not become production feasible, including new diesel engines of two-stroke-cycle design, split-cycle design, free-piston design, rotary design and camless engines with digital valve control such as the Sturman Industries concept. The list of potential technologies also includes oxygen injection into the intake air, hydrogen injection into the intake air, air injection from the air compressor to overcome turbo lag, or the use of fuel-borne catalysts such as platinum and cerium.

Diesel Engine Summary

In summary, to add up all these individual potential reductions to arrive at an overall potential fuel consumption reduction would not be correct because there would be double counting of some effects. The best recent attempts at packaging fuel-saving technologies for engines were in the DOE programs with a goal of demonstrating 50 percent thermal efficiency while meeting 2010 emissions. The National Research Council (2008) review of that program showed a baseline thermal efficiency of 42 percent with a goal of 50 percent, or a 19 percent improvement in thermal efficiency and a 16 percent fuel consumption reduction. Three engine manufacturers—Caterpillar, Cummins, and Detroit Diesel—were funded at a level exceeding $116 million over five years with the following result, according to the report: "These results show that none of the industry partners achieved the goal of measuring 50 percent thermal efficiency at 2010 emissions from a complete engine system."

Cummins has supplied the committee with the following research roadmaps for achieving 49.1 percent thermal efficiency and 52.9 percent thermal efficiency. Figure 4-3 is a research roadmap for 49.1 percent thermal efficiency by 2016, which is an improvement of 17 percent from the 42 percent baseline (14.5 percent reduction in fuel consumption). Figure 4-4 is a research roadmap for 52.9 percent thermal efficiency by 2019, which is an improvement of 26 percent from the baseline 42 percent (20.6 percent reduction in fuel consumption). These roadmaps can be compared to the baseline shown in Figure 4-1. Note that these are plans and goals, not actual development results. Actual results that will be achieved in development may vary from the planned benefits.

In its report for the committee, TIAX (2009, Tables 5-8 and Table 5-9) summarized the diesel engine fuel consumption potential reductions by range of years and by application as shown in Table 4.1.

GASOLINE ENGINE TECHNOLOGIES

Gasoline engines operate with a premixed charge of fuel and air and use spark ignition to start the combustion process. They are used in many Class 2b applications as well as Class 3 to 6 applications. Within the medium-duty truck sector, all gasoline engines operate on the four-stroke-cycle principle and are of an in-line or "vee" configuration. Displacements of these engines typically range from 6 to 8 liters. These engines normally burn gasoline, but with slight changes they can burn natural gas (compressed [CNG] or liquefied [LNG]), propane, hydrogen, ethanol, methanol, and so forth.

The fundamental operating principle for gasoline engines used today relies on creating a well-mixed charge of gasoline and air at the time the spark plug fires. After the combustion process is over, a catalyst in the exhaust system is used to perform the final emissions cleanup. Emissions of NO_x, carbon monoxide, and unburned hydrocarbons are the principal species being treated by the catalyst in the exhaust. The three-way catalyst treats all three of these emissions simultaneously; however, the three-way catalyst will function

[16]Jeff Seger, Cummins, Inc., at committee site visit, May 15, 2009.

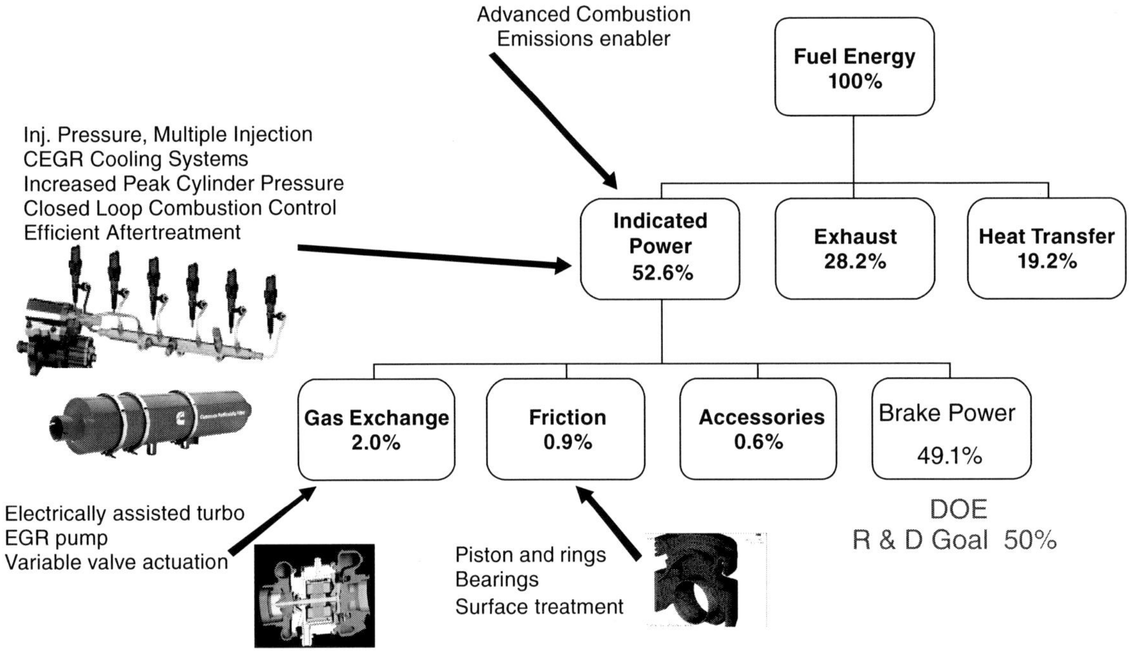

FIGURE 4-3 Research roadmap for 49.1 percent thermal efficiency by 2016. SOURCE: Provided under license by Cummins Inc. Copyright 2009 Cummins Inc. All rights reserved.

properly only if the air/fuel ratio is carefully controlled to be chemically correct, or stoichiometric.[17]

Heavy-duty gasoline engines, and most heavy-duty engines using spark-ignited alternative fuels such as natural gas, use a relatively simple emissions control strategy. The engine operates at a stoichiometric air/fuel ratio across most of the operating range, with relatively high engine-out emissions levels. A three-way catalyst is used both to oxidize hydrocarbon and carbon monoxide emissions and to reduce NO_x. The three-way catalyst can function properly only if the air/fuel ratio is carefully controlled in order to meet current and future emissions requirements. Additional considerations for gasoline engine emissions control include achieving rapid catalyst light-off on startup and controlling evaporative emissions, but the basic emissions control technology for spark-ignited engines is the relatively simple and inexpensive three-way catalyst.

There is a fuel consumption penalty that comes with the three-way catalyst used on gasoline engines. Because the air/fuel ratio must be maintained at stoichiometric all the way down to idle, the pumping losses from throttling are large. Lean operation could provide significant fuel savings but would not allow the NO_x reduction function of the three-way catalyst to work. Many technologies that could be applied to gasoline engines to reduce fuel consumption are not used, primarily because of the need to maintain low NO_x emissions. For example, lean gasoline direct injection (GDI) has the potential to provide double-digit percentage reductions in fuel consumption, but it is not used, because it would result in higher NO_x emissions. The NO_x emissions of a lean GDI engine could be much lower than those of an unregulated engine, but engine makers have not been able to make lean GDI reach the very stringent U.S. NO_x standards applied to new cars and trucks today. GDI has been used in Europe, where NO_x is less stringently regulated.

One consequence of requiring a stoichiometric mixture of air and fuel is that the intake airflow needs to be throttled for lighter load operation. Lighter loads necessitate a lower fuel flow rate into the engine, and since the air/fuel mixture is to be maintained in stoichiometric proportions, the airflow rate needs to be reduced in proportion to the fuel flow rate. The process of throttling the intake airflow results in significant pumping losses that are not present in diesel engines that operate using traditional diesel combustion. This pumping loss is one of the main reasons spark ignition engines are less efficient than diesel engines. The magnitude of the pumping loss depends on the operating duty cycle of the engine. If the engine spends most of its time in light load operation, its throttling losses will be higher than for an engine that spends most of its operation under heavier load. Also, the pumping work will depend on the engine size relative to the vehicle. A smaller engine size in a given vehicle application will spend a higher portion of its operation at a higher load, relative to a

[17]Stoichiometric refers to the chemically balanced reaction of air and fuel. Under stoichiometric conditions there is a certain amount of oxidant (air) such that all of the carbon in the fuel could react to carbon dioxide and all the hydrogen in the fuel could react to water, with no oxidant or fuel left in the products.

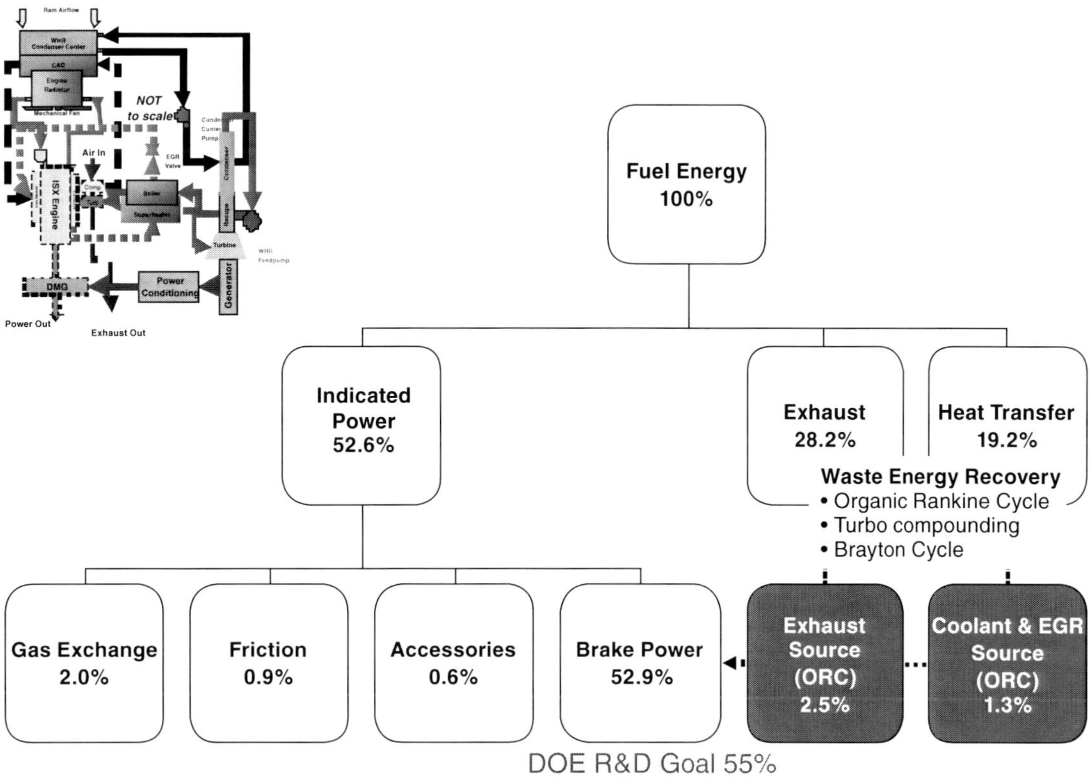

FIGURE 4-4 Research roadmap for 52.9 percent thermal efficiency by 2019. SOURCE: Provided under license by Cummins Inc. Copyright 2009 Cummins Inc. All rights reserved.

TABLE 4-1 Diesel Engine Fuel Consumption (percentage) by Years and Applications

Application	2013-2015	2015-2020
Tractor trailer	10.5	20
Class 6 box truck	9	14
Class 6 bucket truck	7.2	11.2
Refuse truck	10.5	14
Urban bus	9	14
Motor coach	10.5	20
Class 2b pickup and van	14	23

SOURCE: TIAX (2009).

larger engine in the same vehicle, and consequently will have a lower pumping loss than the corresponding larger engine. As an approximate guide, pumping losses might range from 2 to 5 percent of the fuel energy (Patton et al., 2002).

Compared to diesel engines, spark ignition engines are generally simpler and less expensive, they have more effective and lower cost exhaust emissions aftertreatment systems, and they have higher fuel consumption.

The current emphasis in the development of spark ignition engines is on reducing fuel consumption. Figure 4-5 gives a qualitative partitioning of the fuel energy for a typical gasoline-fueled vehicle. This is analogous to Figure 4-1, which gives an energy partitioning for diesel-powered vehicles. Figure 4-5 is illustrative in describing the technologies being considered to reduce gasoline engine fuel consumption.

The proportion of the fuel energy that gets converted into indicated work is a direct measure of the engine's fuel conversion efficiency. Factors that affect an engine's fuel conversion efficiency include irreversibilities[18] in the combustion process, the amount of energy leaving the engine cylinder as heat transfer, and the energy remaining in the exhaust at the end of the expansion process. These losses represent fuel energy that did not get converted into useful shaft work. Not all of the energy that was converted into work in the combustion process makes it to the final shaft output.

[18]Irreversibility is a thermodynamic concept. It is used to describe and quantify the degree of imperfection in any real process. In the context used here it describes the degradation of energy during the combustion process into a form that is less capable of being converted into work. Theoretically it is possible to convert all of the chemical energy contained within the fuel completely into work. Inherent in the chemical reaction of the actual combustion process are irreversibilities that render the resultant thermal energy of the combustion products not completely available to be converted into work, even though the quantity of energy is conserved.

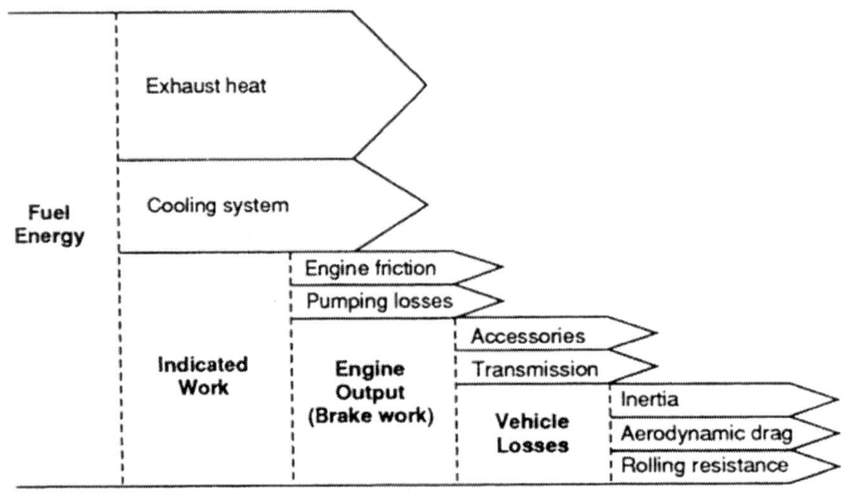

FIGURE 4-5 Partitioning of the fuel energy in a gasoline-fueled engine (proportions vary with vehicle design, type of engine, and operating conditions). SOURCE: NRC (1992).

Some gets used to overcome friction, some is used to pump air and fuel into the engine, and the exhaust gases out of the engine, and some is used to power accessories. The work that makes it to the drive wheels is used to overcome vehicle inertia, aerodynamic drag, and rolling resistance. The relative ranking of these energy uses is highly dependent on the vehicle and the application to which it is being applied.

As indicated in the discussion of the pumping work above, the magnitude of these energy partitions is highly dependent on the engine size, its application in the vehicle, and the duty cycle under which the vehicle is operating. The best way to quantify the partitioning shown in Figure 4-5 would be to take the specific data for the application of interest or through the application of a verified system simulation program.

Opportunities to reduce the fuel consumption of gasoline engines include improving engine efficiency (trying to reduce the proportion of the fuel energy leaving as heat transfer, exhaust energy, and pumping work), reducing the energy lost to friction, and reducing the power required for running accessories. Brief descriptions of different technologies for reducing fuel consumption are given below.

Variable Valve Actuation and Cylinder Deactivation

There are many approaches to VVA. These include cam phasers, variable lift mechanisms, fully flexible valve trains, and cylinder deactivation. The primary loss that the VVA systems are trying to reduce is the pumping, or throttling, loss. A variable compression ratio allows the engine to operate at different compression ratios for different loads in order to maximize the engine efficiency over the widest load range possible. The combination of VVA and variable compression ratio keeps the engine operation closer to its maximum efficiency point, with minimal pumping work, over larger portions of the duty cycle.

Cylinder deactivation is an approach that minimizes pumping losses by varying the total working displacement of the engine. The "smaller" engine operates closer to wide-open throttle at lower loads, which reduces the pumping work.

Cam phasers allow the valve timing to be changed with engine operating conditions. For example, the timing can be shifted with engine speed to optimize the engine breathing with engine speed. VVT can also be used in place of the intake throttle. Either opening the intake valve late or closing it late can regulate the amount of air/fuel mixture captured in the cylinder. This can be done with lower throttling losses than would occur with the conventional intake throttle.

Different engine designs will lend themselves more easily to different valve control technologies. For example, overhead valve systems with the cam in the block versus a single overhead cam versus a double overhead cam design will have to invoke different VVA approaches, which may favor different valve manipulation strategies for the different engine configurations. So decisions as to which valve technology to invoke would be based on the engine configuration, the application's duty cycle, and the incremental cost of implementation. Estimates for the fuel consumption reduction achievable through variable valve lift and/timing range from 1 to 3.5 percent (TIAX, 2009, p. 4-33).

Gasoline Direct Injection Engines

GDI engines refer to an embodiment in which the fuel injector is mounted so that the fuel is directly injected into the cylinder, as opposed to the more common port fuel injection, where the fuel injector is mounted in the intake port. There are two philosophies of engine operation that fall under the classification of GDI engines.

In one case the engine is still operated with a stoichiometric mixture of air and fuel. This approach enables a three-way catalyst to be used in the exhaust, so emissions standards can be met. However, by mandating operation with a stoichiometric mixture, the engine still needs to be throttled. For a direct-injected stoichiometric engine the efficiency improvements come from less fuel being used during transient engine responses and from internal cooling of the cylinder charge from the fuel vaporization. This leads to a higher knock margin, which allows a higher compression ratio to be used. These engines are also more tolerant of EGR, so higher compression ratios can be used without an NO_x penalty. There will be an attendant efficiency improvement with the higher compression ratio. The reduction in fuel consumption achieved via stiochiometric direct injection will be dependent on the extent to which this technology is combined with other technology packages. Based on estimates from a light-duty study (NHTSA, 2009), referenced by TIAX, a fuel consumption reduction with stoichiometric direct injection engines relative to a port-injected engine with VVT described above would range from 2 to 3 percent (TIAX, 2009, p. 4-33).

The other approach to GDI is to attempt to replicate the breathing characteristics of the diesel engine by minimizing the throttling of the intake air and controlling the load by varying the air/fuel ratio. Reduction in fuel consumption with this approach to direct injection comes from reduced pumping losses and higher efficiency from the lean-burning mixture, as well as the potential to increase the compression ratio as with the stoichiometric direct injection. The drawback to this approach is that the three-way catalyst is no longer effective, so the engine has the same aftertreatment challenges and expense as the diesel engine. The application of the lean burn direct injection technology will almost certainly be coupled with using turbocharging (described below) as well, so one must consider the combination of technology packages. Referencing the same NHTSA report referred to above, TIAX reported a potential reduction in fuel consumption by applying turbocharged lean burn GDI technology relative to the VVA nonturbocharged stoichiometric engine of 10 to 14 percent (TIAX, 2009, p. 4-33).

Different Combustion Modes

VVA mechanisms and fuel injection systems open up the possibility of incorporating advanced combustion regimes into the engine operating map. For example LTC, a general classification for auto-igniting combustion modes such as HCCI, partially PCCI, and compression-aided ignition, offers potential for lean low-emissions combustion in which throttling losses are minimized and catalytic converters are not needed (Zhao et al., 2003). Incorporating these alternative modes of combustion into the engine's operating map will require much higher levels of sensing and control than are currently in use in today's engines. For example, it is likely that real-time cylinder pressure sensing would be necessary to activate and control transitions between conventional combustion and LTC-type combustion during engine operation. If these combustion regimes can be incorporated into the operating map of gasoline engines, fuel consumption could be reduced via lean combustion at light loads, with minimal pumping losses, without the need for lean exhaust aftertreatment. Then during higher load operation, where throttling requirements are low, the engine could revert back to stoichiometric operation where the three-way catalyst is effective. It has been estimated that incorporation of LTC operation into the engine could reduce fuel consumption by 10 to 12 percent (TIAX, 2009). See Table 4-2.

Turbocharging and Downsizing

Turbocharging a gasoline engine is similar to turbocharging a diesel engine in that it is motivated by the desire to redirect energy that was leaving the engine in the exhaust gases back into the engine. The turbocharger converts exhaust energy into higher pressure and temperature intake gases using a turbine in the exhaust gas stream and a compressor in the intake air stream. Because of the differences in the fuels and the combustion processes between the two types of engines, there are different constraints that limit the application of turbocharging in a spark ignition engine relative to a compression ignition engine. Because the air and fuel are premixed in a spark ignition engine, as the pressure and temperature of the mixture at the start of compression is increased via turbocharging, the possibility of knocking combustion increases. Consequently it is common for the compression ratio to be decreased when the spark ignition engine is turbocharged. This tends to decrease the thermal efficiency of the engine. It is also likely that more exhaust gas will be recirculated when the engine is turbocharged. However, if successfully implemented, the turbocharged engine will have a higher power density than the nonturbocharged engine so it can be smaller and lighter for the same power output. This reduction of weight and potential reduction in frontal area of the vehicle has an attendant fuel consumption reduction benefit.

Turbocharging can be combined with other technologies, like in-cylinder direct injection and/or VVA to compound the benefits of the various technologies. The direct injection of fuel results in in-cylinder evaporative cooling, which helps counter the increased tendency toward knocking attributable to the turbocharging. Consequently the compression ratio may not need to be lowered to avoid knocking combustion to the same extent it would with turbocharging without direct injection. This coupled with the increased EGR tolerance of the direct injection and turbocharged engines helps keep NO_x emissions from rising. In addition, because the power density of the engine has been increased, it may be possible to run at lower speeds and still generate the power required to drive the vehicle. The lower engine speeds result in lower friction losses, which is beneficial to fuel consumption. So it is possible to combine many of the technologies to compound their benefits. The technology review by TIAX found that a turbocharged downsized engine offers a fuel consumption reduction potential of approximately 2 percent, while a turbocharged downsized direct injection engine with VVA offers a potential 10 to 14 percent reduction in fuel consumption (TIAX, 2009).

Once such technologies as turbocharging, direct injection, and variable valve actuation, and so forth, have been incorporated into the power train, the engine becomes quite flexible. For example, with such flexibility it is possible to use multiple fuels which that are carried separately onboard the vehicle. Approaches such as using gasoline and E85 (Stein et al., 2009) or varying mixtures of gasoline and diesel (Hanson et al., 2010) have been demonstrated. In these demonstrations, fuel consumption levels comparable, or superior to, that of conventional diesel engines have been achieved. The fundamental concept is one of optimizing the fuel characteristics and engine operating conditions for best performance—for example, the lowest fuel consumption and emissions, at each operating point. Such an engine would have characteristics of both gasoline and diesel engines. The duel-fuel concept would require infrastructure changes to address dispensing two separate fuels to a vehicle.

Accessory Loads

As with diesel-powered vehicles, decoupling the direct drive of the accessories from the engine offers the potential to optimally match the accessory use to the duty cycle. Examples of this technology that offer potential to reduce fuel consumption by reducing the amount of engine shaft work that was originally going to the driveline but was tapped to run the accessories instead include electrically driven engine oil and cooling pumps, mechanically clutched or electrically driven radiator cooling fans, electrically driven air-conditioner compressors, and high-efficiency alternators. One benefit of accessory load reduction is that the work saved through accessory optimization goes directly to the vehicle. The fuel-savings potential is highly dependent on the duty cycle of the engine.

Integrated Technologies

It is important to realize that the technologies described above may not be discrete. For example, using VVA for intake throttling reduction along with cylinder deactivation for engine displacement reduction are technologies motivated by the desire to reduce pumping work. Using them together will not result in fuel consumption reductions that are additive. Also, some technologies are facilitators for addressing multiple losses. For example, electric hybrids enable one to downsize the engine, operate the engine over a narrower and optimal portion of it operational map, and reduce accessory load on the engine through accessory electrification.

As discussed above, the fuel savings that comes from reducing any of the losses described above will be highly dependent on the vehicle itself and its duty cycle. To evaluate the potential for the various technologies to reduce the vehicle's fuel consumption, it is best to evaluate the technologies as integrated packages applied to a specific vehicle operating on a representative duty cycle. For example, consider the package cited above of a downsized turbocharged direct injection engine with VVA. Engine and vehicle simulation packages exist that are capable of giving good comparative rankings of the fuel consumption reduction potential of different combinations of technologies.

One such example of a technology assessment program is HEDGE (High Efficiency Dilute Gasoline Engines) at SwRI.[19] In this program a combination of technologies is being evaluated, including a downsized, turbocharged, direct injection, high-EGR, VVA engine, which is fueled with either gasoline or gasoline and diesel (dual fuel), with an emphasis on lower speed operation. Results presented to the committee were encouraging. Laboratory tests showed combinations of technologies that demonstrated gasoline engine thermal efficiencies greater than 40 percent, with fuel consumption reductions on the order of 20 percent, with 35 and 45 percent reductions in particulate matter and NO_x emissions, respectively. It is evident from this work that, as stated above, it is the synergistic combination of various technologies, matched to engine duty cycle, that are critical for achieving reduced fuel consumption with low emissions. Note that introducing such a technology package may require a significant redesign of the engine. An ignition system tolerant of a high-EGR rate is required, and cylinder pressures are higher than in typical gasoline engines.

Shifting to Diesel Engines

Shifting from gasoline to diesel engines offers significant fuel efficiency benefits. This is primarily due to higher compression ratios and reduced gas exchange losses. However, it should be noted that, due to emissions regulations (which have degraded the efficiency of diesels), and advances in spark ignition technology (e.g., HCCI, turbocharging, direct injection), the gap between gasoline and diesels has narrowed considerably. The committee compares diesel and gasoline engines in more detail in the next section.

[19]Chris Chadwell, SwRI, "High Efficiency Dilute Gasoline Engines (HEDGE) for Medium Duty Applications," presentation to the committee, San Antonio, Tex., August 7, 2009.

TABLE 4-2 Technologies for Fuel Consumption Reduction Applicable to Gasoline-Powered Engines for the Medium-Duty Vehicle Class and Estimated Fuel Consumption Reduction and Incremental Costs

		Incremental Cost and Benefit	
Technology	Incremental to...	Percent Fuel Consumption Benefit	Cost ($)
Friction reduction	—	0.5-2.5	110-500[a]
Variable valve timing (VVT)	Friction	1-3	122
Variable valve lift (VVL)[b]	VVT	1-3.5	400-750
Cylinder deactivation[c]	VVL	2.5-3	75
Stoichiometric GDI[d]	Cylinder deactivation	2-3	512-930
Turbocharging and downsizing	S-GDI	2.1-2.2	1,229
Diesel[e]	—	19-24	7,900-9,400
Lean-burn GDI	S-GDI	10-14	750[f]
Gasoline HCCI	S-GDI	10-12	685
Accessory electrification	Applicable to any package	2-4	1,000-2,000 (current) 500 (high volume)

NOTES AND ASSUMPTIONS:
Baselines are explained in the text in the section "Summary of Technologies for Gasoline Engines." The duty cycle is the typical duty cycle of a medium-duty vehicle with a gasoline engine. The diesel fuel consumption benefit includes the 10 percent higher heating value of diesel fuel. "Friction reduction" includes mechanical accessory improvements. Values assume (roughly) constant performance. The time frame covered is 2015 to 2020.

[a]$13 to $49 per cylinder + $5 for lubricants.

[b]$51 per cylinder for discrete VVL; $70-$75 per cylinder for continuous VVL.

[c]Offers marginal benefit on DOHC engines; these benefits reflect SOHC.

[d]$64 to $93 per cylinder.

[e]Diesel fuel consumption incremental to baseline gasoline engine.

[f]Incremental cost over direct injection stoichiometric engine.

SOURCE: TIAX (2009), p. 4-33.

Summary of Technologies for Gasoline Engines

As part of the committee's activities, TIAX was contracted to assemble and categorize different engine technologies, their potential fuel consumption benefit, and their incremental costs. Table 4-2 is a tabulation of the technologies discussed above, their potential fuel consumption benefits, and their incremental costs; the baseline engine on which the estimates of fuel consumption are based is a port fuel-injected, naturally aspirated, fixed-valve-timing engine with 8 to 10 cylinders.

DIESEL ENGINES VERSUS GASOLINE ENGINES

Diesel engines offer significant fuel savings over gasoline engines, when measured on a fuel consumption (gallon per mile) or load-specific fuel consumption (gallons per ton-mile) basis. Depending on the engines and operating conditions, diesel engines can provide 19 to 24 percent lower fuel consumption than gasoline engines. Two factors account for this advantage:

- Diesel fuel has approximately 12 percent higher energy content than gasoline, which allows diesel engines to extract more energy per gallon from the fuel.
- Diesel engines have a higher thermal efficiency than gasoline engines, which allows diesel engines to provide more shaft power for a given amount of fuel energy released.

Several factors contribute to the higher thermal efficiency of diesel engines. One is the lack of a throttle on traditional diesel engines. Pulling air across a closed throttle imposes significant pumping losses, so diesel engines enjoy a significant advantage in pumping loss, especially at light load. Another factor that contributes to the higher thermal efficiency of diesel engines is high compression ratio. This is made possible by the compression ignition cycle of the diesel. The higher expansion ratio that comes with a higher compression ratio allows the diesel to extract more of the combustion energy before the exhaust valve opens near the end of the power stroke. Because of the high cylinder pressures encountered as a result of high compression ratio, turbocharging, lean air/fuel ratios, and the use of EGR, the diesel engine has the disadvantage of being heavier (pounds/cubic inch) and more costly to manufacture. The fuel system of diesel engines also contributes significantly to the cost penalty of diesel engines.

The most common fuels for medium- and heavy-duty engines are gasoline and No. 2 ultralow-sulfur diesel fuel. Both spark-ignited engines and compression ignition engines can burn "other" fuel types as discussed earlier, but most common are renewable fuels (as required by the Renewable Fuels Standard of the Energy Policy Act of 2007), such as ethanol for spark-ignited engines and biodiesel for compression ignition engines. Both of these renewable fuels have lower heating values (Btu/gallon) than their counterpart gasoline and diesel fuel, resulting in higher fuel consumption when measured on a volume basis (liters or gallons). Natural gas is also used in gasoline and converted diesel engines but is low in heating value and used in gaseous form, requiring the expression of fuel consumption data on an equivalent energy basis.

Another factor that may arise if diesel engines are substituted for gasoline engines, or renewable fuels play a bigger role in the mix of fuels used for medium- and heavy-duty vehicles, is the impact on petroleum refineries. Each barrel of petroleum produces various proportions of fuels, such as diesel, gasoline, kerosene, fuel oil, and others, to supply the demand for these fuels. If the fuel market were to shift significantly in the demand for diesel fuel versus gasoline, e.g., refineries would need to modify their processes to change the mix of fuels produced from each barrel of petroleum refined.

Both gasoline and diesel engines incorporate exhaust emissions control systems for hydrocarbons HCs, CO, and NO_x. Gasoline engines also control evaporative emissions and diesel engines control PM. These emissions are controlled by both the engine combustion process (engine out) and the use of catalytic converters, DOCs, DPFs, lean NO_x catalysts, and SCR.

The emissions control approach used with heavy duty diesel engines has been quite different from spark-ignited engines. All strategies used for on-highway trucks through 2009 relied on in-cylinder controls, including EGR, to limit NO_x, rather than on aftertreatment. The 2007 U.S. Environmental Protection Agency (EPA) heavy-duty on-highway standards forced the use of DPFs to control PM. DPFs are the first widely used aftertreatment system on diesel engines, although some diesel engines have also used oxidation catalysts to control HC emissions. For 2010 most diesel truck manufacturers plan to add SCR aftertreatment to meet the new 2010 NO_x requirements.

Volvo is the first company to publicly price heavy-duty diesel vehicles with 2010 emissions control systems. The company plans to charge $9,600 for the SCR system on 2010 model heavy-duty trucks (Fleet Owner Web magazine, March 3, 2009). Volvo's surcharge on 2007 emissions levels is $7,500, which covers the cost of a DPF system for PM control and a cooled EGR system for in-cylinder NO_x control. The total emissions control surcharge of $17,100 for 2010 is not far below the cost of a complete heavy-duty diesel engine. Navistar, maker of international trucks, announced a $6,000 emissions surcharge for 2010 medium-duty trucks, and an $8,000 surcharge for heavy-duty trucks (Reuters, July 28, 2009). Navistar is the only major truck maker that does not plan to use SCR to comply with the 2010 NO_x requirement. The 2010 emissions surcharges are on top of surcharges of $5,000 to $6,000 for medium-duty trucks and $7,000 to $10,000 for heavy-duty trucks that Navistar charged for 2007 emissions (Navistar press release, November 8, 2005). Daimler Trucks North America, makers of Freightliner and Western Star trucks, announced increases of $6,700 to $7,300 for medium-duty trucks and $9,000 for 2010 heavy-duty trucks, compared to 2007 (*Transport Topics*, August 8, 2009).

The cost of meeting new emissions standards with gasoline engines is typically measured in hundreds, rather than thousands, of dollars. Diesel engines start with a significant cost disadvantage compared to gasoline engines, because of their greater strength (to withstand the high-cylinder pressures of compression ignition) and their far more sophisticated fuel systems. Diesel fuel systems have injection pressures of 1,600 to 3,000 bar, while even the expensive (by gasoline engine standards) GDI fuel systems require only 100 to 200 bar. Port injection systems for gasoline engines typically use injection pressures of only a few bar. The need to create and control extreme pressures has a major effect on diesel fuel system cost.

When the higher cost of diesel engines is added to the far higher cost of diesel emissions control aftertreatment, there is a powerful market incentive to move toward gasoline engines, except where the durability of the diesel engine is required. Over the period from 2004 to 2008, diesel engines lost market penetration to gasoline engines in Class 3, 5, and 7 trucks while increasing market penetration in Class 2 and 4 trucks (see Table 4-3).

This trend can be expected to accelerate in 2010 and beyond, when medium-duty diesel engines with aftertreatment may cost $10,000 more than a gasoline engine option. For any operation with relatively low average vehicle miles per year, gasoline engines will make more economic sense. Since gasoline engines are significantly less fuel efficient than diesel engines, this means that fuel consumption of the medium-duty truck fleet will increase as a result of falling diesel engine market share.

TABLE 4-3 Diesel Truck Sales as a Percentage of Total Truck Sales

Class	2004	2005	2006	2007	2008
1	0.10	0.1	0.0	0.0	0.0
2	9.2	9.5	10.1	10.4	12.9
3	68.6	68.6	68.6	42.5	44.1
4	70.6	73.8	75.7	78.5	80.9
5	91.7	92.2	91.6	91.8	92.3
6	75.8	73.4	75.3	52.4	58.0
7	53.6	55.8	58.5	50.4	50.3
8	100.0	100.0	100.0	100.0	99.7
Total	9.1	10.3	11.6	9.3	10.8

SOURCE: DOE, EERE (2009), based on *Ward's Motor Vehicle Facts and Figures*.

TRANSMISSION AND DRIVELINE TECHNOLOGIES

"Transmission and driveline" together refer to the system that connects the propulsion system to the wheels. It includes the transmission, the final drive, and the axle. Options for realizing fuel consumption benefits by optimizing the transmission and driveline generally fall into one of two categories:

- *Improved driveline efficiency.* Strategies that increase the efficiency of the power transfer from the propulsion system to the wheels.
- *Improved system integration.* Strategies that enable the engine to operate at higher average drive cycle efficiency.

Improved integration of the driveline with the power train in a tractor trailer can raise the average efficiency of the engine over an actual real-world drive cycle. The easiest form of this approach entails appropriately matching system gearing to the specific application. This process entails selecting a top gear and rear axle ratio that matches typical cruise speed (ensuring that the engine is in its peak efficiency window). These specifications vary from fleet to fleet; to the extent that a vehicle purchaser knows the specific roads on which a truck will travel, systems can be highly optimized to match their application. Conversely, some purchasers simply use standard specifications, or specify vehicles based on what they purchased in the past. As such, all major original equipment manufacturers (OEMs) make a concerted effort to work with purchasing agents to properly manage the vehicle specification process.

Transmissions

Class 3 through 8 vehicles use three basic types of transmission. The most common transmission type is the manual transmission (MT). Class 3 to 7 trucks typically use 5- to 8-speed transmissions, many of which are synchronized. Class 8 vehicles typically use 9-18 speed transmissions, most of which are not synchronized. Synchronizers are universally used in passenger car transmissions to make shifting easier. The synchronizer is a small clutch that matches the transmission input shaft speed to the speed required on shaft of the gear being engaged. Synchronizers are not used in the main box of heavy-duty transmissions both to reduce cost and to eliminate an expensive wear item. They are used in heavy-duty transmissions to synchronize the main box with the auxiliary box. The auxiliary box is used to multiply the number of speeds of the main box of the transmission. For example, a 5-speed main box can be combined with a 2-speed auxiliary box to form a 10-speed transmission. Another example is the 18-speed transmission, which consists of a 5-speed main box with two 2-speed auxiliary boxes. This makes for a total of 20 available ratios, but two of these are redundant.

With MTs, having more ratios can generally lead to a better match between engine speed and road speed, which reduces fuel consumption. However, there are drawbacks to transmissions with more ratios. They require more work on the part of the driver, and they cost more. They are often larger and heavier than transmissions with fewer ratios, and they may be less efficient because more gears are in mesh at any given time. One additional drawback of a high gear count is the frequent power interruptions caused by the need to shift through a large number of gears.

MTs have the highest market share in long-haul truck applications. According to information supplied by Daimler Trucks North America (DTNA) (Freightliner),[20] the share of MTs in line-haul trucks declined from 90 percent in 2004 to a still dominant 82 percent in 2008. At the other extreme, MTs are rarely used in urban applications such as transit buses and refuse haulers. For long-haul Class 8 trucks, the most common transmission types are 10-speed, 13-speed, and 18-speed. The 13- and 18-speed transmissions have smaller-ratio steps, which allows the driver to better match the engine rpm to road conditions. However, these transmissions also have more gears in mesh at any given time than 10-speed transmissions do, so there is an efficiency penalty for having the additional ratios available. There is also a penalty involved in the more frequent power interruptions for shifting that occur with a higher gear count transmission. The 13- and 18-speed transmissions are most often used by heavy haulers (operators that run over 80,000 pounds gross vehicle weight) and by on/off highway operators. These operators need the flexibility provided by a larger number of gear ratios. Most standard long-haul operators use 10-speed MTs to achieve the best balance between vehicle performance and fuel consumption. In this situation the market has gravitated toward the most fuel-efficient MT available.

The automated manual transmission (AMT) has been gradually gaining market share over the past 10 years. According to DTNA figures, AMTs represented 10 percent of the line-haul transmissions in 2004, increasing to 18 percent in 2008. The AMT is typically based on the platform of a standard MT. Additional actuators and controls are added to allow the transmission control module to take over the shifting activities of the driver. Actuators perform both the shifting and clutch actuation for the driver. The AMT controller can match the shift performance (shift time and smoothness) of a skilled driver, provided the controls are well designed and carefully tuned. The AMT offers several advantages over a conventional manual transmission:

- The requirement for driver skill is lower.
- There is less driver distraction, improving vehicle safety and productivity.

[20]Daimler Trucks North America (DTNA), "Heavy Duty AMT and Automatic Transmission Usage at DTNA LLC," PowerPoint presentation provided to the committee, September 2009.

- The control module decides when to shift, which can be used to reduce fuel consumption compared to an average driver.
- Shifts are always performed smoothly, which can improve transmission durability.

The downsides of the AMT are higher cost and more complexity—more parts that can fail. TIAX (2009, p. 4-70) estimates the cost of AMTs at $4,000 to $5,700 over a comparable MT. There is also a slight weight increase. Donnie Stover, fleet manager for Averitt, reported on December 16, 2008 a 3-percent improvement in average fuel consumption when using AMTs in place of standard MTs in sleeper cab team tractors. These are trucks that use two drivers working together to allow nearly 24-hour operation. This fuel consumption reduction comes from allowing the transmission controller to determine shift points rather than the driver. The best drivers can beat the fuel consumption of an AMT, but average drivers cannot match the results of the AMT.

Other features being added to new AMTs include microprocessor technology to continuously monitor changes in road grade, vehicle speed, acceleration, torque demands, weight, and air resistance.[21] This technology allows the vehicle to select the best gear and fuel setting while minimizing fuel consumption. It also allows the engine and transmission to know when to go into freewheeling when neither power nor engine braking is needed. In these situations, the engine goes back to idle and the transmission slips into neutral when power is not needed. TIAX estimated a 4 to 8 percent reduction in fuel consumption based on site visits with the committee. The fuel savings potential for AMT will vary based on duty cycle and driver training. Long-haul cycles on level ground require little shifting and thus offer little potential for improvement, while duty cycles involving hills, congestion, and urban driving have much more potential for fuel savings, both with AMT and driver training.

The third transmission type seen in Class 3 to 8 vehicles is the traditional torque converter automatic transmission (AT). These transmissions typically have five to seven gear ratios, with a torque converter and a lockup clutch. Many ATs use the torque converter only at low road speed and run in lockup mode in all the higher gears, a feature reduces fuel consumption. Fully automatic transmissions are most popular in urban applications such as transit buses and refuse haulers, but they are also widely used in other applications, including some very heavy on/off highway vehicle applications. Standard 80,000 lb long-haul trucks are one application where ATs are very rare. The DTNA data show that the share of MTs in Class 8 non line-haul applications is lower than for line-haul applications, at 67 percent in 2008. AMT transmissions have about 9 percent share in these non-line haul trucks, while torque converter automatics enjoy a 24 percent share, up from just over 10 percent in 2004. The share of ATs is probably higher in the Class 2b to 7 range.

ATs share the driver skill, productivity, and safety advantages of AMTs. They also offer the ability to complete upshifts under full engine power, something that cannot be done with manual or automated manual transmissions. This can be a significant productivity (trip time) factor in applications with frequent large changes in vehicle speed, such as urban or suburban driving. With an MT or AMT, the engine fueling is shut off during each upshift. This interrupts power generation during the shift, which typically takes about 1 second in lower gears and up to 2 seconds in higher gears. However, after the shift is completed, the engine still requires some time (typically 2 to 3 seconds) to return to full power once the shift is completed.[22] In the future, if the development of heavy-duty dual-clutch transmissions progresses as it has for light-duty vehicles, a dual-clutch transmission will remove the problem of interrupting the power during shifting. There can be a fuel consumption advantage as well as a productivity advantage in performing full-power upshifts, because the engine can continue to operate at an efficient point during and after shifts.

ATs are slightly heavier than MTs, but the use of an AT allows the clutch and flywheel to be replaced by a flexplate and torque convertor. Overall truck weight with an AT is slightly lower than with an MT. TIAX reports the cost of heavy-duty ATs at $15,000 over an MT.

The AT has some fuel consumption penalties compared to a conventional transmission as well. The hydraulic pump required to fill the torque converter and actuate the shift events draws power from the engine. When the transmission operates in converter mode, a significant percentage of the input power from the engine is lost as heat in the torque converter. The AT will operate with an open converter (lockup clutch disengaged) at low engine and vehicle speeds, whereas the MT and AMT use a more efficient closed clutch. The shift schedule of medium- and heavy-duty ATs has evolved over the years to minimize the time spent in converter mode.

The primary disadvantages of the AT compared to other transmission types are much higher cost and more complexity. The warranty period for ATs is much shorter than for MTs, and warranty repairs must be made by the transmission dealership. These factors discourage the use of ATs except in applications where the advantages of using an AT in the vehicle operating cycle are very great. AT applications are typically in urban and suburban operations. There are a number of features under development to improve the efficiency of ATs, including lower friction and lower parasitic hardware, more elaborate shift strategies, a reduced load on the engine during stops, and automatic shift into neutral when the parking brake is applied.[23] TIAX (2009, 4-70) estimated a 0 to 5

[21] Eaton Roadranger Product and Service Update, September 2009.

[22] SwRI, "Heavy Duty Diesel Transient Response," presentation to the committee, San Antonio, Tex., August 6, 2009.

[23] Committee site visit to Allison Transmission, "Allison Fuel Efficiency Improvements," presentation to the committee, May 2009.

percent reduction in fuel consumption from the use of ATs, in tractor-trailer applications, in its report to the committee.

Research has been done on other transmission types, such as continuously variable transmissions (CVTs) and dual-clutch transmissions. Several CVT designs have been proposed for heavy-duty vehicles, and Allison Transmission has signed an agreement with Torotrak to develop its CVT design for heavy-duty vehicle applications. It is not yet clear if any of these alternative transmission types will reach volume production. CVTs tend to have lower mechanical efficiency than MTs or AMTs, but they make up for this by allowing an optimum match between engine speed and road speed under all operating conditions. The main challenges facing the use of CVTs for truck and bus applications are mechanical efficiency, reliability, durability, and cost.

A comparison of transmission fuel consumption performance can be made by considering four driving-cycle components: idle, acceleration, steady-speed cruise, and deceleration. At idle the AT has a disadvantage due to torque converter loss and hydraulic pump parasitic power. The addition of an "auto neutral" function greatly reduces this disadvantage. Under acceleration the AT has an advantage over the other types due to power shifting. Shifts are completed without changing the fueling command, so boost pressure is maintained and engine operation is more efficient. The AMT has an advantage over an average driver with an MT because of computer-controlled shift points. At cruise the MT and AMT have a slight advantage because they do not require a hydraulic pump. It should be noted that cruise fuel consumption is very strongly dependent on speed and the final drive gearing, and these selections are independent of the transmission type. Under deceleration the AT has a slight advantage over other types because there is no need to blip the engine fueling for downshifts. This blip is necessary with both manual and automated manual transmissions to get the engine speed to match the speed of the transmission gears.

Overall, the selection of transmission type has only a relatively small impact on vehicle fuel consumption. The exception is in urban and suburban operation, where the AT may offer a modest reduction in fuel consumption, combined with significantly greater productivity (average trip speed).[24] The higher productivity is a result of avoiding power interruptions during acceleration. Fuel consumption differences due to transmission selection are normally a few percent or less. The shift calibration schedule for AMTs and ATs can have a modest impact on fuel consumption. A driver's shift behavior with MTs can have a significant effect on fuel consumption; a driver with poor habits may suffer a 10 to 20 percent fuel consumption penalty in urban and suburban driving.[25] The line-haul market is likely to move more in the direction of AMTs as costs fall and reliability is proven. This move will provide a modest fuel consumption benefit. A move to ATs for stop-and-go-type operations may also provide a modest benefit. Just as the engine market is very competitive in terms of fuel consumption, so is the transmission market. Long-haul operators will often change transmission type to gain a 1 or 2 percent reduction in fuel consumption, as long as the cost is reasonable and the reliability of the technology is solid.

[24]Eaton, Manual #TCMT0020, June 2009, pp. 1-2; and Dana, Manual #TCMT0021, March 2009, pp. 1-2.

[25]*Heavy Duty Trucking,* January 2009, pp. 62-63.

Rear Axle Ratio

The selection of rear axle ratio is one of the most important decisions in specifying a truck. The axle ratio determines the engine rpm at the vehicle's cruise speed, which is a very important fuel consumption parameter. The axle ratio also determines the grade capability and acceleration performance of the vehicle. A tall (numerically low) axle ratio is typically good for fuel consumption but bad for acceleration performance and grade capability. This means that the choice of axle ratio involves a trade-off between fuel consumption and vehicle performance. Operators select axle ratios based on the type of loads and routes they expect to operate with. Most engine and vehicle OEMs have sophisticated software that can look at specific trucking operations and recommend the best axle ratio for a given application.

Great care must be taken in developing any fuel consumption requirements to avoid a situation where operators are forced to select an axle ratio that is not appropriate for their operation. A logging operation, for example, must use a shorter (numerically higher) axle ratio than a standard long-haul tractor, in order to get adequate off-road performance.

Low-Friction Transmission, Axle, and Wheel Bearing Lubricants

Special lubricants can be used to reduce friction in the transmission and axles. Typically synthetic lubricants are specified to reduce viscosity, especially in low-temperature conditions. Many tests conducted by fleets have documented fuel savings of at least 1 percent when using low-friction lubricants. Because of the relatively low cost, synthetic lubricants are becoming widespread in the industry. Eaton and Dana have recently made low-friction synthetic lubricants standard on all their heavy-duty transmissions and axles.[26] Truck OEMs typically charge $35 to $55 for synthetic lube in tandem-drive axles. Synthetic lube and grease for wheel bearings is also available at relatively low cost.

Rear Axle Types

For Class 3 vehicles several types of axles are used. The standard single-reduction axle with an open differential has the lowest friction and is best suited for highway use. Some

[26]Eaton, Manual #TCMT0020, June 2009, pp. 1-2; Dana, Manual #TCMT0021, March 2009, pp. 1-2.

applications need improved traction for operation in low-friction environments such as snow or off-road conditions. The limited-slip differential uses a clutch pack to provide a limited amount of torque to both ends of the axle when there is a significant speed difference between the two wheels. This is the preferred differential for the front axle of a four-wheel-drive vehicle, since the clutch pack allows some speed differential for easier steering. Some limited-slip axles use electronic actuators to engage the limited-slip clutches.

A range of axle designs and configurations is available for medium- and heavy-duty vehicles. The designs vary based on the weight of the vehicle and the intended application. For drive axles, single, tandem, and tridem (three drive axles) configurations are available. The axle may have no differential, a single-speed differential, a two-speed differential, or a double-reduction differential (two gear sets in series to get a numerically high axle ratio). The most common axle type is the single-speed differential, and more than 17 different ratios are available for this design. The same range of ratios is available in tandem and tridem axles. Axle prices range from \$2,600 to \$15,000, depending on the type selected, the weight rating, and other features.

In general, axle design is selected to match the intended operation of the truck. The selection of axle ratio is the most important factor in determining driveline-related fuel consumption. A ratio that matches proper engine speed at cruise is critical. Other axle features have only a minor effect on fuel consumption. In general, to get the lowest fuel consumption, a vehicle should use the minimum number of drive axles required for the intended application. Each additional drive axle adds friction and weight to the driveline and thus has a fuel consumption penalty. For example, the 6 × 2 arrangement for tandem axle trucks is estimated to provide about a 1 percent fuel consumption reduction compared to the standard 6 × 4 arrangement, at the expense of limited traction (TIAX, 2009, p. 4-69). A 6 × 4 arrangement refers to a truck with six wheels (dual wheels count as one), with one steer axle and two drive axles. This is the standard arrangement for long-haul tractors. A 6 × 2 tractor has only one drive axle, with the second rear axle used only to carry load. To make a 6 × 2 design practical, a method of unloading the nondriven axle must be used for situations where traction is critical. The 6× 2 layout is popular in Europe, but not in the U.S. market. Resale value is a major factor keeping fleets from specifying 6 × 2 tractors. Buyers are concerned about running into low- or no-traction difficulties due to the driven axle being lifted to the point of no traction when crossing uneven road or ground. This issue can be avoided by using air bags and a driver control valve to lower the drive axle back onto the road surface, but this adds cost and complexity to the vehicle.

Transmission and Driveline Summary

In its report for the committee, TIAX (2009) summarized the transmission and driveline fuel consumption potential reduction by range of years and by application as shown in Table 4-4.

TABLE 4-4 TIAX Summary of Transmission and Driveline Potential Fuel Consumption Reduction (percentage) by Range of Years and by Application

	2013-2015	2015-2020
Tractor trailer	5.0	7
Class 6 box truck	1.5	4
Class 6 bucket truck	1.2	3.2
Refuse truck	1.5	4
Urban bus	1.5	4
Motor coach	2.0	4.5
Class 2b pickup and van	4.5	7.5

The committee believes that the claims for reduced fuel consumption of tractor trailers would apply only to those not properly specified today. Many tractors are already well specified for their application, and the savings is likely to be less than shown above.

HYBRID POWER TRAINS

A hybrid vehicle (HV) combines at least two energy converters, such as internal combustion engines (ICEs), electric drives, and hydraulic drives. The ultimate goal of the HV is to provide the equivalent power, range, and safety as a conventional vehicle while reducing fuel consumption and harmful emissions. HVs have the potential to realize several advantages, including the following:

- *Regenerative braking.* A regenerative brake is an energy mechanism that reduces vehicle speed by converting some of its kinetic energy into a storable form for future use instead of dissipating it as heat as with a conventional brake. The significance of regeneration becomes apparent when one considers that approximately 60 percent of the total energy spent in the U.S. Federal Urban Driving Schedule is used to overcome the effect of inertia and that, theoretically, up to 50 percent of this energy could be recovered. This "maximum theoretical percentage of recoverable energy" will vary with duty cycle and vehicle characteristics, since it depends on the deceleration level and vehicle aerodynamic drag and rolling resistance. Regenerative braking can also reduce brake wear and the resulting fine particulate dust.
- *Higher electric machine efficiency.* In comparison with the ICE, the electric machine is a simpler and more efficient machine. For instance, the moving parts of an electrical machine consist primarily of the armature (DC motor) or rotor (AC motor) and bearings.

- *Improved torque characteristics.* Electric machines are more suited to vehicle applications, with high torque at low speed and less torque at cruising speed. When the electric machine torque dips, the ICE must be engaged to harness torque.
- *Reduced emissions.* Reduced emissions occur through smoothening of transients and idle elimination.
- *Operate at best efficiency.* For selected configurations, optimal engine operation. operate the engine in its "sweet spot," staying close to its best efficiency line (also called E-line).
- *Downsize engine.* Engine downsizing might be possible to accommodate average load (not peak load) and consequently reduce engine and power train weight.
- *Engine shutoff.* Engine shutoff is possible, thereby reducing fuel consumption, emissions and noise vibration and harshness. In the case of line-haul vehicles, engine shutoff can be achieved through electrification of overnight hotel load; and in the case of service trucks such as utility vehicles, the power take-off can be electrified.
- *Accessory electrification.* Accessory electrification allows parasitic loads to run on an as-needed basis. Electrified accessories are often more efficient than belt-driven ones.
- *Better drivability.* An electric machine reacts faster to a throttle input than an ICE; furthermore, torque from the ICE and the electric drive train can be added up whenever needed in certain configurations.
- *Robustness.* For some configurations, such as the parallel, the vehicle may be operable with either of the power sources when one fails.
- *Plug-in hybrids.* Plug-in hybrids can help absorb excess electricity from the grid at night and improve energy security and diversity.
- *Electrification.* Electrification can enable waste heat recovery, thus bringing on board systems to generate and store electricity, and use it when needed.

HV disadvantages include the following:

- Increased power train and electronic complexity.
- Increased vehicle mass due to addition of components.
- Increased cost due to additional components and complexity of the power management.
- Overall system reliability can be lower due to increased complexity.
- If not optimized for the appropriate drive cycle, benefits may not be fully realized, or fuel consumption may even increase.

The two major types of hybrids are electrical and hydraulic:

- A hybrid electric vehicle (HEV) combines electric and mechanical power devices. The main components of an HEV that differentiate it from a standard ICE vehicle are the electric machine (motor and generator), energy storage (e.g., battery or ultracapacitors), and power electronics. The electric machine absorbs braking energy, stores it in the energy storage system, and uses it to meet acceleration and peak power demands. HEVs are widely used in almost all vehicle weight classes—light- medium- and heavy-duty vehicles.
- A hydraulic hybrid vehicle (HHV) combines hydraulic and mechanical components. The four main components of a hydraulic hybrid power train are the working fluid, fluid reservoir, hydraulic pump/motor (in a parallel hybrid system) or in-wheel motors and pumps (in a series hybrid system), and an accumulator. The hydraulic accumulator stores the energy (as highly compressed nitrogen gas) and a variable displacement pump acts as a motor while driving the wheels and as a generator while absorbing regenerative braking energy. This system suits medium- and heavy-duty vehicles operating with high-power, low-energy requirements, including stop-and-go driving profiles (e.g., refuse vehicles, inner-city buses, and delivery vans).

The task of achieving fuel savings using a hybrid architecture depends on the type of power train selected as well as the component sizes and technology, the vehicle control strategy, and the driving cycle. When approached as a system, an HV is an integrated propulsion system. Fuel consumption reductions can be realized only after sensible optimization of power management based on a suitable driving cycle. The vehicle's power demand is met by the different power sources on board. For instance, a simple acceleration, cruising, and braking cycle for an HEV demonstrates the best use of different power sources based on the vehicle's power demand: during small accelerations, only the energy storage power is used (electric vehicle mode), and during braking some of the energy is absorbed and stored. The ICE does not start to operate during low-power demands due to its poor efficiency compared to the electric system. The ICE is used only during medium- and high-power demands where its efficiency is higher. More discussion on vehicle-level management strategies is given in a subsequent section.

Comparison of Energy Storage Devices

Generally light-duty vehicles, small trucks, and transit buses typically make use of electric systems, whereas heavy-duty vehicles make use of both electrical and hydraulic systems (except transit buses). Truck applications enable regenerating and reusing significant amounts of braking energy. Consequently, power flows through the hybrid subsystem can be very high. This makes both ultracapacitors and hydraulic

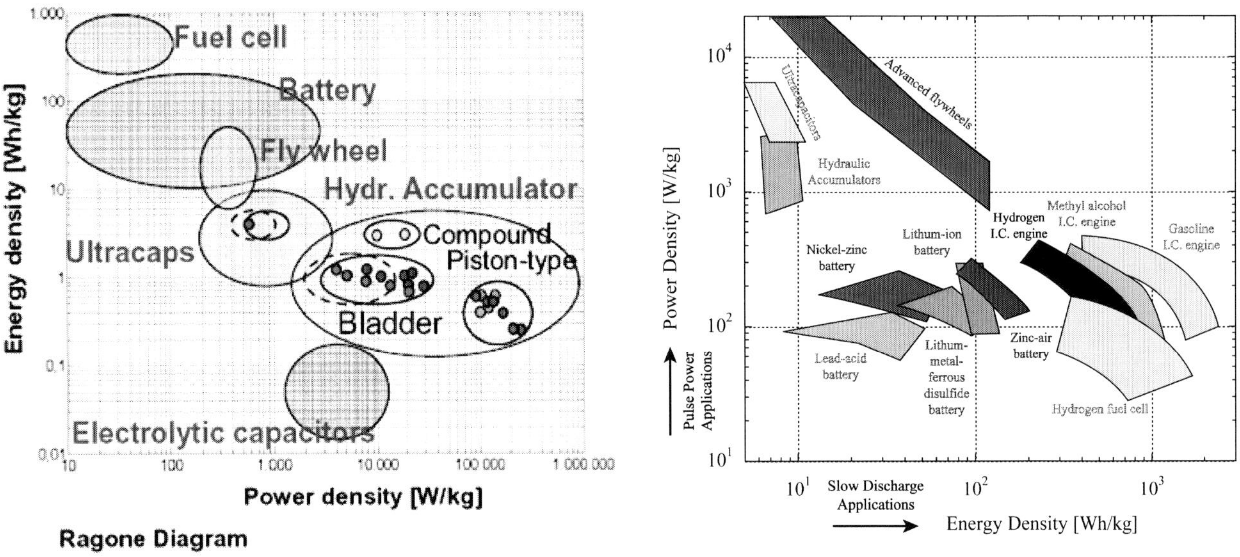

FIGURE 4-6 Power density versus energy density of various technologies. SOURCE: Baseley et al. (2007). Reprinted with permission from SAE Paper 2007-10-4150. © 2007 SAE International.

storage very attractive since they are characterized by very high power density levels. However, the energy density of both ultracapacitors and hydraulic systems is lower than that of batteries and, hence, energy cannot be supplied over a long duration. The high-power systems are well suited for a driving cycle with several start-stops, as energy can be captured and released quickly. A battery, on the other hand, has greater energy density and can be used for long energy storage and supply. The Ragone diagrams shown in Figure 4-6 compare the power density versus the energy density of different energy storage systems. Note that the ultracapacitors, hydraulic accumulators, and advanced flywheels have the highest power density among present-day storage systems, but their energy density is limited significantly.

Hybrid Vehicle Architecture

A number of different system architectures are being considered to meet different applications. They are broadly classified as series, parallel, and power split. The selection of system architecture depends mainly on the application. The following sections describe some of the possible power train configurations under each architecture.

Series Hybrids

Series Hybrid Electric

In a series HEV, as illustrated in Figure 4-7, an electric generator, coupled to an ICE, supplies electricity to the electric machine to propel the truck and to the energy storage system when it needs to be recharged. Generally for batteries the engine/generator set keeps the energy storage system

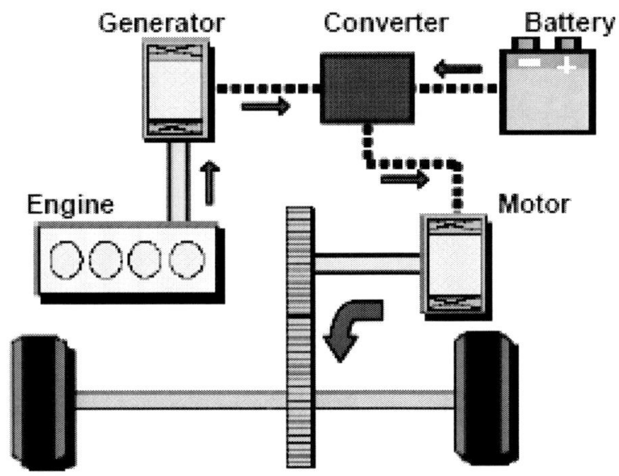

FIGURE 4-7 Series hybrid electric vehicle. Courtesy of University of Michigan.

charged between 50 and 70 percent for charge sustaining HEVs.[27] (For PHEVs, the energy storage system is used in a wider operating range (e.g., 80 to 30 percent for the GM Volt). In this configuration the ICE and vehicle speeds are decoupled and only the electric machine is connected to the wheels. The ICE does not need to speed up or slow down as

[27]"Charge sustaining" means that any energy used from the battery has to be provided back either through regenerative braking or from the engine. The battery state of charge (SOC) usually operates in a small window (around 15%) to protect the battery life.

the load varies. As a consequence, the engine can run at its optimum performance (best engine efficiency zones), greatly reducing fuel consumption. Moreover, the ICE never idles, thus reducing overall emissions. However, because the electric machine is the only one connected to the wheels and the engine/generator set is sized for sustained grade ability, this configuration requires a large energy storage system pack, electric machine, and engine. For this system to be viable, it must possess an overall high efficiency in total power processing. While the added mass and the component inefficiencies make that configuration unlikely for small trucks, it is more viable for large vehicles, such as buses, which are less sensitive to increases in overall weight. It should be pointed out that the elimination of the mechanical driveshaft can be an important advantage of adopting the series configuration. One example is the transit buses where elimination of the mechanical driveshaft makes it possible to lower the bus floor for improved wheelchair access.

Several variations of the series configurations have been considered. One of the important considerations in the design of a series HEV is related to the use of a single-gear ratio versus a two-speed transmission. Using a single-gear ratio usually leads to low maximum vehicle speed and poor performance at high speed due to the low electric machine torque at that regime. When applications require better performance at high speeds, a two-speed transmission is considered. If electric machines are used at each of the wheels, instead of a single electric machine, torque vectoring is possible, improving vehicle stability.

Torque vectoring can be defined as distributing the majority of the power to the wheels that have traction. If the front wheels have better traction than the rear wheels (rear wheels might be running on icy patches for some particular time), the power is transmitted to the front wheels and not the rear wheels. Torque vectoring can also be done between the front left and front right wheels. As each wheel can be powered independently, the number of degrees of freedom to control the vehicle traction is increased, and in turn vehicle stability can be improved. Torque vectoring is achieved by using redesigned differentials, which means that wheels don't need to be stopped, and even better, the vehicle won't suffer from a sudden loss of power as it is negotiating an unexpected loss in traction.

Series Hybrid Hydraulic

HHVs employ the same basic architecture as HEVs. However, in an HHV the battery is replaced with a hydraulic accumulator, and electric machines are replaced with hydraulic motors/pumps. As in the HEV architecture, kinetic energy from braking can be recovered and stored: in an HHV, braking energy is used to drive a hydraulic motor that pumps fluid from a low-pressure fluid reservoir to a high-pressure accumulator. This energy can then be used to supplement engine power by releasing the fluid in the high-pressure accumulator back to the low-pressure reservoir, driving the motor in the process. The series configuration of the HHV can be implemented using a high-pressure accumulator along with a low-pressure reservoir, as shown in Figure 4-8.

Current hydraulic hybrids are capable of capturing on the order of 70 percent of kinetic energy from heavy-duty vehicles. This is due to both the high rates at which power can be recovered and the fact that the energy storage system is virtually lossless, i.e., there is very little energy lost (Gotting, 2007). Hydraulic hybrids are targeted for power-driven applications—that is, duty cycles that have high regenerative braking requirements but relatively low energy requirements, such as refuse trucks and house-to-house delivery vehicles. Series hydraulic hybrids are still in the prototype stage. They are being evaluated for pickup and delivery vehicles and for refuse haulers, for which they have demonstrated fuel consumption reductions on the order of 50 percent (J. Kargul, U.S. EPA, presentation to the committee, April 6-7, 2009).

Parallel Hybrids

Parallel Hybrid Electric

Parallel hybrids have mechanical connections to the wheels from both the electric machine and the engine, as il-

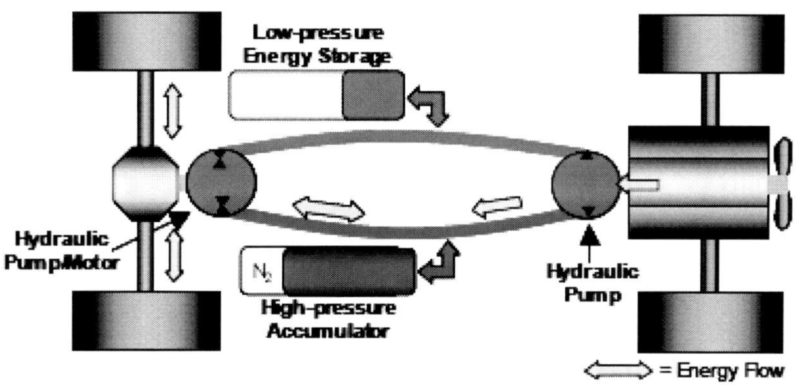

FIGURE 4-8 Series engine hybrid hydraulic vehicle. SOURCE: Eaton-HTUF (2009). Courtesy of Eaton.

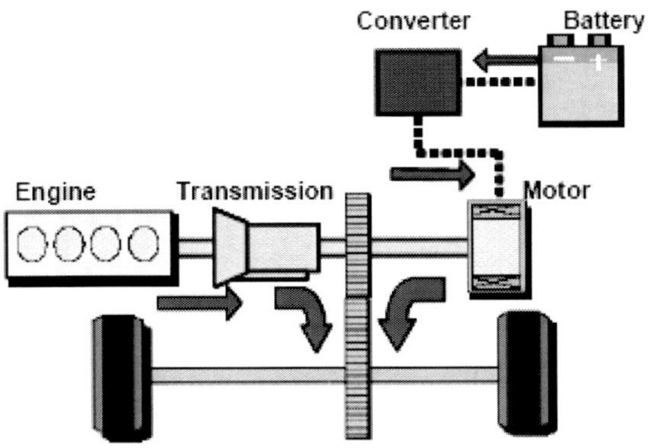

FIGURE 4-9. Parallel hybrid electric vehicle. Courtesy of University of Michigan.

lustrated in Figure 4-9. The electric machine can be located anywhere between the output engine shaft and the wheels. These vehicles do not need a dedicated generator; the electric machine can be used as a generator to recharge the batteries. In a parallel HEV the electric machine can assist the engine during startup and acceleration.

Because the electric machine and the engine are both coupled directly to the wheels, they can share the power during acceleration. Therefore, it is possible to downsize both the engine and the electric machine compared to series hybrids (the vehicle mass is then reduced compared to the series hybrid architecture). It is also possible to increase the hybridization degree by downsizing the engine and upsizing the electric machine; note, though, that downsizing the ICE is not practical in applications that require extended high-power operation, such as long-haul trucks. For some configurations the ICE can operate close to its best efficiency curve, with the electric machine assisting it or recharging the battery. However, it should be noted that the ICE speed is not independent of the vehicle speed even though there is some degree of freedom over the engine load.

Numerous options have been studied for parallel architectures, ranging from micro or mild versions of the starter-alternator type to full-featured pre- and posttransmission hybrids, as described below.

Starter-Alternator Type

In a starter-alternator configuration the electric machine is connected to the engine crankshaft directly on the shaft, as shown in Figure 4-9, or coupled through a belt, as shown in Figure 4-10. The main advantage of this configuration is the ability to turn the ICE off during idling. Since the electric machine speed is linked to the engine, the vehicle cannot operate in electric mode other than for extremely low speeds (e.g., creep). In addition, the electric machine is used to smooth the engine torque by providing power during high transient events to reduce emissions. Finally, some regenerative braking energy is recaptured but only a small amount due to the limitation in size of the electric machine. Two families are defined, based on the battery voltage: micro and mild HEV. This system requires minimal modifications and cost.

Pretransmission Type

For pretransmission parallel HEVs, the electric machine is located in between the clutch and the transmission, as illustrated in Figure 4-11. This configuration allows operation in electric mode during low- and medium-power demands, in addition to the ICE on/off operation. The location of the electric machine allows torque multiplication through the transmission, allowing a small maximum speed and providing good assist during high-power demands.

Post-transmission Type

In a post-transmission configuration, shown in Figure 4-12, the electric machine does not have the benefit of gear ratio changes. As a result, it must operate on a very broad vehicle speed range, requiring high torque over a wide speed range. The main disadvantage of the configuration is the need for high torque requirements for electric machines,

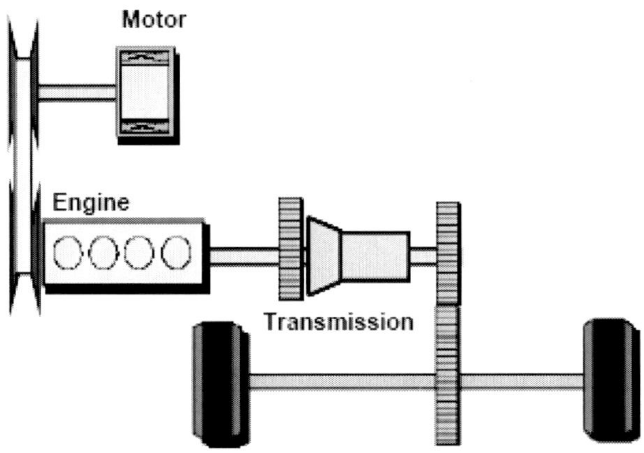

FIGURE 4-10 Example of integrated starter generator configuration coupled through a belt. Courtesy of University of Michigan.

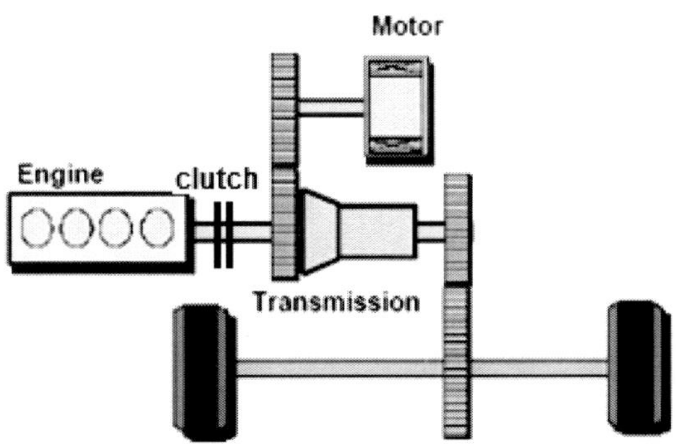

FIGURE 4-11 Example of pre-transmission parallel configuration. Courtesy of University of Michigan.

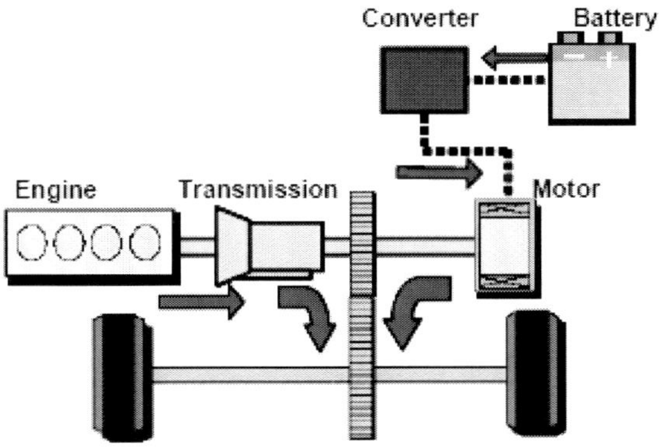

FIGURE 4-12 Example of post-transmission configuration. Courtesy of University of Michigan.

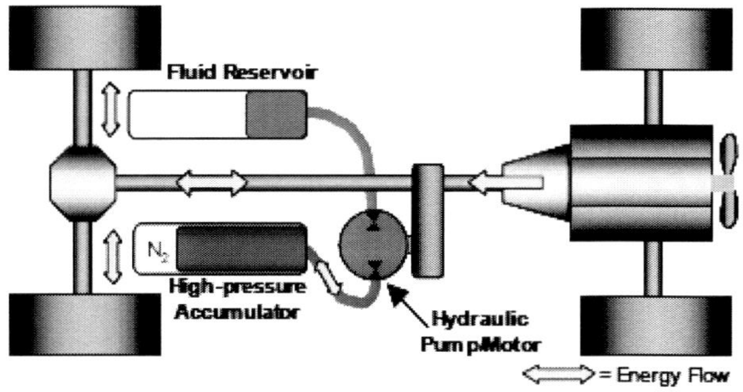

FIGURE 4-13 Parallel hydraulic launch assist hybrid architecture. SOURCE: Eaton-HTUF (2009). Courtesy of Eaton.

leading to an electric machine that is both larger and heavier than the corresponding machine in the pre-transmission configuration. However, compared to the pre-transmission type, it allows a higher efficiency path from the energy storage to the wheel, increasing the amount of regenerative braking captured at the energy storage system.

Parallel Hydraulic Hybrid

A parallel hydraulic hybrid utilizes two power sources, the ICE and the hydraulic motor for launch assist (HLA) to improve acceleration and reduce fuel consumption. The hydraulic pump/motor is located behind the transmission for more effective regeneration during braking. The hydraulic pump/motor is coupled to a propeller shaft via a transfer case, as shown in Figure 4-13. The HLA power source is an axial piston pump/motor with variable displacement. The hydraulic displacement per revolution can be adjusted via inclination of the swash plate to absorb or to produce desired torque. When pumping, hydraulic fluid flows from the low-pressure reservoir to the high-pressure accumulator; when motoring, hydraulic fluid flows in the reverse direction. The accumulator contains the hydraulic fluid and inert gas such as nitrogen, separated by a piston. When hydraulic fluid flows in, the gas is compressed, and its internal energy is increased. When discharging, fluid flows out through the motor and into the reservoir. The reservoir can be regarded as an accumulator working at much lower pressure (e.g., 8.5 to 12.5 bar). The size of hydraulic components is configured to absorb sufficient braking energy.

Power Split Hybrids

Power Split Hybrid Electric

Power split hybrids combine the best aspects of both series and parallel hybrids to create an extremely efficient system. As shown in Figure 4-14, this system divides the engine power along two paths: one goes to the generator to produce electricity, and one goes through a mechanical gear system to drive the wheels. In addition, a regenerative system uses the kinetic energy of deceleration and braking to produce electricity, which is stored in the energy storage system.

The most common configuration, called an input split, is composed of the following components: a power-split device (transmission), two electric machines, and an engine. According to the situation, all these elements operate differently. Indeed, the engine is not always *on* and the electricity from the generator may go directly to the wheel to help propel the truck or through an inverter to be stored in the battery. The different possibilities of an input split configuration are as follows:

- When starting out, when moving slowly, or when the state of the battery charge is high enough, the ICE is not as efficient as electric drive, so the ICE is turned *off* and the electric machine alone propels the truck.
- During normal operation, the ICE power is split, with part going to drive the vehicle and part being used to generate electricity. The electricity goes to the electric machine, which assists in propelling the truck. The generator acts as a starter to activate the engine.
- During full-throttle acceleration, the energy storage (e.g., battery) provides extra energy.

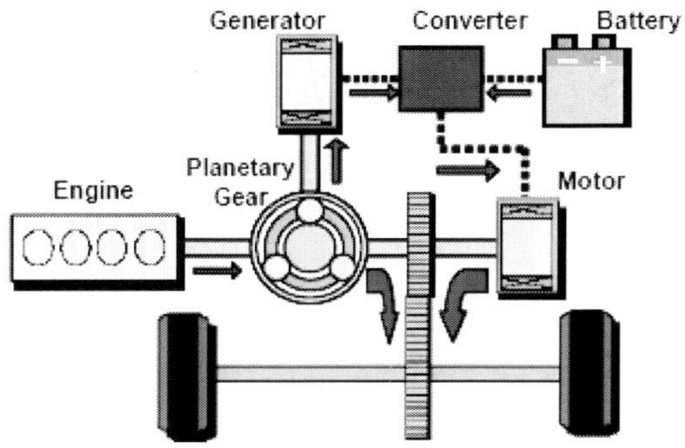

FIGURE 4-14 Power-split hybrid electric vehicle. Courtesy of University of Michigan.

- During deceleration or braking, the electric machine acts as a generator, transforming the kinetic energy of the wheels into electricity.

Note that the two electric machines and the planetary gear behave as a continuously variable transmission (CVT) whose instantaneous gear ration depends on the amount of power fed from the generator to the motor.

Several variations of the power split have been implemented, each providing different advantages. In a single-mode power-split hybrid, the first electric machine is used to control the engine speed while the second one provides the remainder of the power required to follow the vehicle trace. In a dual-mode power system, composed of a compound mode in addition to the input mode, the size of the electric machine can be minimized as each motor is used to control the engine speed in different conditions. A dual-mode power system with several fixed gears minimizes the electric machine power requirements, and the system efficiency can be further improved by reducing the energy recirculation through the use of the fixed gears. The dual-split typically requires at least two planetary gears and one or more clutches, in contrast to the single-split, which requires only one planetary gear and no internal clutches.

The multimode power-split effects are as follows:

- Smaller electric machine peak power and sizes.
- Addition of clutches to transmission increases spin and pump losses.
- ICE may not be at its optimum point during the fixed-gear mode.
- Higher tractive capability during fixed-gear mode.

Power-Split Hybrid Hydraulic

In principle, hydraulic power-split architectures similar to the electric power-split architectures can be implemented as well. For instance, in Tavares et al. (2009), a hydraulic power-split system similar to the Toyota hybrid system in a Prius was chosen. Instead of using electric components, the proposed system is a hydraulic hybrid, comprising of two pump/motors coupled to a planetary gear set and a hydro-pneumatic accumulator for energy storage.

Plug-in Hybrids

Plug-in hybrids differ from HEVs by their ability to recharge the energy storage system through the electric grid. Since the vehicle is designed with high energy storage, batteries are usually used for this application. All the HEV configurations described above can be used for plug-in hybrids. In most cases, due to the electric energy focus, the electric machine power is increased compared to an HEV.

The vehicles can be recharged during the night when electricity prices are low in terms of demand and cost. Commercial vehicles, made by Smith Newton, are being delivered in the United States to AT&T, Frito-Lay, Coca-Cola, Staples, Kansas City Power & Light, and Pacific Gas and Electric Company. Odyne develops and manufactures plug-in hybrid electric drive systems for medium- and heavy-duty trucks. Odyne has developed three major systems: 10 kWh, 18 kWh, and 35 kWh. These systems can be used on a wide variety of truck applications, including bucket trucks, digger derricks, and underground utility vehicles. Eaton Corporation, Ford Motor Company, and the Electric Power Research Institute (EPRI) have developed both diesel and gasoline versions of a plug-in hybrid system for trucks (EPRI, 2008; Eaton, 2009). The system has been demonstrated on the medium duty, Ford F550, "trouble" truck platform used to repair and maintain the transmission and distribution infrastructure of utilities. By using grid electricity stored in batteries for part of the vehicle's daily duty cycle, the plug-in vehicle can operate at the job site for several hours continuously, running the bucket, power tools, lights, and accessories without the need to run the gasoline or diesel engine. The plug-in hybrid truck, which is estimated to deliver fuel economy improvements of up to 70 percent compared with a conventionally powered truck, with corresponding reductions in harmful emissions, was developed for Southern California Edison. Eaton's current hybrid systems used widely in work trucks and delivery vehicles, can reduce fuel consumption by 30 to 60 percent, with similar percentages in emission reductions, extended brake life, and idle time reductions up to 87 percent during work-site operations.

Plug-in hybrids will benefit considerably from intelligent vehicle technologies, especially if the algorithm knows how much farther the vehicle is going to be used and at what rate. In addition to drive-cycle characteristics, such as vehicle speed or acceleration, distance is a critical parameter to minimize plug-in vehicles fuel consumption (Karbowski, 2007).

Knowing what each type of hybrid demands in terms of electric power will give an immediate picture of the size and type of battery needed and in turn an idea about the weight and cost of implementing it.

For heavy-duty applications where the driver sleeps in the vehicle with the engine running while it is parked, a significant source of fuel consumption is the approximately 8 hours of engine idling time used a day to operate the air conditioning, heat, or on-board appliances (such as a television or microwave) and also to keep the fuel warm in cold weather. While hybrids in general aim for idle elimination, the ICE in a plug-in or conventional hybrid vehicle can be run at a specified and efficient power range to store energy in the battery packs. The ICE can be switched on and off as the state-of-charge requires. For plug-in hybrid vehicles, these so-called hotel loads can be powered with grid electricity at rest stops, if the required infrastructure is in place.

Batteries for Hybrid Vehicles

With the increased use of electrified systems in vehicles for both drivelines and accessories, it is important to consider the choice of the right battery technology. The battery's contribution to the vehicle's overall power is growing at a rapid pace, and the choices of battery affects the cost, reliability and service life, packaging space and weight, recyclable and "green" issues of the vehicle itself. As a rule of thumb, the battery system represents one-third of the overall increase in the cost of hybridization (Alamgir and Sastry, 2008). While hybrids and electric vehicles technologies are very promising, batteries are the Achille's heel. Some of the several battery technologies available include lead-acid, zinc bromine, nickel cadmium, nickel metal hydride, lithium ion, and lithium polymer. Lead-acid, zinc bromine and nickel metal hydride (Ni-MH) were used in the initial era of EVs and HVs, but current vehicles rely on Ni-MH and lithium ion (Li-ion) battery technologies, with the latter seemingly having superior attributes for a vehicle system.

Figure 4-15 compares the relative gravimetric energy and power capabilities for the battery types used or being considered for automotive applications. The figure shows (as rectangular domains) the approximate ranges of energy and power densities required for the batteries of the various advanced-technology vehicles, including full HEV, plug-in hybrid electric vehicle (PHEV), and full performance battery electric vehicle (FPBEV); for details on vehicle types, refer to Kalhammer et al. (2007). These so-called Ragone plots show that, for each type, batteries designed for high power densities have substantially lower energy densities than batteries optimized for high energy (FPBEV designs). Whenever the performance domain for a specific vehicle type is below and to the left of the Ragone performance characteristic for a particular battery type, properly engineered versions of that battery type can be expected to meet vehicle power and energy requirements. It appears that Li-ion has good potential to be configured for superior power and energy density to meet various applications.

Figure 4-16 depicts the Li-ion performance status versus the targets set by the DOE FreedomCAR program (NREL, 2004). While Li-ion meets most of the requirements, but concerns remain regarding cost and life, as well as abuse tolerance to extreme operating temperatures and rapid charging and discharging rates within a safety level to be used in cars. In particular, the performance of Li-ion batteries erodes drastically at extreme temperatures (above 65 °C or below 0 °C). Therefore, in order to maintain battery life and performance, expensive and complex thermal management systems might be required. Also, under abusive conditions such as inadvertent overcharge, short circuit, or over-heating, a Li-ion cell will generate gas and experience an increase of internal pressure (Snyder et al., 2009). Hence, Li-ion technology might present a number of new system design and validation challenges that ensures robustness and durability for vehicle applications compared to NiMH technology. Accordingly, there is a need to carefully consider the trade-offs between Li-ion and NiMH technology, in terms of various factors, including system design, validation implications, and performance, among others.

Li-ion developments with iron, manganese, and nickel instead of cobalt have made them cheaper and safer. Modification of the electrode nanostructure has increased charge and discharge rate and cycle life. Current prototypes built by SAFT, LG, and A123, among others, are demonstrating the potential for lower-cost, longer-life Li-ion battery systems with less need for complex thermal management.

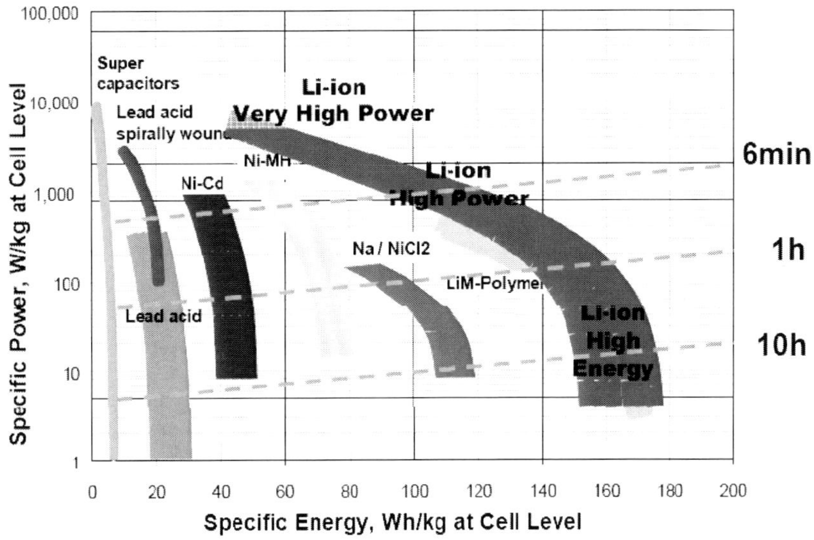

FIGURE 4-15 Battery type versus specific power and energy. SOURCE: Kalhammer et al. (2007).

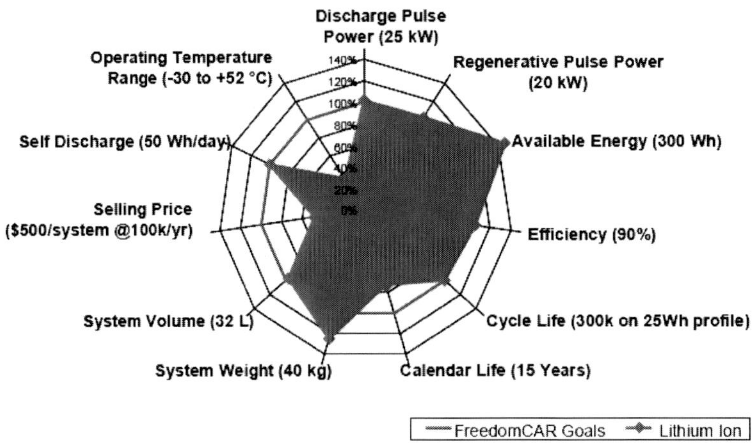

FIGURE 4-16 Li-ion status versus targets (for power-assist HEV). SOURCE: DOE (2005).

This is expected to help make HEVs more competitive in the marketplace and enable consumers to receive a faster payback. For instance, Ford has tested in the Escape hybrid model a Li-ion battery system that is said to be 20-30 percent smaller and 50 percent lighter than today's Ni-MH technology, at 30 percent lower cost.[28] Also, Argonne National Laboratory reports that the Enerdel/Argonne lithium-ion battery promises high reliability, light weight, and potential to meet the U.S. Advanced Battery Consortium's $500 manufacturing price criterion for a 25-kilowatt battery (ANL, 2008). A recent NRC report has estimated costs for Li-ion batteries for light-duty plug-in electric hybrid vehicles, including projections of costs per kWh, which present a range for what such batteries may cost in the future (NRC, 2010).

Adding Ultracapacitors to Battery Packs (Dual Energy Storage)

Ultracapacitors (UCs) have their own merits and demerits compared to batteries. They can be charged and discharged *very* quickly, they have a longer life, and their performance does not degrade appreciably with use. On the other hand, for an equal size/weight, their energy-storing capacity (energy density) is extremely low. Despite this shortcoming, UCs can be supplemented by batteries. UCs offer a way to extend the life of a hybrid vehicle's power source, reducing the need to oversize its battery packs. If UCs were paired with batteries, they could protect batteries from intense bursts of power such as those needed for acceleration, thereby reducing the need to oversize battery packs and extending the life of the batteries (NREL, 2004). UCs could also ensure that a vehicle can accelerate just as well at the end of its life as at the beginning. UCs are shown to improve the performance of batteries especially for start-stop ISA hybrids and mild hybrid systems. UCs can enable redesigning of batteries to hold more energy. Paired with UCs, batteries would not need to deliver bursts of power and so could be made with just a few layers of very thick electrodes, reducing the amount of supporting material needed. That could enable storing twice as much energy in the same space. The combined system has a better low-temperature performance and improved power and energy storage abilities. Disadvantages include increased energy storage cost and need for power converters.

Hybrid Technology Status

Table 4-5 presents a snapshot of the technology status and applications of the various hybrid system architectures currently available in the market.

Hybrid Electric Vehicles

Class 3 to 6 Straight Box Trucks

Hybrid type: Parallel heavy-duty hybrid system with no electric power takeoff (ePTO).

Fuel savings 20 to 25 percent were demonstrated depending on duty cycle. Table 4-6 shows the configuration used by three different manufacturers and the benefit obtained.

There are also a number of opportunities to continue to optimize the system. These include full electrification of accessories, which will allow engine shutdown at idle; integration of the hybrid system with emissions control;

[28]"Ford's Accelerated Battery Research Drives Development of Vehicle Electrification Plans." Available on the Ford Motor Company website at http://media.ford.com/article_display.cfm?article_id=30221.

TABLE 4-5 Different Vehicle Architectures, Their Status as of Today and Primary Applications

Architecture	Technology Status	Primary Applications
Medium-duty/heavy-duty parallel HEV	Available now: Eaton, Azure, Volvo	Refuse, urban pickup, and delivery (P&D)
Medium-duty/heavy-duty parallel HEV w/e PTO	Available now: Eaton, Volvo	Bucket truck
Parallel gasoline or diesel HEV bus	Available now: ISE, Enova, BAE	Transit bus
Two-mode diesel HEV bus	Allison	Transit bus
Series gasoline or diesel HEV bus	Available now: ISE	Transit bus
Parallel hydraulic hybrid	Introduced in 2009: Eaton, Parker Hannifin, Crane Carrier	Refuse, urban P&D
Series hydraulic hybrid	Demo vehicles	Refuse, urban P&D
Parallel Class 2b	Demo vehicles	Class 2b pickups and vans
Two-mode Class 2b	Demo vehicles	Class 2b pickups and vans
Line-haul dual-mode HEV	Demo vehicles	Line-haul tractor trailer,
Line-haul parallel HEV	Demo vehicles	Line-haul tractor trailer, motor coach

SOURCE: TIAX (2009).

TABLE 4-6 Production-Intent Medium-Duty and Heavy-Duty HEV Systems, No ePTO

System Attribute	Eaton	Volvo	Azure
Motor (peak)	44 kW	120 kW	100 kW
Battery	Li-ion, 2 kWh	Li-ion	Ni-MH 2.4 kWH 288 V
Percentage FC	20-25	20	23-25
No idle?	No	Yes	Yes
Electric launch	Yes	Yes—12 mph	Yes
Applications	Class 3-7 delivery; refuse soon	Refuse; delivery soon	Small delivery, shuttle

SOURCE: TIAX (2009).

engine downsizing in certain applications;[29] and improved integration and packaging. In combination these enhancements could improve fuel consumption benefits by another 5 to 10 percent (TIAX, 2009, page 4-81). Table 4-7 shows the predicted improvement that can be achieved with the parallel HEV of the future compared with the present.

[29]Note that both the scope for engine downsizing and the benefits are very limited in heavy-duty applications and apply more to light-duty cases. For example, in-line haul fuel consumption with an 11- to 13-liter engine is the same as with a 15-liter engine. Since the durability of the larger engine is better, the market has gone mostly to the larger engine, except in very weight sensitive applications (bulk haulers, tankers, etc.).

TABLE 4-7 Hybrid Technology, Benefits and Added Weight for Class 3 to Class 6 Box Trucks

Architecture	Percent Benefit (FC)	Weight Added (lb)
Parallel HEV	20-25	450
Parallel HEV, future	25-35	350

SOURCE: TIAX (2009).

TABLE 4-8 Hybrid Technology, Benefits and Added Weight for Class 3 to Class 6 Bucket Trucks

Architecture	Percent Benefit (FC)	Weight Added (lb)
Parallel HEV w/ePTO	30-40	~650
Future Parallel w/ePTO	35-45	~500

SOURCE: TIAX (2009).

TABLE 4-9 Hybrid Technology, Benefits and Added Weight for Refuse Haulers

Architecture	% Benefit (FC)	Weight Added (lb)
Parallel HEV	20	450
Parallel HEV (future)	25-30	~350
Parallel HEV, ePTO	25	~650
Parallel HEV, ePTO (future)	30-35	~500
Parallel HHV	20-25	1,000
Series HHV	40-50	~1,500

SOURCE: TIAX (2009).

Class: 3-6 Bucket Trucks

Hybrid type: Parallel hybrid electric with ePTO.

HEVs with an ePTO system have demonstrated a 30 to 40 percent reduction in fuel consumption and a nearly 90 percent reduction in idle time.[30] Table 4-8 compares this system with the predicted future of this type of vehicles.

Refuse Haulers

Hybrid type: Parallel hybrid electric with and without ePTO.

Fleet tests have demonstrated fuel consumption benefits on the order of 20 percent (TIAX, 2009, page 4-83) (see Table 4-9). These systems can also be tuned to maximize acceleration, which increases productivity by allowing a single vehicle to visit more houses per day. Systems tuned to maximize productivity have been shown to increase productivity by 11.5 percent while decreasing fuel use by 14 percent, for a total fuel savings of 26 percent. Both HEVs and HHVs also

[30]Freightliner, Run Smart, Business Class M2e Hybrid, AFVi Technology Showcase, May 2008. Available at www.oregonsae.org/Meetings/M2Hybrid/M2e_SAE.ppt.

offer a significant operation and maintenance (O&M) savings by more than doubling brake life. The grid-charged compactor option (where the compactor used in a refuse hauler can be operated using electricity obtained from the grid) reduces fuel use by 25 percent.

Tractor Trailers

Hybrid type: Dual mode and parallel hybrid electric.

A line-haul duty cycle limits the amount of energy that can be recovered from regenerative braking. Rather, the advantages of hybridizing in this segment come from electrifying accessories and hotel loads, as well as limited gains from energy recovery and launch assist.

A dual-mode HEV (Arvin Meritor's system; TIAX, 2009), and a parallel HEV (Eaton's system; Coryell, 2008) are considered. Over-the-road benefits (i.e., not including idle reduction) are estimated to range from 5 to 7 percent for the parallel system (NESCCAF, 2009; Coryell, 2008) and 6 to 9 percent for the dual-mode system.

These on-road benefits are supplemented by overnight hotel road reduction, which reduces fuel use by an additional 5 to 8 percent, similar to that observed for auxiliary power unit idle reduction systems. Hotel loads are met by running the engine for a few minutes (4 minutes per hour, according to an Eaton/PACCAR demonstration) to recharge the battery (Carpenter, 2007).

Transit Buses

Hybrid type: Series, parallel, and dual-mode hybrid electric.

Fuel consumption savings range from 12 to 50 percent depending on the severity of the duty cycle (see FTA, 2005, page 2). The Central Business District cycle fuel consumption is decreased some 50 percent compared to the nonhybrid diesel baseline. The benefit decreases to 18 and 12 percent for the arterial and commuter cycles, respectively. This decrease is expected since the hybrid system's benefit is increased for cycles with more stops and starts. In general, the fuel efficiency figures can be represented as shown in Table 4-10.

TABLE 4-10 Hybrid Technology, Benefits and Added Weight for Transit Buses

Architecture	Percent Benefit (FC)	Incremental Weight (lb)
Gasoline series	25-35[a]	2,000
Diesel series	30-40	2,600
Diesel parallel and dual-mode	22-35[b]	940-2,840

[a]Site visits.
[b]FTA (2005).
SOURCE: TIAX (2009).

Motor Coaches

Hybrid type: Parallel hybrid electric.

The Northeast Advanced Vehicle Consortium showed on-road fuel savings for an MCI motor coach of 5 to 40 percent on a high-speed suburban duty cycle (FTA, 2005).

Class 2b Pickup Trucks

Hybrid type: Parallel and dual-mode hybrid electric.

The fuel consumption benefit for a Class 2b dual-mode hybrid is estimated to range from 20 to 30 percent, based on estimates from Vyas et al. (2002) and EPA (2008).

Simulation-Based Assessment and Interpretation of HEV Potential for Fuel Savings

The previous section demonstrates that reported fuel-saving benefits resulting from the adoption of various HEV architectures can vary over a considerable range, even for the same truck class. This is attributed to numerous factors, including differences in component characteristics and sizes, power management strategies, duty cycles, and driver behavior. High-fidelity modeling and simulation can play an important role in exploring potential benefits and assessing alternatives using common boundary conditions, as well as parametrically assessing the effects of various critical variables. It is crucial to stress the importance of accurate, real-world driving cycles to evaluate the potential performance of different hybrid vehicle architectures and their performance compared to traditional vehicles. As an illustration of the power of modeling and simulation when applied to HV fuel consumption assessments, select architectures are compared under prescribed driving cycles. Further, this section reports on the impact of using real-world driving cycles, including the effect of removing the breaks from the highway cycle to more accurately reflect real-world conditions and the effect of the grade of the driving cycle on fuel consumption.

Argonne National Laboratory (ANL) has conducted a case study on mild- and full-hybrid versions of a parallel pre-transmission hybrid (see Figure 4-17) for a Class 8 long-haul truck and compared predicted fuel consumption with the conventional power train configuration. The "mild-hybrid" configuration augmented the baseline engine with a 50-kW motor and a 5-kWh battery, thus enabling engine idle off, torque assist, and regenerative braking. The "full-hybrid" configuration featured the baseline engine plus a 200-kW motor (traction), a 50-kW motor (starter + generator), and a 25-kWh battery, thus enabling all features of a mild hybrid plus operation in electric-only mode and long idle, (i.e., battery capable of meeting energy demands for 10 hours for full hotel stop).

The three versions of the truck (conventional, mild and full hybrid) were simulated on three highway cycles (HHDDT [Heavy Heavy-Duty Diesel Truck Schedule] 65, HHDDT Cruise, HHDDT High Speed) and two transient/urban cycles (HHDDT Transient, UDDS Truck). Tests were conducted

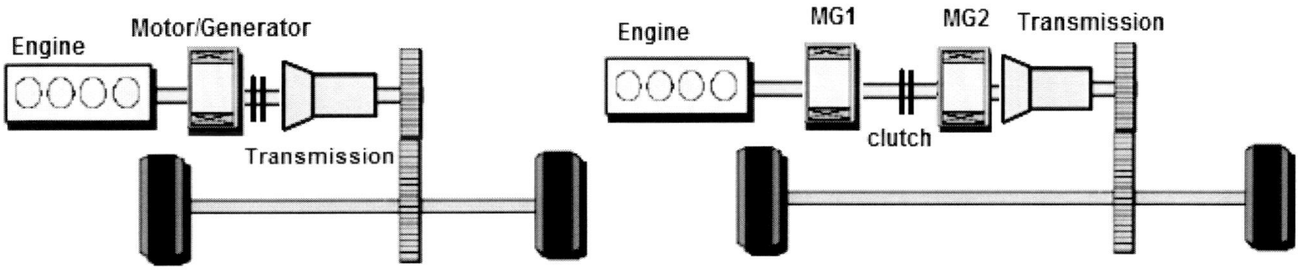

FIGURE 4-17 Hybrid configurations considered in ANL study. SOURCE: ANL (2009).

TABLE 4-11 Characteristics of Primary Drive Cycles

	Average Speed (mph)	Maximum Speed (mph)	Maximum Acceleration (m/s^2)	Maximum Deceleration (m/s^2)	Distance (miles)	Duration (s)	Time Stopped (%)
HHDT65	50	66.7	2	−2.8	26.5	1,904	5
HHDT Cruise	39.9	59.1	0.42	−0.59	23.1	2,083	6
HHDDT High Speed	50.2	66.1	0.69	−1.2	10.5	757	6
HHDDT Transient	15.3	47.5	1.32	−2.4	2.8	668	17
UDDS Truck	18.7	57.7	1.9	−2.1	5.5	1,060	33

SOURCE: ANL (2009).

at 50 percent load and full load. Table 4-11 gives important information about the drive cycles, which significantly affect the performance of the hybrid system, as will be shown in the results. The drive cycles used in the tests are shown in Table 4-12.

Figure 4-18 illustrates the predicted fuel savings of mild and full parallel hybrid configurations compared to a conventional power train for a Class 8 truck. For both configurations, the fuel savings are lower on the highway cycles, which is to be expected since the hybrid system does not help much at cruising speeds where the engine already operates efficiently. It should be noted that neither hybrid system has enough electrical storage to contribute to cruise power demand for any significant length of time. The mild-hybrid configuration shows fewer savings than the full hybrid, peaking at 11 percent, while the full hybrid can save up to 40 percent on an urban cycle. The fuel savings also tend to be lower with added mass, and this can be seen by the lower percentage benefit for the 100 percent load compared to the 50 percent load case.

Figure 4-19 shows the fraction of the total braking energy that is recovered at the wheel—meaning not including the driveline and electric machine losses involved in the channeling of that energy into the battery. The recovery rate depends on the cycle aggressiveness during deceleration. A heavier truck is more likely to reach its regenerative braking torque limitation sooner than a lighter one—hence the lower predictions for the fully loaded truck.

Figure 4-20 compares engine efficiencies for the conventional vehicle with the two hybrids over the drive cycles at two different loads. The mild hybrid does not show signifi-

TABLE 4-12 Profiles of Primary Drive Cycles

Drive Cycle	Profile
HHDDT 65	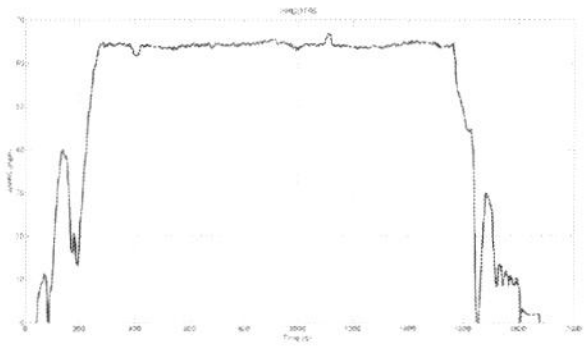
HHDDT Cruise	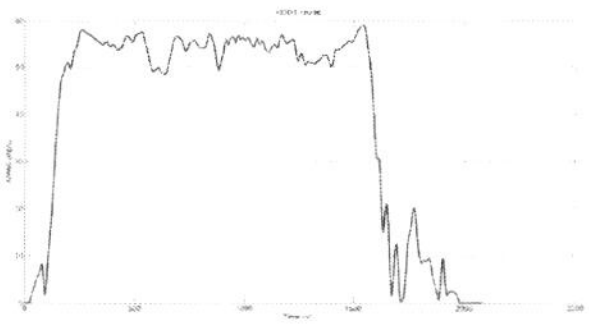
HHDDT High Speed	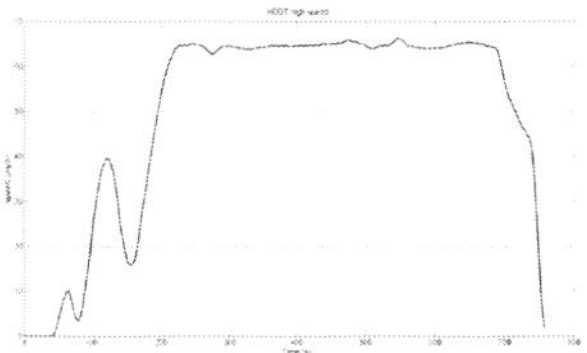
HHDDT Transient	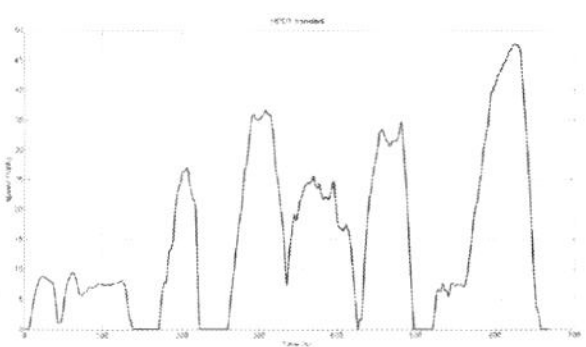
UDDS Truck	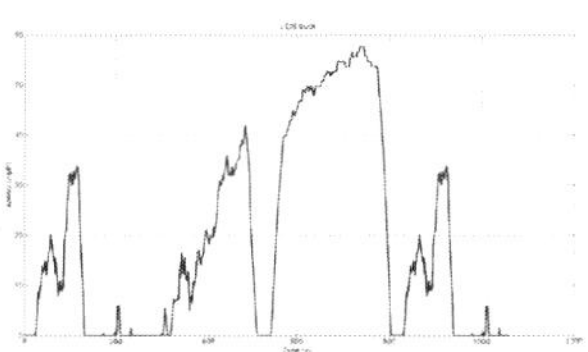

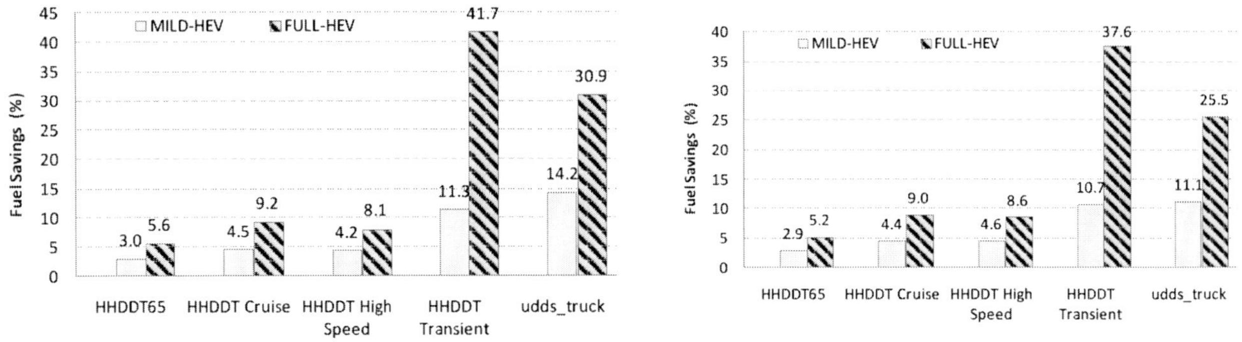

FIGURE 4-18 Fuel savings with respect to conventional cycles on standard drive cycles under (left) a 50 percent load and (right) a 100 percent load. SOURCE: ANL (2009).

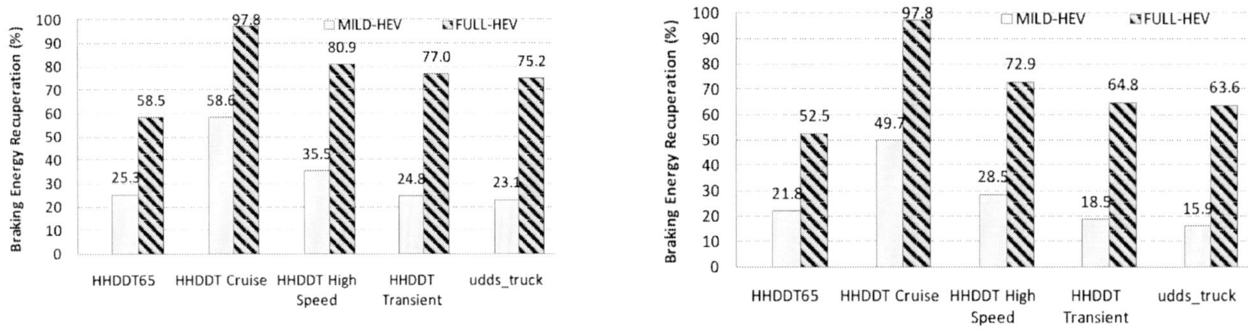

FIGURE 4-19 Percentage of braking energy recovered at the wheels under (left) a 50 percent load and (right) a 100 percent load. SOURCE: ANL (2009).

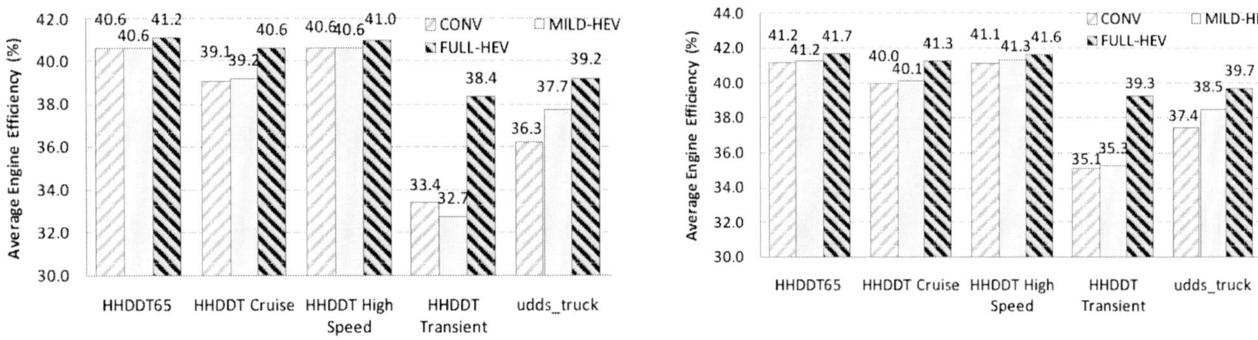

FIGURE 4-20 Percentage average engine efficiency of conventional and hybrid trucks for (left) a 50 percent load and (right) a 100 percent load on standard cycles. SOURCE: ANL (2009).

cant improvement since start-stop is the only main feature in it to aid in engine efficiency. The full hybrid gains in the transient and urban cycles as the engine can be completely switched off in electric-only mode.

Effect of Drive Cycle on Hybrid Performance

The drive cycle or duty cycle plays an important role in determining the following: type of hybrid technology to be used, level of hybridization and sizing of components, and power management strategy.

Effect of Removing Breaks from Highway Cycle

Since the HHDDT cycle is short and does not represent the real highway cycle, a new cycle was formulated with original acceleration followed by a cruising part and finally a deceleration part. Figure 4-21 shows how the new drive cycle was obtained from the HHDDT cycle. The results of removing stops from the HHDDT cycle have been grouped in Figure 4-22. In every other bar the hybrids are compared with the conventional vehicle, hence the values are greater.

When the breaks are removed from the highway cycle, there is a significant drop in fuel savings in the hybrid con-

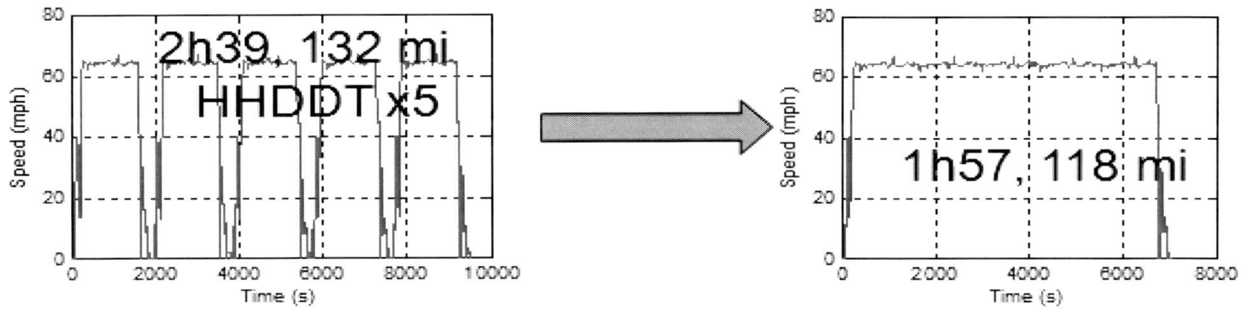

FIGURE 4-21 HHDDT 65 cycle repeated five times with stops (left) and without stops (right). SOURCE: ANL (2009).

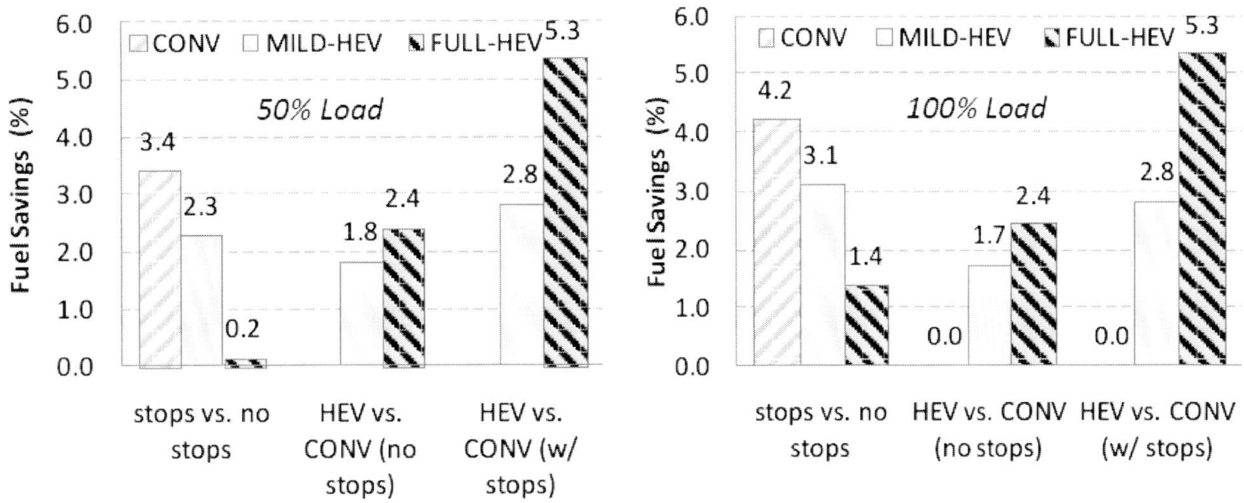

FIGURE 4-22 Fuel consumption reduction due to stop removal, with respect to conventional vehicles without stops, and with respect to conventional vehicles with stops (50 percent load on the left, 100 percent load on the right). SOURCE: ANL (2009).

figuration. However, the conventional configuration benefits the most when the stops are removed as the hybrids could have recovered part of the kinetic energy while braking. In the case of the full hybrid, the savings are more than halved (5.3 percent fuel saved on a cycle with stops, 2.4 percent fuel saved on a cycle without stops). In general, the hybrids still outperform the conventional vehicles in all cases as there are still some gains using the hybrid system even when the stops are removed.

Fuel Savings in Grades for Hybrid Configurations

Due to the lack of real-world drive cycles that include grades and to illustrate the potential benefits of hybridization in a "hilly" terrain, idealized sinusoidal road profiles were created. The elevation of such a road is a sinusoidal function of the horizontal distance, with a "hill" period varying between 1 and 3 km. Maximum grades also vary from 0 to 4 percent. All combinations of maximum grade and period were analyzed. Figure 4-23 shows the profile created.

For the mild-hybrid truck, the motor reaches its rated power when braking for grades 3 percent and higher when half-loaded and at or above 2.5 percent when fully loaded. The full hybrid hits its regenerative braking limit only when fully loaded at or above 3.5 percent grade. Thus, the full hybrid can capture more kinetic energy while braking, as expected.

The simulation results suggest that the mild hybrid has no advantage over the conventional vehicle when the grade is less than 2 percent, as there is not enough energy generated for accessories (see Figure 4-24).[31] Charge balancing is hard to achieve, so the engine might be used to charge and hence may result in higher than expected fuel consumption. Furthermore, there is not much reduction in fuel consumption with grades greater than 3 percent, as the electric machine will reach its maximum potential.

The results also indicate that the available fuel savings with the full-HEV configuration can increase by as much as

[31]Note that a simulation of vehicle performance on bus routes in San Francisco, where the grades can be demanding for conventional buses, found that the hybrid bus performed the best on fuel savings as well as on emissions of NO_x and particulate matter (SAE, 2004).

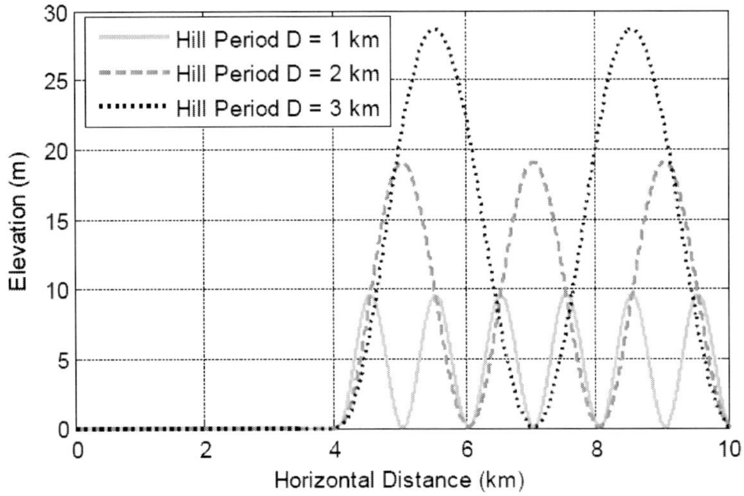

FIGURE 4-23 Representation of the grades considered. SOURCE: ANL (2009).

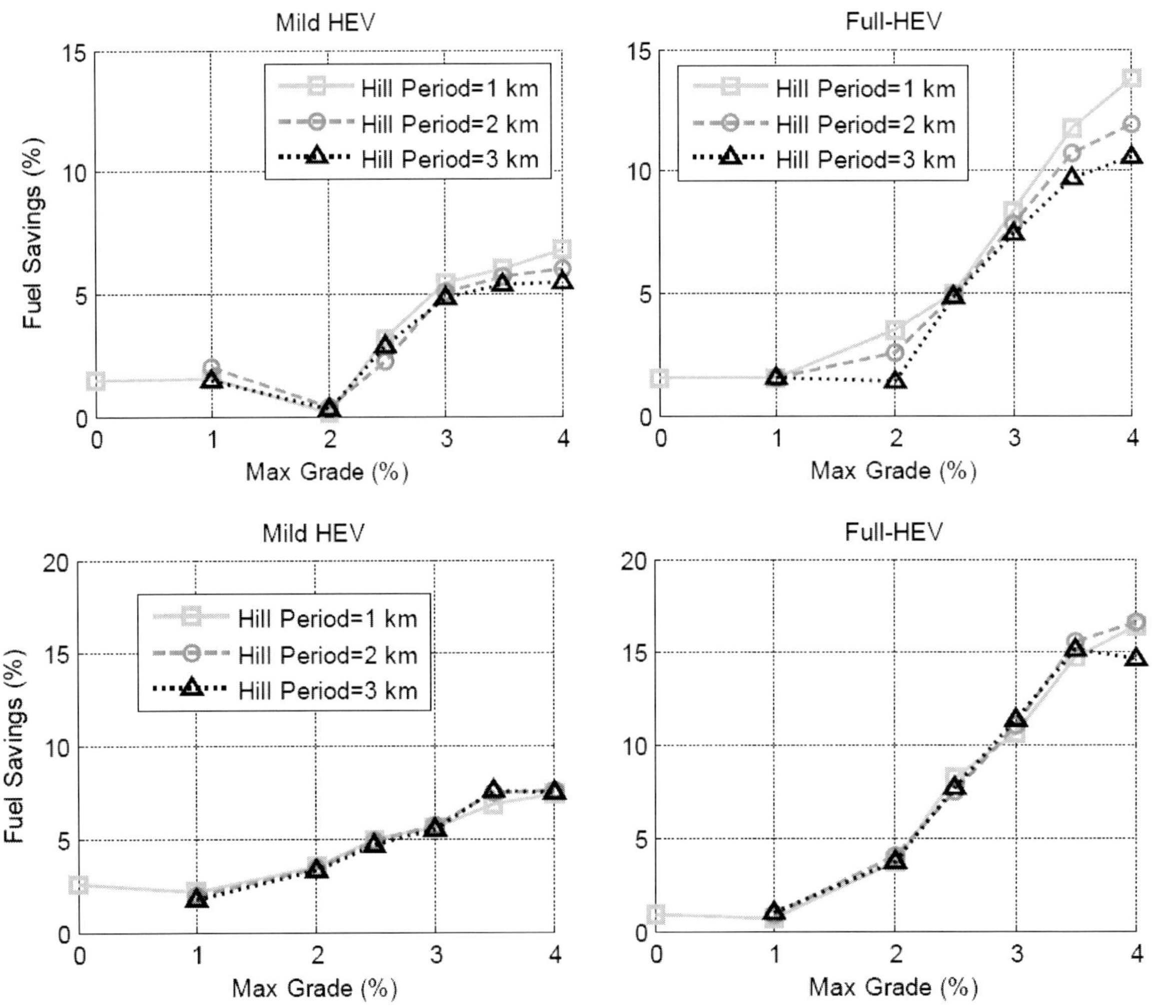

FIGURE 4-24 Fuel savings of hybrid trucks with respect to conventional trucks as a function of maximum grade for various hill periods; (left) 50 percent load and (right) 100 percent load. SOURCE: ANL (2009).

14 percent for a 4 percent maximum grade with a hill period of 1 km. This suggests that the value of hybridization in tractor-trailer trucks may be significant in hilly terrains.

Hydraulic Hybrid Vehicles

Hydraulic hybrids have demonstrated fuel savings for medium- and heavy-duty applications. The energy savings can be attributed to optimization of engine operation and regenerative braking energy absorption.

EPA has been actively involved in vehicle-level demonstrations of this technology by using hydraulic launch assist in retrofitted urban delivery trucks. A hydraulic package of a reversible hydraulic pump/motor and accumulators was added to the vehicle to reduce fuel consumption, while keeping the vehicle's conventional engine and transmission. Fuel savings of 25 and 45 percent were realized during city driving. For instance, Ford Motor Company, working jointly with EPA, demonstrated hydrualic power assist on a full-size sport utility vehicle platform fitted with a hydraulic pump/motor and valve block provided by infield technologies and carbon fiber accumulators developed by EPA. The pump/motor was connected to the vehicle driveshaft in parallel with the conventional power train. The vehicle demonstrated the ability to improve fuel economy by close to 24 percent on a start/stop, city-typical driving cycle (Kepner, 2002). Emissions were reduced by 20 to 30 percent. Better acceleration, reduced brake maintenance, and reduced operating costs (consumer payback of 4 to 5 years for city driving) were among other benefits (EPA, 2004). If the transmission and transfer case were replaced by a complete hydraulic drive train, larger fuel consumption benefits could be realized. In June 2006, EPA and United Parcel Service (UPS) demonstrated the world's first full hydraulic hybrid delivery truck, which realized 60 to 70 percent reductions in fuel consumption under urban driving conditions and up to 40 percent reductions in greenhouse gas emissions, with an estimated payback shorter than 3 years.

Currently, EPA is focusing more on full-series hydraulic hybrids, along with improved vehicle aerodynamics, tires and advanced ICEs, including HCCI gasoline engines, free piston engines, completely variable displacement engines, alcohol engines, and exhaust heat recovery systems. Eaton Corporation, in collaboration with the EPA, has developed a series hydraulic hybrid power system that combines a high-efficiency diesel engine and a unique hydraulic propulsion system to replace the conventional drive train and transmission (Eaton, 2009). The engine operates at its "sweet spot facilitated by the continuously variable transmission (CVT) functionality of the series hybrid hydraulic system and by regenerative braking." Fuel savings of 50 to 70 percent have been achieved, corresponding to a 40 percent reduction in greenhouse gases, 50 percent reduction in unburned hydrocarbons, and 60 percent reduction in particulate matter (EPA, 2009).

Simulation-Based Assessments of Hydraulic Hybrids

EPA predicted that the UPS prototype hydraulic hybrid vehicles would be able to capture and reuse 70 to 80 percent of the otherwise wasted braking energy (EPA, 2009).

Kim and Filipi (2007) from the University of Michigan conducted a simulation study on a series hydraulic hybrid light truck (mass = 5,112 kg). Approximately a 68 percent reduction in fuel consumption can be achieved in city driving and about 12 percent in highway conditions. The energy savings can be attributed to regeneration and engine shutdowns and a smaller fraction to optimization of engine operation. Design optimization over the complete driving cycle enabled right-sizing of all hydraulic pumps/motors, the accumulator volume, and the gear ratio of the two-step transmission. Downsizing the engine to roughly 75 percent of the baseline matched the acceleration of the conventional vehicle. In addition, it was found that having two smaller propulsion motors for each axle reduced fuel economy consumption by an additional 10 percent compared to a single propulsion motor architecture. The advantage is that the rear electric machine could be used mainly for acceleration and the front electric machine for regenerative braking, where both electric machines operate on high loads and in turn yield greater efficiency. Wu et al. (2004), also from the University of Michigan, developed a power management algorithm for a parallel hybrid and predicted fuel savings ranging from 28 to 48 percent depending on the types of pumps used.

Anderson et al. (2005) conducted a simulation study comparing hydraulic hybrid and electric hybrid vehicles. Fuel savings over a variety of driving cycles were found to be 39 percent for parallel hydraulic hybrids, compared to 31 and 34 percent improvements for parallel and series electric hybrids, respectively. Gotting (2007) estimated the fuel consumption benefit for the parallel HHV to range from 20 to 25 percent for Class 3 to 6 box trucks.

While indicative of the range of potential benefits, it should be noted that simulations are carried out under ideal conditions—hence results typically represent best-case scenarios. Real-world savings in fuel consumption are likely to be lower, because of off-design duty cycles and practical production vehicle constraints.

Power Management in Hybrid Vehicles

Power management is key in obtaining maximum performance and fuel savings from a hybrid vehicle. The objective of a power management algorithm in a hybrid vehicle is to compute the optimum operating point of the overall system for any amount of power demanded from the driver. The cost criteria are usually fuel economy and emissions. The way the power is managed will depend a lot on the sizing and characteristics of each of the components and their instantaneous state of operation. Mechanical efficiencies of the components (i.e., transmission, torque converter, differential), rolling resistance, aero drag coefficients, and instantaneous operating point of components, among others, are of great importance with respect to fuel consumption but are considered as given. The hybrid power train architecture is also assumed to be given. There are three general power management algorithm types, as outlined below.

Heuristic Rule-Based

This type of control is constituted by heuristic rules—for example, if/then statements to split the power demanded from the driver into the electrical and mechanical subsystems. The majority of these rules are thermostatic—that is, actions are triggered when certain conditions are met. They are simple and easy to implement and less computationally expensive to develop, but they need a significant amount of tuning in order to achieve better results compared to a conventional vehicle. Their basic simplicity means the system will not operate at its best potential at all times and there is significant room for improvement. A rule-based algorithm for a particular vehicle can never be readily used for another. Nevertheless, this type of control is often used due to its ease of implementation and lower cost.

Table 4-13 shows the difference in fuel economy figures obtained during simulation of a hybrid electric transit bus using three different rule-based control algorithms.

Chu et al. (2003) developed a series HEV military bus (15,000 kg) prototype model and formulated an energy management strategy enabling the ICE to operate in its peak efficiency zones for reduced fuel consumption. A parametric design methodology also was established. Simulation results on four different driving cycles show that, on average, fuel savings improve by about 17 percent and, acceleration times by 25 percent, with top speed and gradability being the same compared to a conventional vehicle.

Real Time

Lately, there has been a growing research effort toward developing real-time power management algorithms of the power split between the thermal and electrical paths of HEVs. The main aspects of this approach are concerned with (1) the self-sustainability of the electrical path, which must be guaranteed for the entire driving cycle since the storage system cannot be expected to be recharged by an external source (fuel converter primarily, brake regeneration secondarily) and (2) the fact that no, or only limited, a priori knowledge of future driving conditions is available.

Such algorithms consist of an instantaneous optimization (Sciarretta et al., 2004; Rodatz et al., 2005; Pisu and Rizzoni, 2007); the objective function in this optimization is fuel consumption and emissions. Decisions on the energy flow path (engine path and electrical path) can be evaluated based on an Equivalence Consumption Minimization Strategy. The equivalence between electrical energy and fuel energy can be evaluated by comparing the cost of energy produced at any instant. An instantaneous objective function combines the weighted sum of electrical energy and fuel energy and is evaluated with regard to selection of a proper equivalence factor value at any instant. If the engine can produce the required power more efficiently than the electrical system (sweet-spot operating points), the engine energy path is favored, and when the motor can produce power more efficiently than the engine (low speed/low load), the electrical path is favored.

The key in a real-time power management system is simultaneous optimization of the operating points of the entire system as a whole and not just the engine alone. It aims for the best overall efficiency from the engine to the wheels or from the battery pack to the wheels as the case may be. However, owing to greater interactions among the involved subsystems (engine, motor, battery), the control complexity can rise rapidly with the number of agents or their behavioral sophistication. This increasing complexity has motivated continuing research on computational learning methods toward making autonomous intelligent systems that can learn how to improve their performance over time while interacting with the driver. These propulsion systems need to be able to sense their environment and also integrate information from the environment into all decision making.

This challenge can be effectively addressed by applying principles of cognitive optimization techniques. The problem can be formulated as sequential decision making under uncertainty in which an intelligent system (e.g., hybrid-electric vehicle, power train system) learns how to select control actions so as to reduce fuel consumption over time for any different driving cycle (Malikopoulos, 2009).

Dynamic Programming

Power management control algorithms employing dynamic programming (Scordia et al., 2005; Lin et al., 2003; Perez et al., 2006; Ogawa et al., 2008; Karbowski et al., 2009) rely on computing off-line the optimal control policy with respect to the available power train variables—that is, the power split between the thermal (engine) and electrical

TABLE 4-12 Fuel Economy and Exhaust Emissions of Hybrid Electric Transit Bus with Various Control Strategies, Taipei City Bus Cycle

Hybrid Electric Bus	Fuel Economy (km/L)	CO (g/km)	HC (g/km)	NO_x (g/km)	PM (g/km)
Speed control	1.82	1.14	0.31	28.52	0.29
Torque control	2.02	0.71	0.27	20.50	0.26
Power control	2.15	0.70	0.24	18.98	0.23
Diesel bus					
Conventional control	1.29	4.12	0.59	55.56	0.61

SOURCE: Wu et al. (2008).

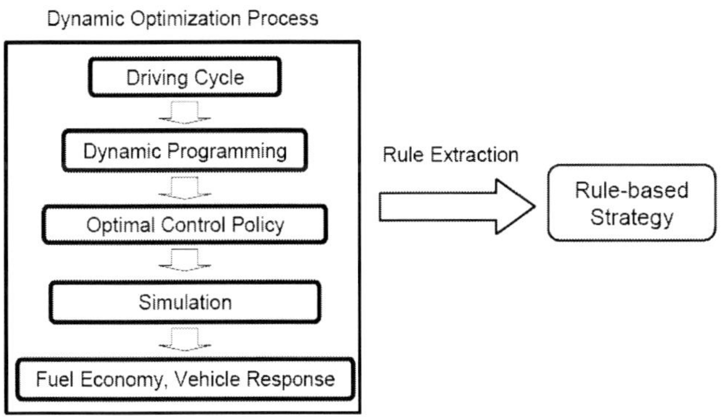

FIGURE 4-25 Dynamic programming process and rule extraction from the result. SOURCE: Lin et al. (2003).

paths (motor, generator, and battery), gear selection, etc., over a given driving cycle or a family of given driving cycles. Even though dynamic programming is not directly implementable, its results can be used to form an efficient rule-based control algorithm (as shown in Figure 4-25) or to develop vehicle-level control using neural networks (Delprat et al., 2001). The derived optimum policy is then approximated with simple rules and shift logic functions in order to be implemented in real time.

Figure 4-26 shows an example of a rule-based algorithm used in real time on an engine map to determine which of the power devices are being used depending on visitation points during the duty cycle. The main shortcoming of this approach is that it is efficient only for the driving cycles used in deriving the optimal policy. In addition, due to the high computational cost of dynamic programming, only simplified models of HEVs can be used. As a result, the extracted optimum policy omits a significant number of HEV dynamics that affects the efficiency of the derived policy even for the given driving cycle derived. Despite the aforementioned shortcomings, power management results based on deterministic dynamic programming methods are useful to serve as the benchmark of possible performance. Table 4-14 compares the predicted fuel consumption of a conventional vehicle against a hybrid power management optimized using dynamic programming and rule-based algorithms.

Crosscutting Issues and Future Outlook for Hybrids

During the committee's discussions with manufacturers and suppliers, a number of overarching themes emerged with respect to hybrid technology discussed.

Brake O&M Benefits

In addition to saving fuel, hybridization significantly reduces brake costs. Suppliers and OEMs expect that hybrids will more than double brake life. For some applications these savings can outweigh the fuel savings.

TABLE 4-14 Predicted Fuel Consumption Comparison: Conventional (nonhybrid), Dynamic Programming (DP), and Rule-Based (RB)

	DP	RB	Conventional
Mpg	13.85	12.65	10.39
Fuel (gallon)	0.5259	0.5757	0.7005

SOURCE: Lin et al. (2003).

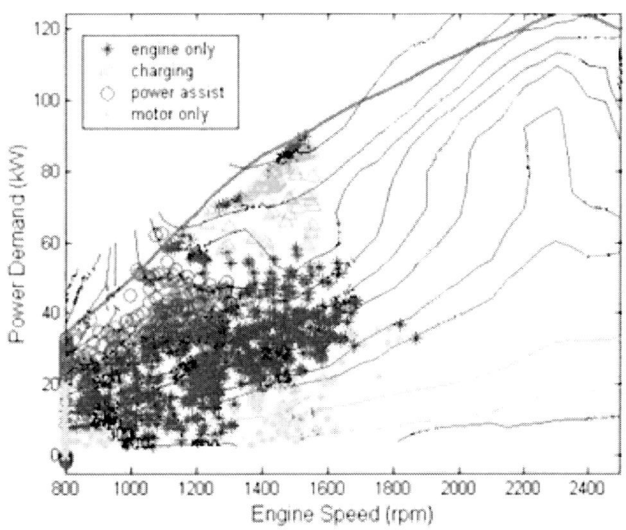

FIGURE 4-26 Implementing dynamic programming as a rule-based algorithm in SIMULINK. SOURCE: Lin et al. (2003).

System Integration

As hybrid systems become more fully integrated into overall vehicle architecture, there are a number of opportunities to further optimize the system:

- Depending on the application, hybridizing a vehicle can enable engine downsizing. Currently, this is not widely done. Part of the reason is that industry is still figuring out which applications can use a downsized engine without sacrificing mission performance. Another important factor is that smaller engines that meet medium-/heavy-duty warranty requirements may be unavailable. One supplier suggested that they would be able to offer optimized, right-sized power trains at volumes of 5,000 to 10,000 vehicles per year.
- There is opportunity for substantial integration between hybrid systems and exhaust aftertreatment systems. For example, a system designer could vary engine load to manage exhaust gas temperature, thereby offering emissions benefits. However, current standards measure engine-out emissions. To get credit for any benefit, EPA would need to switch to measuring emissions on a vehicle basis or develop a method to credit the lower vehicle emissions.
- Improved system power density, component efficiency, and design integration are expected to offer additional fuel-saving opportunities.
- Electrification is viewed as an enabler for more efficient waste-heat recovery systems, such as electric turbo-compounding or electric bottoming cycles. In a hybrid vehicle, electric waste-heat systems can offer an additional 1 to 2 percent efficiency benefit at neutral cost compared to an equivalent mechanical waste-heat system.
- By narrowing the design window, hybrid systems enable the engine to be optimized to deliver peak fuel economy within a narrow operating band. This opportunity applies primarily to dual-mode or series systems.

Taken as a whole, improved systems integration can offer an additional 5 to 10 percent improvement in fuel efficiency in future systems in the years 2015-2020. In tandem, higher sales volume can reduce costs by a factor of 2.

Hybrid Power Train Summary

In its report for the committee, TIAX (2009) summarized the hybrid fuel consumption potential reductions by range of years and by application, as shown in Table 4-15.

Based on work discussed in this chapter as well as on the TIAX summary for hybrid power trains, the committee estimated potential fuel consumption reduction as shown in Table 4-16.

TABLE 4-15 Hybrid Fuel Consumption Reduction Potential (percentage) Compared to a Baseline Vehicle Without a Hybrid Power Train, by Range of Years and Application

	2013-2015	2015-2020
Tractor trailer	NA	10
Class 6 box truck	22	30
Class 6 bucket truck	35	40
Refuse truck	20	25
Urban bus	30	35
Motor coach	NA	NA
Class 2b pickup and van	NA	18

NOTE: NA, not applicable.
SOURCE: TIAX (2009).

TABLE 4-16 Estimated Fuel Consumption Reduction Potential (percentage) for Hybrid Power Trains

Tractor trailer	5-10[a]
Class 6 box truck	20-35
Class 6 bucket truck	30-45
Refuse truck	20-35
Urban bus	12-50
Motor coach	5-40
Class 2b pickup and van	18-30

[a]Includes some reduction in hotel load, some idle reduction, and some electrification of accessories.

FINDINGS AND RECOMMENDATIONS

Diesel Engine Technologies

Finding 4-1. Many individual technologies for reducing load-specific fuel consumption of diesel engines were identified. Some technologies are being used in 2010 by nearly all manufacturers (common rail fuel injection and selective catalytic reduction, SCR), and some are being used by a limited number of manufacturers (turbocompounding and multiple turbochargers). One manufacturer, Cummins, has shown a roadmap for 49.1 percent thermal efficiency by 2016 and 52.9 percent by 2019, which are 14.5 and 20.6 percent reductions in fuel consumption, respectively, from a 2008 baseline, compared to current diesel fuel consumption. Significant technical challenges remain to be overcome before many of the fuel-saving technologies described in this section can be successfully implemented in production.

Gasoline Engine Technologies

Finding 4-2. Technologies exist today, or are under development, that offer the potential to reduce the fuel consumption of gasoline-powered vehicles operating in the medium-duty vehicle sector. The most beneficial technologies and the magnitude of fuel savings will be dependent on the configuration of the engine and the duty cycle of its application. Under optimal matching of technology and duty cycle, fuel consumption reductions of up to 20 percent appear to be possible compared to 2008 gasoline engines in the 2015 to 2020 time frame. The economic merit of integrating different fuel-saving technologies will be an important consideration for operators and owners in choosing whether to implement these technologies.

Recommendation 4-1. Development of fuel-saving engine technologies and their effective integration into the engine/power train is critical for reducing fuel use by medium- and heavy-duty vehicles and helping the nation to meet national goals related to energy security and the environment. The federal government should continue to support such programs in industries, national labs, private consulting companies, and universities.

Diesel Engines Versus Gasoline Engines

Finding 4-3. Diesel engines can provide fuel consumption advantages, compared to gasoline engines, of 6 to 24 percent depending on application, duty cycle, and baseline gasoline engines.

Finding 4-4. Diesel engines are increasing in cost primarily due to emissions aftertreatment equipment (DPF and SCR), which can cost over $17,000. Because of this cost increase (and diesel fuel prices), dieselization of Class 6 trucks in the new sales fleet went from 75.8 percent in 2004 to 58.0 percent in 2008. The effect of 2010 emission regulations has yet to be felt, but it is expected to accelerate the trend toward gasoline engines in medium-duty trucks.

Recommendation 4-2. Because the potential for fuel consumption reduction through dieselization of Class 2b to 7 vehicles is high, the U.S. Department of Transportation/National Highway Traffic Safety Administration (NHTSA) should conduct a study of Class 2b to 7 vehicles regarding gasoline versus diesel engines considering the incremental fuel consumption reduction of diesels, the price of diesel versus gasoline engines in 2010-2011, especially considering the high cost of diesel emission control systems, and the diesel advantage in durability, with a focus on the costs and benefits of the dieselization of this fleet of vehicles.

Transmission and Driveline Technologies

Finding 4-5. The transmission ratio and axle ratio affect fuel consumption by determining the engine speed versus road speed of the vehicle. A properly specified transmission and axle will allow the engine to run at its best fuel consumption operating range for a given road speed.

Finding 4-6. Manual transmissions have the least mechanical losses. An automated manual transmission can reduce fuel consumption by reducing driver variability (4 to 8 percent benefit). The fully automatic transmission can improve productivity by reducing the shift time (power shift) and by avoiding engine transient response delays and can reduce fuel consumption (up to 5 percent) by reducing driver variability, but the AT has higher parasitic losses.

Recommendation 4-3. The industry should continue its practice of training dealers and provide training materials for truck specifications affecting fuel consumption, such as transmission ratios, axle ratios, and tire size.

Hybrid Power Trains

Finding 4-7. Fuel consumption reductions on hybrid vehicles 5 to 50 percent have been reported by enabling optimum engine operation, downsizing in certain cases, braking energy recovery, accessory electrification, and engine shutdown at idle.

Finding 4-8. A wide range of hybrid electric and hydraulic architectures have been demonstrated. The selection of a particular system architecture depends mainly on application, duty cycle, and cost-benefit trade-offs.

Finding 4-9. The realized fuel consumption benefits of a particular hybrid technology and architecture implementation are strongly dependent on application and duty cycle. Optimization of component sizing and power management are keys to maximizing the potential for fuel consumption reductions while satisfying performance and emission constraints.

Finding 4-10. Computer simulation of medium- and heavy-duty vehicles is an effective way to predict fuel consumption reductions considering the additional variables in a hybrid vehicle system, but such systems are not standardized, leading to a wide variety of results and unpredictability.

Recommendation 4-4. NHSTA should support the formation of an expert working group charged with evaluating available computer simulation tools for predicting fuel consumption reduction in medium- and heavy-duty vehicles and developing standards for further use and integration of these simulation tools.

BIBLIOGRAPHY

Alamgir, M., and A.M. Sastry. 2008. Efficient batteries for transportation applications. Paper No. 08CNVG-0036. Presented at SAE International, Convergence Transportation Electronics Conference 2008, Detroit, Mich. Available at http://amsl.engin.umich.edu/publications/08SAE_Alamgir_Sastry.pdf.

Allison Transmission. 2008. Superior Fuel Efficiency. Allison Transmission Brochure SA5704EN. Indianapolis, Ind.

Allison Transmission. 2009. Allison fuel efficiency improvements. PowerPoint presentation provided to the committee by Allison Transmission. July.

Allison Transmission. 2009. Field Test Data on Class 8 Distribution Tractors. Excel spreadsheet summary provided to the committee by Allison Transmission. July.

Altoona Bus Research and Testing Center, Online Bus Database. Available at http://www.altoonabustest.com/.

Anderson, S.R., D.M. Lamberson, T.J. Blohm, and W. Turner. 2005. Hybrid route vehicle fuel economy. Advanced Hybrid Vehicle Powertrains 2005. SAE Paper 2005-01-1164. Warrendale, Pa.: SAE International.

Andrew, M. 2006. Lithium-ion: Enabling a spectrum of alternate fuel vehicles. Presentation by Johnson Controls-Saft to the Zero-Emission Vehicles (ZEV) Technology Symposium, Sacramento, Calif., Sept. 25-27.

ANL (Argonne National Laboratory). 2008. EnerDel/Argonne Advanced High-Power Battery for Hybrid Electric Vehicles. August. Available at http://www.transportation.anl.gov/batteries/enerdel.html.

ANL. 2009. Evaluation of Fuel Consumption Potential of Medium and Heavy Duty Vehicles Through Modeling and Simulation.

Baseley, S.J., C. Ehret, E. Greif, and M. Kliffken. 2007. Hydraulic hybrid systems for commercial vehicles. SAE Paper 2007-01-4150. Warrendale, Pa.: SAE International.

Chu, L., W. Wang, and Q. Wang. 2003 Energy management strategy and parametric design for hybrid electric military vehicle. SAE Paper 2003-01-0086. Warrendale, Pa.: SAE International.

Coryell, R. 2008. PACCAR—Eaton Hybrid Collaboration: Taking HD Hybrid from Concept to Commercialization. Presented at the NESCCAF/ICCT Workshop on Improving the Fuel Economy of Heavy-Duty Fleets II, San Diego, Calif. Feb. 20.

Delprat, S., T.M. Guerra, and J.Rimaux. 2001. Optimal Control of a Parallel Powertrain. Presented at the International Electric Vehicle Symposium (EVS18), Berlin. October.

DOE, EERE. 2009. 2008 Vehicle Technologies Market Report. Golden, Colo.: National Renewable Energy Laboratory. July. Available at http://www.nrel.gov/docs/fy09osti/46018.pdf.

Eaton EPRI press release. Available at http://www.reuters.com/article/pressRelease/idUS195396+19-Mar-2009+BW20090319.

Eaton-HTUF (Hybrid Truck User's Forum) Conference. 2009. Eaton hybrid hydraulic technology & development. Slide 3. Atlanta. October 27-29.

Eaton. 2009. Series Hybrid Hydraulic. Available at http://www.eaton.com/EatonCom/ProductsServices/Hybrid/SystemsOverview/Series Hydraulic/index.htm.

EPA (U.S. Environmental Protection Agency). 2004. Hydraulic Hybrid Technology—A Proven Approach. EPA brochure 420-F-04-024. March. Available at http://www.epa.gov/oms/technology/420f04024.pdf.

EPA. 2009. Region 9: Air Programs. Hydraulic Hybrid Vehicles. Available at http://www.epa.gov/region09/air/hydraulic-hybrid/.

EPRI (Electric Power Research Institute). 2008. Plug-In Hybrid Trouble Truck: An EPRI/Utility Alliance with Eaton Corporation and Ford Motor Company. Palo Alto, Calif.: EPRI. Available at http://mydocs.epri.com/docs/public/000000000001016496.pdf.

FTA (Federal Transit Administration). 2005. Analysis of Electric Drive Technologies for Transit Applications: Battery-Electric, Hybrid-Electric, and Fuel Cells. U.S. Department of Transportation, Report FTA-MA-26-7100-05.1. August. Available at http://www.navc.org/Electric_Drive_Bus_Analysis.pdf.

Freightliner vehicle specification information provided by the Columbia Truck Center, a Freightliner dealership in Columbia, S.C., July 2009.

Freightliner Business Class Data Book PRL-72E.013

GM BMW DaimlerChrysler. 2006. Hybrid Kooperation. Internationales Wiener Motorensymposium.

Gotting, G. 2007. Hydraulic Launch Assist HLA System. Eaton. Presented to the Houston Advanced Research Center.

Hanson, R., R. Reitz, D. Splitter, and S. Kokjohn. 2010. An experimental investigation of fuel reactivity controlled PCCI combustion in a heavy-duty engine. SAE Paper 2009-01-0864. Warrendale, Pa.: SAE International.

ICCT (International Council on Clean Transportation). 2008. Japan's "Top Runner" Fuel Economy Standards for Heavy-Duty Vehicles. Washington, D.C.: ICCT. February 11.

Kalhammer, F., B. Kopf, D. Swan, V. Roan, and M. Walsh. 2007. Status and Prospects for Zero Emissions Vehicle Technology. Report of the ARB Independent Expert Panel 2007. Prepared for the State of California Air Resources Board (ARB), Sacramento, Calif. April 13.

Karbowski, D. 2007. Plug-in Vehicle Control Strategy: From Global Optimization to Real Time Application. Presented at the International Electric Vehicle Symposium, EVS23, Anaheim, Calif. December 2-5.

Karbowski, K., F. von Pechmann, S. Pagerit, J. Kwon, and A. Rousseau. 2009. Fair comparison of powertrain configurations for plug-in hybrid operation using global optimization. SAE Paper 2009-01-1383. Presented at the SAE World Congress, Detroit, April.

Kepner, R.P. 2002. Hydraulic power assist—A demonstration of hydraulic hybrid vehicle regenerative braking in a road vehicle application. SAE Paper 2002-01-3128. Warrendale, Pa.: SAE International.

Kim, Y.J., and Z.S. Filipi. 2007. Simulation study of a series hydraulic hybrid propulsion system for a light truck. SAE Paper 2007-01-4151. Warrendale, Pa.: SAE International.

Kruiswyk, R. 2008. Annual Progress Report 2008 Vehicle Program: Advanced Combustion Engine Technologies. Prepared for the U.S. Department of Energy by Caterpillar. Washington, D.C.: U.S. Department of Energy.

Lin, C.-C., H. Peng, J.W. Grizzle, and J.-M. Kang. 2003. Power management strategy for a parallel hybrid electric truck. *IEEE Transactions on Control Systems Technology*, 11(6): 839-849.

Lin, C., et al. 2003. Control system development for an advanced-technology medium-duty hybrid electric truck. SAE Paper 2003-01-3369. Warrendale, Pa.: SAE International.

Liu, J., and H. Peng. 2008. Modeling and control of power-split vehicle, *IEEE Transactions on Control Systems Technology*, 16(6): 1242-1251.

Malikopoulos, A.A. 2009. Convergence properties of a computational learning model for unknown Markov chains. *Journal of Dynamic Systems, Measurement and Control*, 131(4).

NESCCAF/ICCT, Southwest Research Institute, and TIAX. 2009. Reducing Heavy-Duty Long Haul Combination Truck Fuel Consumption and CO_2 Emissions. ICCT. October.

NHTSA (National Highway Traffic Safety Administration) 2009. Average Fuel Economy Standards Passenger Cars and Light Trucks Model Year 2011. Docket No. NHTSA-2009-0062. Washington, D.C.: U.S. Department of Transportation. March.

NRC (National Research Council). 1992. Automotive Fuel Economy: How Far Should We Go? Washington D.C.: National Academy Press.

NRC. 2008. Review of the 21st Century Truck Partnership. Washington, D.C.: The National Academies Press.

NRC. 2010. Transitions to Alternative Transportation Technologies: Plug-in Hybrid Electric Vehicles. Washington, D.C.: The National Academies Press.

NREL (National Renewable Energy Laboratory). 2004. Milestone Report 20040901. Golden, Colo.: NREL.

Ogawa, T., H. Yoshihara, S. Wakao, K. Kondo, and M. Kondo. 2008. Design estimation of the hybrid power source railway vehicle based on the multiobjective optimization by the dynamic programming. *IEEE Transactions on Electrical and Electronic Engineering*, 3(1): 48-55.

Patton, K.J., A.M. Sullivan, R.B. Rask, and M.A. Theobald. 2002. General Motors Corporation aggregating technologies for reduced fuel consumption: A review of the technical content in the 2002 National Research Council report on CAFE. SAE Paper 2002-01-0628. Warrendale, Pa.: SAE International.

Perez, L.V., G.R. Bossio, D. Moitre, and G.O. Garcia. 2006. Optimization of power management in an hybrid electric vehicle using dynamic programming. *Mathematics and Computers in Simulation*, Vol. 73, Special Issue, pp. 244-254.

Pesaran, A., T. Markel, M. Zolot, and S. Sprik. 2005. Ultracapacitors and Batteries in Hybrid Vehicles. Presented at Advanced Capacitors, San Diego, Calif. July 11-13. National Renewable Energy Laboratory document NREL/PR-540-40630. Available at http://www.nrel.gov/vehiclesandfuels/energystorage/pdfs/38484.pdf.

Pisu, P., and G. Rizzoni. 2007. A comparative study of supervisory control strategies for hybrid electric vehicles, *IEEE Transactions on Control Systems Technology*,15(3): 506-518.

Rodatz, P., G. Paganelli, A. Sciarretta, and L.Guzzella. 2005. Optimal power management of an experimental fuel cell/supercapacitor-powered hybrid vehicle. *Control Engineering Practice*, 13(1): 41-53.

SAE (Society of Automotive Engineers). 2004. Analysis of the performance and emissions of different bus technologies on the city of San Francisco Routes. SAE Paper 2004-01-2605. Warrendale, Pa.: SAE International.

Sciarretta, A., M. Back, and L. Guzzella. 2004. Optimal control of parallel hybrid electric vehicles. *IEEE Transactions on Control Systems Technology*, 12(3): 352-363.

Scordia, J., M. Desbois-Renaudin, R. Trigui, B. Jeanneret, F. Badin, and C. Plasse. 2005. Global optimisation of energy management laws in hybrid vehicles using dynamic programming. *International Journal of Vehicle Design*, 39(4): 349-367.

Snyder, K.A., X.G. Yang, and T.J. Miller. 2009 Hybrid vehicle battery technology—The transition from NiMH to Li-Ion. SAE Paper 2009-01-1385. Warrendale, Pa.: SAE International.

Stein, R., C. House, and T. Leone. 2009. Optimal use of E85 in a turbocharged direct injection engine. SAE Paper 2009-01-01490. Warrendale, Pa.: SAE International.

Tavares, Fernando, Rajit Johri, and Zoran Filipi. 2009. Simulation study of advanced variable displacement engine coupled to power-split hydraulic hybrid powertrain. ICES2009-76113. Proceedings of the ASME Internal Combustion Engine Division 2009 Spring Technical Conference ICES2009 May 3-6, Milwaukee, Wisc.

TIAX. 2009. Assessment of Fuel Economy Technologies for Medium- and Heavy-Duty Vehicles. Report prepared for the National Academy of Sciences by TIAX LLC, Cupertino, Calif. July 31.

Vuk, C. 2006. Electric Turbo Compounding. A Technology Whose Time Has Come. Presented at the Diesel Engine-Efficiency and Emissions Research (DEER) Conference, August 20-24, Detroit, Mich.

Vyas, A, C. Saricks, and F. Stodolsky. 2002. The Potential Effect of Future Energy-Efficiency and Emissions-Improving Technologies on Fuel Consumption of Heavy Trucks. Argonne National Laboratory (ANL) Report ANL/ESD/02-4. Argonne, Ill.: ANL.

Wu, B., C.-C. Lin, Z. Filipi, H. Peng, and D. Assanis. 2004. Optimal Power Management for a Hydraulic Hybrid Delivery Truck. Available at http://www-personal.umich.edu/~hpeng/VSD_from_AVEC_HHV.pdf.

Wu, Y.-Y., B.-C. Chen, and K.D. Huang. 2008. The effects of control strategy and driving pattern on the fuel economy and exhaust emissions of a hybrid electric bus. SAE Paper 2008-01-0306. Warrendale, Pa.: SAE International.

Zhao et al. (eds.) 2003. *Homogeneous Charge Compression Ignition (HCCI) Engines: Key Research and Development Issues.* Warrendale, Pa.: SAE International.

5

Vehicle Technologies for Reducing Load-Specific Fuel Consumption

The technologies that can be used to reduce fuel consumption in medium- and heavy-duty vehicles vary by vehicle type, duty cycle, and the year the technology becomes available. For instance, a Class 8 tractor trailer operating on the interstate will benefit from technologies that improve aerodynamic performance and reduce rolling resistance, whereas a Class 2b pickup truck will benefit little from these technologies. This chapter first reviews the ways in which energy is lost in the operation of medium- and heavy-duty vehicles. It then reviews technologies and techniques for reducing the fuel consumption of these vehicles, including technologies that improve aerodynamic performance and that reduce rolling resistance, auxiliary loads, and idle. It also covers mass and weight reduction, and intelligent vehicle technologies.

VEHICLE ENERGY BALANCES

The potential efficiency improvements being considered in this study can be illustrated by reviewing the energy losses for the various vehicle classes. The U.S. Department of Energy (DOE) 21st Century Truck Partnership Technology Roadmap (DOE, 2006) provides the following tables for energy losses. The engine losses were calculated from a typical accounting of fuel energy usage, such as that shown in Figure 5-1. The engine losses are primarily a result of heat transfer to the coolant and heat loss through the exhaust. The remaining energy is used to power the vehicle and auxiliaries under the conditions set forth in the tables below. In Figure 5-1, engine accessories are components essential to engine operation, such as the fuel pump, water

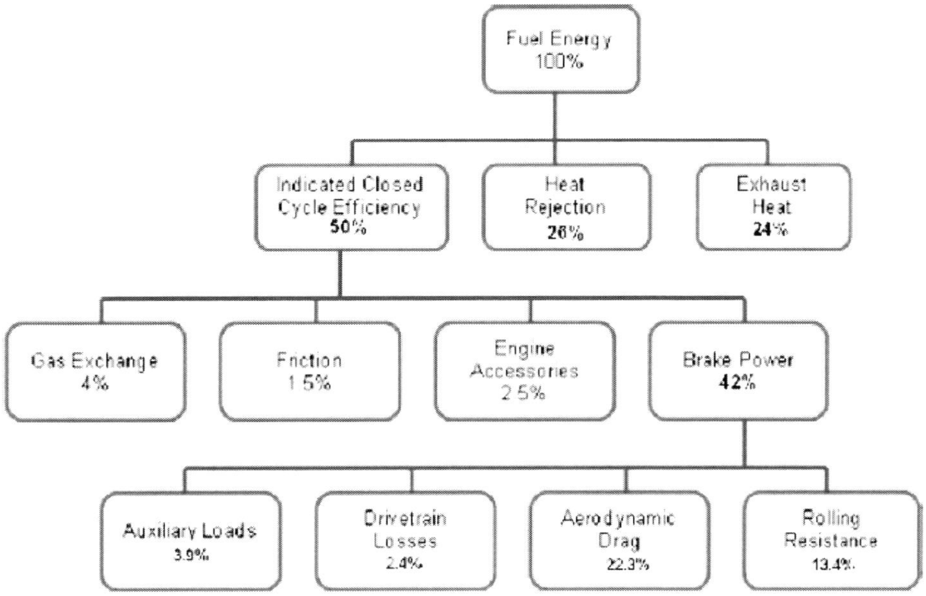

FIGURE 5-1 Energy balance of a fully loaded Class 8 tractor-trailer on a level road at 65 mph, representing the losses shown in Table 5-1. SOURCE: TIAX (2009).

pump, and oil pump, while auxiliary loads are accessories used in a vehicle's operation, such as the power steering, air compressor, cooling fan, and air-conditioning compressor.

The energy losses for a Class 8 tractor with a 53-ft van trailer, fully loaded to 80,000 lb gross vehicle weight (GVW) and operating on a level road at 65 mph, for one hour, are shown in Table 5-1. The energy losses for a Class 3 to 6 medium duty truck, loaded to 26,000 lb GVW and operating on a level road at 40 mph for 1 hour, are shown in Table 5-2. (Note that this steady-state operating point is not typical of the duty cycle for a Class 3 to 6 medium-duty truck.) The energy losses for a 40-ft transit bus with one-half seated load (32,000 lb) and the air conditioning on, operating over the central business district cycle, for 1 hour, are shown in Table 5-3. Note the high percentage of energy devoted to auxiliary loads.

Vehicle energy balances such as those described in Tables 5-1 through 5-3 identify the energy required to propel a vehicle down the road at a specific speed and with a specific load. The following sections break down these areas of losses as follows: aerodynamics, auxiliary loads, rolling resistance, vehicle mass (weight), and idle reduction.

TABLE 5-1 Energy Balance for a Fully Loaded Class 8 Vehicle Operating on a Level Road at 65 mph for One Hour

Energy Sources	Baseline (kWh)	Baseline (hph)
Engine losses per hour	240	321.8
Auxiliary loads	15	20.1
Drivetrain energy	9	12.1
Aerodynamic energy	85	114.0
Rolling resistance energy	51	68.4
Total energy used per hour	400	536.4

NOTE: hph, horsepower-hour.
SOURCE: TIAX (2009), p. 2-3.

TABLE 5-2 Energy Balance for a Fully Loaded Class 3 to 6 Medium-Duty Truck (26,000 lb) Operating on a Level Road at 40 mph for One Hour

Energy Sources	Baseline (kWh)	Baseline (hph)
Engine losses per hour	73.1	98.0
Auxiliary loads	1.5	2.0
Drivetrain energy	3.3	4.4
Aerodynamic energy	18.9	25.3
Rolling Resistance energy	23.0	30.8
Total energy used per hour	119.8	160.6

NOTE: hph, horsepower-hour.
SOURCE: TIAX (2009), p. 2-5.

TABLE 5-3 Energy Balance for a 40-ft Transit Bus Operating over the Central Business District Cycle for One Hour

Energy Sources	Baseline (kWh)	Baseline (hph)
Engine losses per hour	86.8	116.4
Auxiliary loads	36.4	48.8
Drivetrain energy	13.4	18.0
Aerodynamic energy	1.3	1.7
Rolling resistance energy	7.2	9.7
Total energy used per hour	145.1	194.6

NOTE: Transit bus with one-half seated load (32,000 lb) and air conditioning on. [Baseline] hph, horsepower-hour.
SOURCE: TIAX (2009), p. 2-6.

AERODYNAMICS

Truck Aerodynamics

Anyone comparing the commanding size of a tractor-van trailer combination to a small sedan or even a full-size sport utility vehicle (SUV) understands that the aerodynamic drag[1] of these large vehicles exceeds that of any light-duty vehicle. Some quantitative comparisons of those differences will be made later. For now, consider again the energy summary for the tractor-van trailer given in Table 5-1 and Table 5-4. Clearly, for this class of truck, aerodynamic load reduction is a key for successful fuel consumption reduction.

Early Studies

K.R. Cooper of the Canadian National Research Council summarized some of the earliest heavy truck wind tunnel testing performed in 1953 at the University of Maryland (Cooper, 2004). Many of the aerodynamic design solutions now available or being developed for Class 8 tractors and box van trailers were evaluated in that 1953 study. Those devices were shown to reduce aerodynamic drag by about 50 percent as compared to the predominant truck configurations of the 1950s (the cab-over-engine tractor). The "near-practical" streamlined result is shown in Figure 5-2.

Airshield introduced a commercial cab roof-top air deflector in about 1965. This device received some trucking company interest, especially after the 1973 petroleum crisis.

TABLE 5-4 Operational Losses from Class 8 Tractor with Sleeper Cab-Van Trailer at 65 mph and GVW of 80,000 lb

Operating Load	Power Consumed (hp)	Power Consumed (%)
Aerodynamic	114	53
Rolling resistance	68	32
Auxiliaries	20	9
Drivetrain	12	6
Braking	0	0
Total	214	100

SOURCE: DOE (2008).

[1]Aerodynamic drag refers to forces that oppose the motion of a vehicle through air.

FIGURE 5-2 University of Maryland, streamlined tractor, closed gap, three-quarter trailer skirt, full boat tail. SOURCE: Cooper (2004), p. 15, Fig. 4, Case 8. Reprinted with kind permission of Springer Science and Business Media.

Industry interest was encouraged by the graphic illustrations produced by the National Research Council of Canada, as shown in Figure 5-3.

Recent History

The introduction of the Kenworth T-600 in 1985 marked the industry's first serious attempt to incorporate aerodynamic improvements in truck tractors (see Figure 5-4). The T-600 included features such as a streamlined hood and fenders, an aerodynamic faired bumper, fuel tank fairings, and air filters mounted under the hood. These changes resulted in a significant reduction in aerodynamic drag compared to contemporary tractor models.

Continuing development led to additional improvements such as cab extenders to reduce the gap between tractor and trailer, more aerodynamic mirrors, and full-length side fairings. By 1990 all major truck/tractor manufacturers had introduced aerodynamic models, although "traditional" models continue to be available. See Figure 5-5 for identification of the common aerodynamic features.

FIGURE 5-3 National Research Council of Canada: Smoke pictures, cab with deflector (right). SOURCE: Cooper (2004), p. 11, Fig. 2. Reprinted with kind permission of Springer Science and Business Media.

FIGURE 5-4 Kenworth 1985 T600 aerodynamic tractor. SOURCE: Photo courtesy of Kenworth Truck Company.

Early Efforts Toward Engineering Measurements

The development of aerodynamic features led to a need for test procedures that could quantify the performance of these features. Most industry standards were developed through a consensus process by the Society of Automotive Engineers (SAE). Some of the widely used standards are as follows:

- *Coast-down tests on a track.* By 1976 SAE issued a procedure to quantify and standardize this test, SAE J1263. This standard allows estimation of both the coefficient of aerodynamic drag, C_d, and the rolling resistance coefficient, C_{rr}. SAE J1263 was intended primarily for passenger car applications, and the accuracy and repeatability of this procedure may be inadequate for heavy duty vehicles.
- *Wind tunnel tests.* Truck companies began performing wind tunnel evaluations. Since a limited number of wind tunnels are available that can handle full-size trucks, scale models were widely used. By 1981, SAE had developed a recommended practice, SAE J1252,

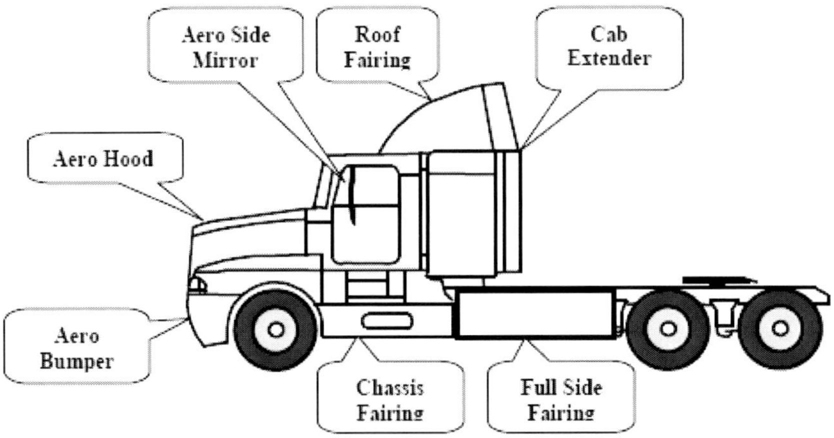

FIGURE 5-5 Aerodynamic sleeper tractor aerodynamic feature identification.

for wind tunnel tests. This procedure is the only accurate method of determining wind-averaged drag, by accounting for the effects of side wind. See Chapter 2 for a description of the SAE J1252 test procedure.
- *Fuel consumption measurement with full-size trucks.* In 1986 the SAE and the Truck Maintenance Council (TMC) introduced the joint TMC/SAE Fuel Consumption Test Procedure—Type II, SAE J1321, which provides the in-service fuel consumption of one test vehicle compared to a control truck. This test allows for correction of parameters outside the control of the researchers performing the test, such as ambient conditions and wind.

SmartWay Partnership of the U.S. Environmental Protection Agency

The U.S. Environmental Protection Agency's (EPA) partnership with the truck industry to reduce emissions (especially greenhouse gases) and fuel consumption alike is described in Chapter 3. The EPA's SmartWay truck specification has a significant dependence on improved aerodynamic performance. It requires certain fuel consumption reducing aerodynamic design features to be applied on tractor-van trailer combinations (Figure 5-6):

- The SmartWay tractor must be a high-roof aerodynamic sleeper cab, with aerodynamic bumper, mirrors, side truck fairings, side extender fairings, and roof fairings. All six domestic tractor manufacturers now supply SmartWay compliant tractors.
- The SmartWay-certified van trailer must be equipped with side skirt fairings plus either a trailer boat tail or a tractor-trailer gap fairing. The combined trailer aerodynamic treatment is estimated to achieve at least a 5 percent fuel consumption reduction compared to a standard trailer. At least eight trailer manufacturers supply SmartWay-compliant trailers (EPA, 2009).

As noted in Chapter 3, California adopted the SmartWay specification and validation processes as integral to its Global Warming Solutions Act in December 2008. That adoption provided for a mandatory introduction schedule of these fuel-saving specifications for both tractors and trailers in the 2011-2013 period (CARB, 2008).

Technology of Aerodynamic Improvements

The standard metric for comparing aerodynamic losses is the drag coefficient, C_d (see in Chapter 2 the section "Truck Tractive Forces and Energy Inventory"). Vehicle designers seek to minimize the drag coefficient in order to reduce fuel consumption at higher vehicle speeds, where aerodynamic drag represents a substantial fraction of the energy needed to keep the vehicle moving.

Drag coefficients for current aerodynamically designed tractors with smooth-sided van trailers (T-T) are about 0.6 to 0.65, which is higher than the values normally found in light-duty vehicles. Most current automobile sedans achieve a C_d of about 0.3 to 0.4, and the C_d of SUVs is typically 0.4 to 0.5. The higher C_d values for tractor trailers is primarily due to the fact that they are essentially large boxes optimized for the movement of freight.

Drag coefficient values are usually measured in wind tunnel tests. The C_d of a vehicle directly facilitates the computation of aerodynamic energy loss. Further, the wind tunnel test can provide a relatively simple and precise method to evaluate the wind yaw[2] effects on a vehicle's C_d (Cooper, 2004). An accurate and repeatable process for establishing C_d values is essential for the successful application of whole-truck computer modeling to evaluate fuel consumption effectiveness of various drag-reducing devices.

Each truck tractor manufacturer has developed in-house processes to validate the technical performance of their aerodynamic solutions. Manufacturers do not publish C_d values, evidently because the procedures used by different manufacturers are not known to be directly comparable. Manufacturers have not agreed to use a common standard such as SAE J-1252.

Aerodynamic Energy Loss

As described in the Road Load Power paragraph in Chapter 2, the resisting aerodynamic horsepower is proportional to $C_d \times A \times V^3$, where A = frontal area and V = forward velocity. This illustrates the important role of vehicle speed on aerodynamic horsepower loss. It is helpful to graphically display aerodynamic power consumption as a function of road speed. Consider Figure 5-7, where the blue ($C_d = 0.625$) curve is typical of today's tractor-trailer combination. The green curve represents a 20 percent reduction in the C_d, and therefore in the aerodynamic power loss. A 20 percent reduction in C_d results in a fuel consumption reduction of about 10 percent at 65 mph (TMA, 2007, p. 10).

Figure 5-7 also shows the curve for power consumed by tire rolling resistance. As described in the Chapter 2 paragraph on road load power, the tires' rolling resistance power loss is proportional to: $C_{rr} \times W \times V$, where W = vehicle gross weight and V = forward velocity. The C_{rr} value used in the figure is typical of those on current tractor trailers but not of the lower value required for SmartWay certification.

Heavy-duty tractors not only have higher C_d values than light-duty vehicles, they also have a frontal area that is 3 to 3.7 times larger than cars and SUVs. The large frontal area is driven by the need to package a large payload capacity. As a result, the C_d*A values (drag coefficient times frontal area) of heavy trucks are roughly 4.7 to 7.7 times higher than

[2]"Yaw" refers to a wind whose direction is not directly in line with the forward motion of the vehicle (i.e., a side wind).

FIGURE 5-6 2009 model year Mack Pinnacle (left) and Freightliner Cascadia (right) SmartWay specification trucks. SOURCE: Courtesy of Mack and Freightliner Cascadia.

light-duty vehicles, which results in an aerodynamic power consumption 4.7 to 7.7 times higher. The benefits of aerodynamic features are a strong function of both operating speed and annual vehicle miles traveled (VMT). As Figure 5-7 shows, aerodynamic drag is larger than rolling resistance at speeds above 48 mph for a typical current truck. At 32 mph, however, aerodynamic drag is only half of tire rolling resistance, and aerodynamic drag becomes insignificant at low speeds. The sensitivity to VMT applies to any fuel-saving feature: the more miles a vehicle travels, the larger the potential fuel savings becomes.

In determining whether to apply aerodynamic features to a vehicle, it may be appropriate to consider a duty cycle average road speed hurdle. A method to quantify a weighted aerodynamic-average speed (WAAS) has been established that provides for an average of the mileage-weighted velocity^3 (V^3). If it is deemed that a speed hurdle is appropriate, a numerical value for the hurdle speed must be established, and the WAAS must be verifiable. For example, will a tractor-container/trailer chassis operate at a low average mph by virtue of its operation over short distances between ports and rail terminals? CARB has taken this issue into account in its greenhouse gas (GHG) regulation, where drayage tractors are exempt if operated within 100 miles of the port. If this approach is applied in the general case, it would likely require use of electronic onboard data recorders to substantiate the short distance and/or below-speed-hurdle reality. The required record keeping and oversight could become very burdensome.

Details of Aerodynamic Solutions

There are four regions of the tractor-van trailer combination truck that are amenable to aerodynamic design improvements. These regions include the various tractor-related details, the tractor-trailer gap, the trailer skirt, and the trailer

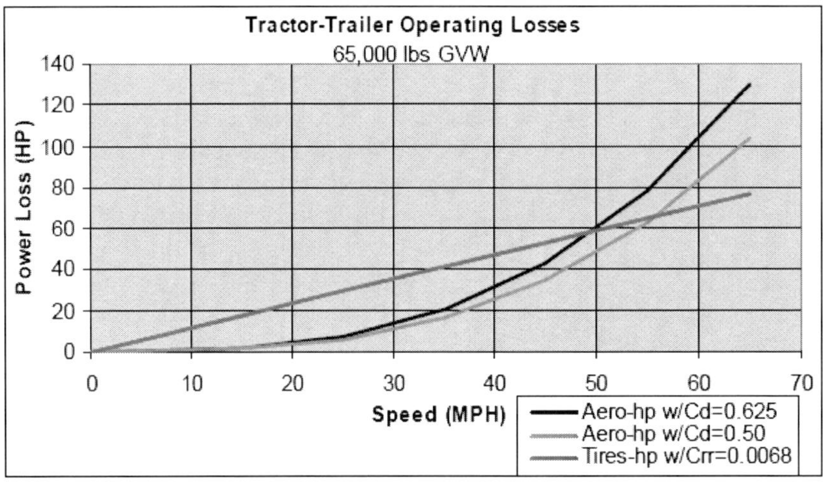

FIGURE 5-7 Aerodynamic and tire power losses for tractor-van trailer combination.

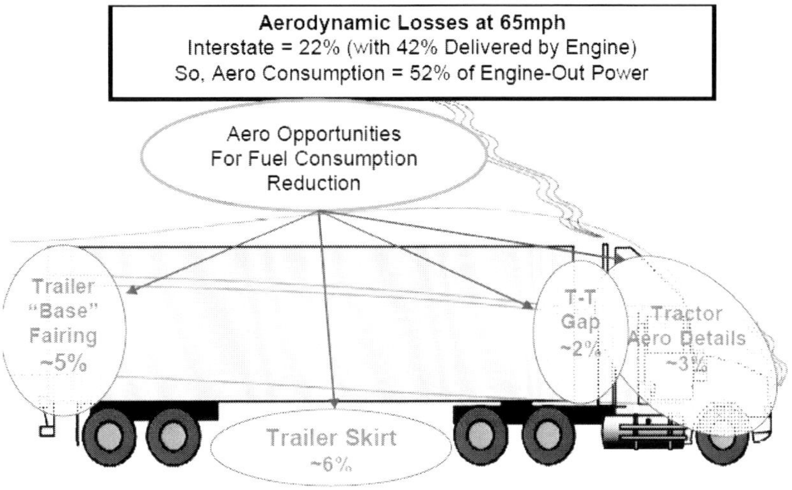

FIGURE 5-8 Tractor-trailer combination truck showing aerodynamic losses and areas of energy-saving opportunities. Percent changes refer to fuel consumption. SOURCE: Based on Wood (2006). Courtesy of Richard Wood.

"base" fairing, which are all illustrated in Figure 5-8, along with the approximate fuel consumption reductions that seem to be achievable in the near term.

Aerodynamics of the Truck Tractor

The contemporary image of aerodynamically optimized tractors is that of sleeper cab tractors equipped with many fairings, such as the SmartWay trucks shown in Figure 5-6 and the left-side image shown in Figure 5-9. These tractors are typically used in long-haul applications where the ability to provide "hotel" accommodations is important. However, many long-haul operators use a terminal-to-terminal system that does not require sleeper tractors. These operators use day cab tractors such as that shown on the right in Figure 5-9. Shorter-haul operators tend to avoid aerodynamic fairings because they provide limited fuel savings and are prone to damage in urban operations and during frequent stops at loading docks. Day cab tractors often are equipped with only a roof fairing, and for nonvan applications they may carry no fairings at all. Day cab tractors make up about one-third of all tractor sales, and so they are a significant portion of the market.

Most tractor manufacturers introduced tractor offerings in the 2003 to 2008 period that included purposeful, major improvements in their aerodynamic performance accomplished by attention to many details and utilizing most of the evaluation tools noted earlier. Reports of fuel consumption reductions of up to 6 percent were received during committee site visits.

FIGURE 5-9 Volvo full sleeper cab (left) and day cab (right). SOURCE: Courtesy of Volvo.

TABLE 5-5 Class 8 Tractor Aerodynamics Technologies, Considering the 2012 Time Frame

Technology	Fuel Consumption Reduction (%)	C_d Improvement (%)	Cost ($)	Industry Adoption Rate (%)
Day cab roof deflector	4-7	13	1,000-1,300	Most
Sleeper roof fairing	7-10	15-20	500-1,000	Standard
Chassis skirt	3-4	4-7	1,500-2,000	50-60
Cab extender	2-3	4-5	300-500	80-90
Next-generation package	3-4	6-8	2,750	2012 Introduction

SOURCE: TIAX (2009), p. 4-35.

Table 5-5 shows the market shares of various aerodynamic features on aerodynamic-style tractors. Roof shields have a very high market share, while the relatively damage-prone chassis skirts have a lower share. The side-of-cab extender works with the roof shield to minimize the gap between tractors and trailers. This gap has a major role in determining the overall vehicle C_d. The recommended maximum gap[3] is typically 30 in. from the rearmost feature on the tractor to the trailer face. Smaller gaps do have drawbacks, however, in that they limit the ability of the vehicle to operate in tight spaces.

The next-generation aerodynamics package shown in Table 5-5 represents the forecast of tractor manufacturers for sleeper cab tractor aerodynamics improvement in the 2012 time frame. These features will be designed and optimized for long-haul applications. It is expected that many of these features may not be compatible with short-haul operations, and thus their application on day cab tractors will be limited. Characteristics of short haul operation include curb encounters, severe road-crossing humps, backing maneuvers, and tight street-side clearances. All of these combine to damage many of the aerodynamic surfaces that could be successful in long-haul duty.

There are two consequences of the fragility of tractor aerodynamic features that must be considered. One is that trucks specified with many aerodynamic features will not be attractive or cost-effective in short-haul operations, because of the fragility of aerodynamic features and because of restricted maneuverability. The other issue is that tractors specified for short-haul operations will be less efficient if they are pressed into long-haul service for any reason. Excessive specialization of tractors can lead to logistics problems for operators, as well as to lower used tractor values in cases where the original operating intent does not match the second buyer's application. An example of this is sleeper cab tractors with full-height air deflectors pulling flat bed trailers. In this case the aerodynamic feature actually costs fuel rather than saves fuel, because the high roof sleeper increases the frontal area of the truck beyond what the trailer requires. In many cases the application of a high roof sleeper with a flat-bed trailer is just temporary, but in some cases this fuel-wasting combination may be long term.

Another marketplace factor in tractor design that delays full implementation of aerodynamic features is the preference for traditional styling. This preference for a traditional look is prevalent among owner-operators, and many small fleets use traditional styling as a driver retention feature. Notice the differences in traditional styling compared to aerodynamic styling in Figure 5-10. Traditional features known to have high drag-inducing effects include the large, flat bumper, along with features protruding into the airstream such as head lamps, air cleaners, and dual exhaust stacks, as well as "west coast" side mirrors. While manufacturers have made useful aerodynamic improvements to traditional models of years gone by, these traditional features are believed to invoke a fuel consumption increase of at least 5 percent compared to the aerodynamic model. (TIAX, 2009, Table 4-24). Some operators are well aware of the fuel consumption penalty, while others are likely to underestimate it.

Day cab tractors constitute roughly one-third of Class 8 tractors. So far it has not been possible to match the aerodynamic performance of the best sleeper models with day cabs. Fortunately, it appears that many of the day cab and other short-haul tractors accumulate fewer miles and thus consume less fuel than over-the-highway tractors. More specific data gathering is needed to quantify the fuel consumed by various applications of tractor trailers.

A 2-year collaborative study of a variety of design improvements that would reduce aerodynamic drag on tractor trailers was completed in 2007 by four members of the Truck Manufacturers Association (TMA) and DOE. Their research evaluated the effect of post-SmartWay designs on combination tractor-trailer aerodynamics. A number of potential tractor features were evaluated, including alternative rearview mirror designs, treatments of the tractor-trailer gap such as gap fillers and trailer gap flow control devices, and features to manage airflow under the vehicle and between the tractor and trailer. In addition to the tractor features, a number of trailer features also were evaluated.

The 2007 TMA/DOE study started with a computational fluid dynamics (CFD) modeling evaluation of potential aerodynamic feature concepts. The CFD models allowed the researchers to explore the effect of many design parameters on C_d (see in Chapter 2 the section "Computational Fluid Dynamics"). The most promising concepts from the analyti-

[3]"Gap" refers to the distance from the rearmost vertical cab feature to the front of the trailer face; where a cab extender is employed, it is this rearmost feature.

FIGURE 5-10 Peterbilt Traditional Model 389 (left) and Aerodynamic Model 387 2 (right) (SmartWay). SOURCE: Courtesy of Peterbuilt Motors Company.

cal study were then tested in either scale model or full-scale wind tunnels to validate the modeling, quantify the C_d improvements and further refine the concepts. Finally, full-scale prototype hardware was created for vehicle testing using the SAE J-1321 protocol (TMA, 2007, p. 5).

The 2007 TMA/DOE study successfully revealed improved aerodynamic design features, but it also served to spotlight the difficulties of achieving major new reductions in tractor and tractor-trailer C_d. This last point also confirms that today's modern tractor designs already perform well aerodynamically, within constraints such as the need to provide payload capacity and to be compatible with existing infrastructure such as loading docks.

The heat management requirements present another vehicle design factor that influences the aerodynamic characteristics. Larger cooling packages limit the ability of designers to reduce vehicle C_d. The total heat rejection from the vehicle determines the size of the cooling package required. The heat rejection includes engine-related heat rejection from the engine radiator, charge air cooler, and exhaust gas recirculation (EGR) coolers, but there are other sources of heat rejection as well. The transmission and steering systems may have coolers, along with the air-conditioner condenser. The addition of charge air cooling in the early 1990s, and the addition of EGR in 2002, has led to increased cooling system size and heat rejection requirements. Certain engine efficiency design features are mentioned in Chapter 4 that might reduce the size of current truck cooling system components and/or basic engine size itself. For example, "light-hybrid" systems have been mentioned as very useful to reduce light-load engine operation in the tractor-trailer class. Unfortunately, a corresponding reduction of engine size and heat rejection does not seem likely, since peak power demand will still be fulfilled only from the engine under many conditions. Similarly, most "bottoming-cycle" concepts will demand additional underhood space and will greatly increase overall power train heat rejection, as opposed to heat rejected in the exhaust gases, further challenging frontal styling and possibly increasing C_d values.

Aerodynamics of the Truck Trailer (with Focus on Typical 53-Ft-Length Box Vans)

Significant progress has been made with aerodynamic components added to the trailer (see Table 5-6). Unfortunately, there are three major impediments to widespread incorporation of aerodynamic trailer features. One is that in many operations the tractor and trailer owners are not the same. As a result, the trailer owner does not benefit from the fuel consumption reduction achieved by pulling an aerodynamic trailer. The second issue is that there are many more trailers than tractors, since trailers are widely used as temporary storage. As a result, the investment in aerodynamic improvements must be amortized over many fewer miles than is the case for tractors. The third impediment is the reality that there are no aerodynamic-system integrators

TABLE 5-6 Current Van Trailer Aero-Component Performance

Trailer Aerodynamic Technology	Skirts	Boat Tails	Nose Cone	Vortex Stabilizer	Bogie Cover
Range of fuel economy improvement (% mpg)	5.6-7.5	2.9-5.0	2.0->4.0	1.0	1.0
Range of costs	\$1,600-\$2,400	n/a	\$800-\$1,260	\$500	n/a

SOURCES: Based on responses to committee questionnaire and information on manufacturers' websites.

in the medium- and heavy-duty trucking industries. Trailer manufacturers are not owned by or related with tractor companies, and trailer aerodynamic-device manufacturers constitute yet a third layer of unaligned companies. For the most part it falls to the carriers themselves to sort through the emerging aerodynamic devices to find the most cost effective solutions.

Table 5-6 gives a partial summary of supplier-reported information obtained from responses to a committee questionnaire[4] (see Figures 5-11 to 5-14). Data are typically reported from SAE J1321 full-vehicle tests, in mpg improvement. However, individual testing procedures are not consistent, average test speeds differ, and it is not known whether the statistical requirements of the test procedure are consistently adhered to. Also, the data are not adequate to conclude that benefits achieved by combining aerodynamic devices would be completely additive. In a section below, results from a combination of devices are presented, and these results show that simple addition of individual results does not provide the correct result.

Note that the combined effects of several aerodynamic features (Full Package, in Figure 5-15) provide an average fuel consumption reduction of 9.3 percent. The full package includes the partial gap filler, full or partial trailer skirt, and base flaps (base fairings and boat tails). Further, if it is assumed that individual performances at the 75th percentile would eventually be achieved by 2015-2020, then the combined full package (of partial gap, full skirt, and base flaps) would be 12.1 percent. This result is derived through a method of multiplication of fuel consumptions reductions.[5] Although there are more than 10 independent manufacturers of trailer aerodynamic devices, at least eight trailer manufacturers have certified 53-ft van trailers in the EPA's SmartWay Partnership (EPA, 2009). One trailer manufacturer, Wabash National Corp., announced in July 2009 the production availability of a trailer skirt of its design.

[4]Trailer Aerodynamic Component Performance. Private survey, NRC Committee on Medium and Heavy-Duty Vehicles, May to August, 2009. Respondents included AdamWorks, ATDynamics, Air Tab, Freight Wing, Inc., Laydon Composites Ltd., Nose Cone Manufacturing Co., Wabash National Corp., and Windyne, Inc. (remarks only). The committee also obtained material from the websites (accessed August 2009) of manufacturers that did not return the survey questionnaire, including Aerodynamic Trailer Systems (http://fuelsaverbyats.com/ats_company_info.htm), Nose Cone Manufacturing Co.(http://www.nosecone.com/apvan.htm), Transtex Composites (http://www.transtexcomposite.com/); and Windyne Inc. (http://www.windyne.com/).

[5]The fuel consumption reduction of the combined technology packages is calculated multiplicatively (*not additively*) according to the following equation:

$$\%\ FC_{Package} = 1 - (1 - \%\ FC_{Tech\ 1})(1 - \%\ FC_{Tech\ 2})(1 - \%\ FC_{Tech\ N})$$

where $\%\ FC_{Tech\ x}$ is the percent reduction of an individual technology, and therefore $(1 - \%\ FC_{Tech\ x})$ is the consumption associated with the reduction (personal communication between TIAX consultant Matt Kromer and C. Salter).

Pneumatic Blowing to Reduce Trailer Drag

One researcher has extensively studied this concept, where low-pressurized air is discharged across curved trailing surfaces used on the trailer's rear face. Reported results indicate that a fuel consumption reduction of about 8 percent at 65 mph may be achievable with this process. (The 8 percent is net savings, after accounting for the energy to pressurize the required plenum, Englar, 2005, p. 12). Such a system would need to be integrated into the trailer design for effective packaging, including compatibility with loading docks. One trailer aerodynamics manufacturer, AT Dynamics, has recently initiated development for a production-viable active flow control system for trailer rear edges. The cost and complexity of a pneumatic blowing system would be a substantial challenge to production implementation.

Cost-Effectiveness

DOE remarked in a December 4, 2008, research solicitation that "there has not been a strong pull from fleets due to concerns about cost, return on investment, durability and maintenance requirements." Much of this lack of demand stems from the reality of trailer quantities. Currently, the trailer-to-tractor ratio is about 2.8. Most larger trucking companies report individual ratios ranging from about 1.1 to 4.0. One large private carrier reported a trailer-to-tractor ratio exceeding 8 (*Transport Topics*, 2009).

This reality adds to the difficulty of adding aerodynamic devices within normal and favorable capital acquisition metrics, such as net present value (NPV). The NPV for trailer skirts is strongly dependent on fuel prices and can easily exceed a 3-year zero-cost hurdle at a 2.8 trailer-to-tractor ratio. The retrofit of trailer aerodynamic devices might be a useful fuel-savings strategy, due to the long lifetime of highway trailers, which is 20 years or more, as a result of their low on-road utilization (compared to tractors).

Another cost-effectiveness issue is the fact that trailer aerodynamic improvements in duty cycles with low average speeds must be judged as a particularly poor value. For example, the fuel consumption benefit decreases by nearly 90 percent if the average speed is 30 mph rather than 60 mph (see Figure 5-7).

Safety Issues for Trailer Aerodynamic Devices

Since damage in normal vehicle operation is a major issue with aerodynamic features, virtually all reporting skirt developers have placed a high priority on ensuring that their designs are significantly road damage tolerant. Many have video clips on their Web sites showing resistance to railroad grade crossing humps, steep loading dock accesses, and snow accumulation or snow piles, as well as low-speed collisions with equipment such as fork-lift trucks. Nevertheless, caution is appropriate, as a proliferation of such low-hanging devices may create a new source of on-road hazards similar to tire tread sections today. One manufacturer, AdamWorks,

FIGURE 5-11 ATDynamics trailer tail (left) and FreightWing trailer skirt (right). SOURCE: Courtesy of Freight Wing.

FIGURE 5-12 Nose cone trailer "eyebrow." SOURCE: Photo provided with permission by FitzGerald Corporation, national marketer for Nose Cone Mfg. Co. Nose Cone is a registered trademark.

FIGURE 5-13 Laydon vortex stabilizer (left) and nose fairing (right). SOURCE: Courtesy of Laydon Composites.

FIGURE 5-14 Trailer bogie cover. SOURCE: TMA (2007).

automatically lowers its trailer skirt as a function of road speed from 16 to 6 in. This strategy permits close proximity of the bottom of the skirt to the road which enhances its high-speed performance while avoiding deployment under more hazardous low-speed conditions.

Aerodynamic features may also provide safety benefits unrelated to their primary purpose of saving fuel. There are numerous testimonials that trailer skirts and boat tails substantially improve trailer tracking stability, especially in crosswinds, and significantly reduce road spray from trailer tires. One trailer skirt manufacturer, Laydon, claims that its devices have been verified to comply with the European heavy truck side under-ride regulation. European Council Directive 89/297/EEC (ECD, 1989) mandates side under-ride

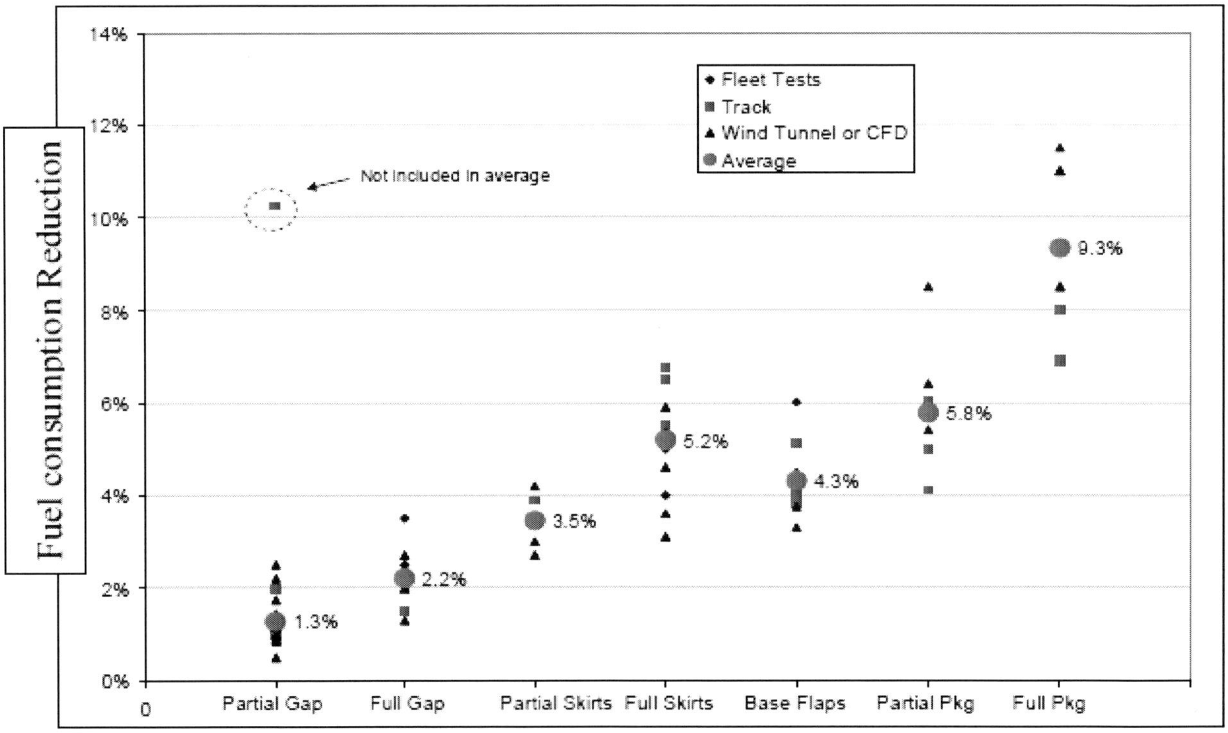

FIGURE 5-15 Summary of trailer aerodynamic device fuel consumption reduction. SOURCE: TIAX (2009).

protection on trucks to prevent pedestrians, bicycle riders, and motorcyclists from falling under the wheels of a vehicle when it turns.

Weight, Length, and Width Issues for Trailer Aerodynamic Devices

Weight

Trailer skirts can add 200 lb or more to a 53-ft trailer's weight. Boat tails can add up to 200 lb to the trailer. Likewise, nose cones and/or vortex stabilizers can add up to 100 lb weight. Simulations have shown that about 1,000 lb of incremental truck weight incurs about 0.5 percent fuel consumption increase for a tractor trailer in over-the-road duty (NESCCAF/ICCT, 2009, p. 50). For operators running at the legal weight limit, payload is reduced by any empty weight increase, causing a load-specific fuel consumption increase of 2.2 percent per 1,000 lb added to the vehicle's empty weight. Consideration should be given to allowing carriers to be permitted this increase in weight without penalty.

Length

Boat tails will also impose trailer length increases; again carriers should be permitted such length increase without penalty, even if the boat tail is structural. One boat tail manufacturer has advised receiving a U.S. Department of Transportation (DOT) length exemption under 23 CFR 658.16, Exclusion from Length and Width Determinations, for an additional 4 ft.

Width

Trailer skirts are often installed under the van, not exceeding a 102-in. width limit. Some manufacturers have experimented with side skirts that cover the trailer bogie (and also the tractor bogie). As these are usually non-structural components, it is believed they are permitted within the current width regulation, up to 106 in. width, also according to 23 CFR 658.16. Potential issues include docking in narrow confines and drivers who are unaccustomed to a truck that is potentially 2 in. wider on each side.

Aerodynamics of Tractor-Trailer Interface (Gap)

The gap between the tractor rear face and the trailer front face is filled with large vortices at high road speed. This air motion creates a low-pressure drag on the tractor's rear face. The conditions worsen with oblique wind direction, which causes more air to get into the gap between the tractor and trailer, increasing trailer drag. The average wind velocity throughout the 48 contiguous states is 7 mph. This results in prevailing effective yaw angles of 7 to 11 degrees on the East and West coasts, and up to 14 degrees in the Midwest. Such conditions combine to increase drag by 30 to 55 percent on tractor trailers (Wood, 2009, pp. 2, 3).

The considerable air turbulence in the gap has been ameliorated by the use of cab extenders. Manufacturers typically recommend that a gap not to exceed 30 inches, as measured from the trailing edge of the extender to the trailer face. Indeed, cab extenders are integral to the SmartWay specification.

Several developers offer a "nose" fairing and vortex stabilizer for installation on the trailer front face, as noted earlier. Several developers have prototyped partial and even complete gap closure devices, with somewhat limited improvement in system performance (1 percent fuel savings for full gap closure on an SAE J1321 test at 65 mph; TMA, 2007, pp. 56, 64).

One developer combined three design elements to manage the gap flows: a smoothed tractor underside (from bumper to back of engine), a vertical airflow blocker on the front face of the trailer, and a significantly increased cab extension. Together, these features achieved a 1.3 percent fuel consumption reduction (J1321 at 65 mph; TMA, 2007, p. 78). However, the effect of these features in the presence of a side wind (which was not reported) may be greater.

Aerodynamics of Trailers, Double Vans, and Trailers with Shorter Than 53-ft Lengths

There are a variety of current and former industry standard van-length trailers: 28-ft trailers (also 27-ft), which are usually used as doubles, plus 45-ft and 48-ft trailers.

It is expected that gap treatments and boat tails will perform on any length single trailer equally as well as with 53-ft trailers. Likewise, 45-ft and 48-ft trailers equipped with skirts should have benefits only slightly smaller than the 53-ft results. While single 28-ft trailers surely can be equipped with skirts, data are not available to quantify that benefit. Further, it is expected that on double trailers some form of gap treatment between the two trailers would be effective, but again no data are available.

There is some aerodynamic data for multiple trailers. Cooper has reported results for standard trailers without aerodynamic treatments. These results, shown in Figure 5-16, compare a single 27-ft trailer with 27-ft doubles and a 45-ft single. Note that these are C_d data. The 27-ft doubles have a C_d that is 33 percent greater than the single 27-ft trailer, but only 17 percent greater than the single 45-ft. Finally, the 45-ft trailer's drag is 12 percent greater than the 27-ft trailer's (these are zero yaw results; Cooper, 2004, p. 17). These data suggest that there is a significant aerodynamic drag penalty for doubles, which is likely to increase under the effect of crosswinds.

Cooper clarifies that, even though the double trailers have 33 percent higher drag than a single trailer, the freight capacity increases by 100 percent (both in terms of weight and cubic volume). So the freight-hauling efficiency provides a net 38 percent drag reduction per unit freight quantity carried and thus a nearly 20 percent fuel consumption reduction per unit freight quantity (Cooper, 2004, p. 17). Cooper's drag

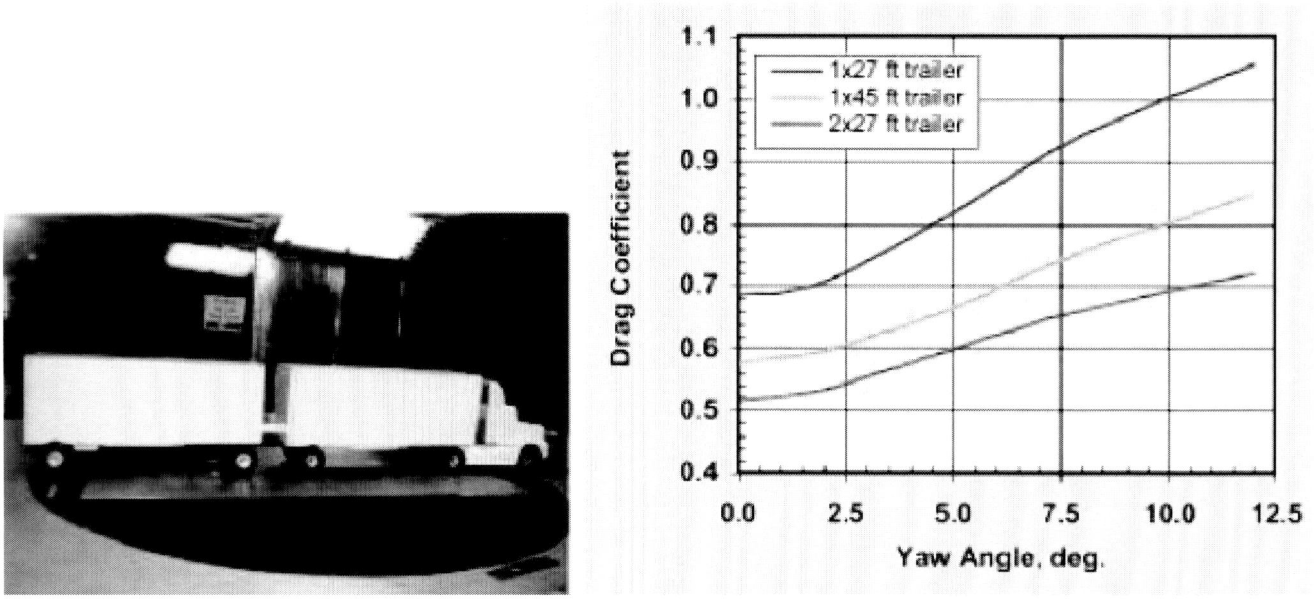

FIGURE 5-16 Drag coefficient for aerodynamic tractor with single or double trailers. SOURCE: Cooper (2004), p. 17. Reprinted with kind permission of Springer Science and Business Media.

and fuel consumption results bolster arguments that favor the increased use of long combination vehicles (LCVs), as perhaps the most cost-effective vehicle design feature available to reduce fuel consumption in the heavy-duty truck industry.

Laydon Composites Ltd. has recently completed scale-model wind tunnel tests on double 28-ft trailers equipped with forward-looking skirts, plus a vortex generator on the trailer's faces. These skirts included trailer and tractor axle skirts. The fuel consumption reduction of the combination shown in Figure 5-17 is 9.9 percent at 60 mph, compared to a standard 28-ft double vehicle configuration. These results are based on yaw averaged and weighted winds, using the National Aeronautics and Space Administration procedure for estimating consumption from the drag results.

Aerodynamics of Trailers Other Than Dry Vans

While the dry van constitutes the largest portion of body styles in the U.S. fleet of trailers, there is a proliferation of

FIGURE 5-17 Laydon double trailer arrangement with trailer skirts and vortex stabilizers on both trailers; SOURCE: Courtesy of Laydon Composites.

other styles. The design of many of these non-dry vans is far less amenable to incorporation of all or even some of the aerodynamic devices found helpful on the dry van. Polk has provided a dataset for all trailers by body style for the state of Florida (Table 5-7). The results are summarized here as a surrogate for the national averages.

Aerodynamics of Van Refrigerated

The trailer-face-mounted refrigeration unit will dictate different gap treatment solutions than those used with the dry van. The refrigeration unit requires an airflow to provide engine combustion air, engine cooling air, and refrigeration condenser cooling air. On the other hand, the refrigeration unit itself may provide an aerodynamically significant reduction in drag, by virtue of providing some vortex control and associated pressure increase on the tractor rear face. Of course, the trailer skirt and trailer base areas are similar to those on a dry van, so similar aerodynamic features can be used. Freight Wing, Inc. has fitted skirts to a refrigerated van (see Figure 5-18), but test results for this trailer type are not yet available.

Aerodynamics of Flatbed Trailers

This trailer body style is among the most problematic of the semitrailer family because the space above the deck carries an endless assortment of products and implements, many with grossly unsymmetrical geometries. The trailer skirt area is judged to be as amenable to aerodynamic treatment as the dry van. One trailer aerodynamics supplier, Freight Wing, has fitted a flatbed with a skirt but has not yet tested the configuration. Note that the trailer model in Figure 5-19 also has a spread axle, which somewhat complicates skirt addition and may limit its potential effectiveness.

TABLE 5-7 Florida Trailer Population by Body Style

Body Style	Population (%)
Auto transporter	0.4
Beverage	1.4
Container chassis	4.3
Dump	2.3
Flatbed	10.6
Grain	0.2
Livestock	0.3
Lowbed	2.0
Tank	4.1
Transfer	1.4
Van	60.9
Van refrigerated	11.8
Other	0.3
Total	100

SOURCE: Personal communication between L. Hart and C. Salter, committee member, June 2, 2009.

Aerodynamics of Container Chassis

This body style appears substantially similar to a van trailer. A removable container box is attached to a skeleton chassis consisting of a frame, a king pin, and an axle bogie. The difference from an aerodynamic point of view is that this style is equipped with square corners and many external ribs. These ribs provide the requisite strength to the container box to deal with the handling forces when loaded containers are lifted on and off the chassis frame but also can add to the trailer C_d (see Figure 5-20). The trailer skirt area is judged to be amenable to aerodynamic treatments, similar to those of the dry van. One trailer aerodynamics supplier, Freight Wing, has fitted a container chassis with a skirt but has not yet tested the configuration (see Figure 5-21). Unfortunately, typical empty-chassis handling and stowage practices will put trailer skirts at high risk of damage.

Aerodynamics of Tank Trailer

Since tanker trailers are normally operated at the maximum legal weight limit, the design is constrained to achieve minimum tare weight within the structural demands of the unit. While the tank itself is typically cylindrical, often with a hemispherical or somewhat rounded front face, the functional needs often result in an external skeleton of pipes, tubes, and so forth, to facilitate product loading/discharge and personnel protection when accessing certain operational devices. Further, the rear shape of the leading tractor poorly matches the trailer face in aerodynamic terms. Standard high-roof sleepers or day cab air deflectors are too high for tanker-trailer applications. Trailer skirts can be fitted as the dry van. One trailer aerodynamics supplier, Freight Wing, has fitted a tank trailer with a skirt, but has not yet tested the configuration (see Figure 5-22). Because any weight added for aerodynamics will reduce the load that a tanker can carry, these features are not likely to be financially attractive to any operator unless regulatory allowances are made.

Aerodynamics of Auto Transporter Trucks

The population of this body style is relatively small. This design style also has a complex nonaerodynamic array of structural tubes deploying the moving floors. Interestingly, one transport company, Precision Motor Transport Group (PMTG), has created a variety of trailer configurations incorporating curtain sides, rounded noses, and boat tails. This carrier specializes in transport of upscale sedans, and the auto capacity is inferior to the more standard designs. PMGT's trailer system can hold six to eight sedans, compared to nine to eleven for a typical transporter trailer. The aerodynamic performance must be substantially superior to the traditional design, but no data are available at this time.

PMGT's solutions raise the question of what could be done to improve other current body styles that have poor

FIGURE 5-18 Refrigerated van trailer with Freight Wing skirts. SOURCE: Courtesy of Freight Wing.

FIGURE 5-19 Freight Wing skirts on flatbed trailer. SOURCE: Courtesy of Freight Wing.

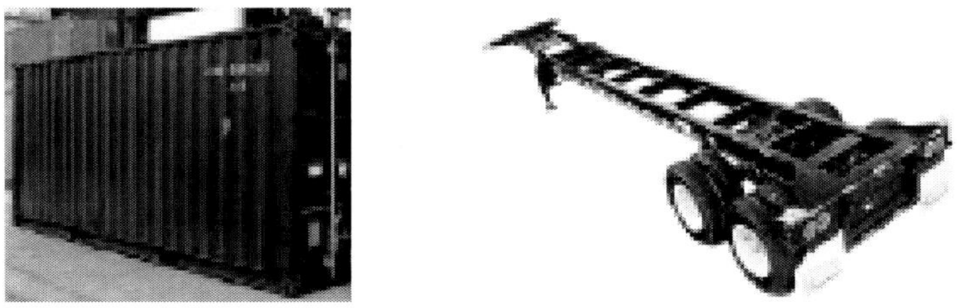

FIGURE 5-20 New 40-ft-long container built by TRS Containers (left) and container chassis (right). SOURCE: Courtesy of TRS Containers, Avenel, New Jersey.

FIGURE 5-21 Container chassis with Freight Wing trailer skirt. SOURCE: Courtesy of Freight Wing.

FIGURE 5-22 Tank trailer with Freight Wing skirts. SOURCE: Courtesy of Freight Wing.

FIGURE 5-23 Sturdy-Lite curtain side design for flatbed trailers. SOURCE: Courtesy of Sturdy-Lite.

aerodynamics. For example, could many of the normal flatbed/lowboy styles accommodate a curtain side/roof frame to achieve a major improvement in aerodynamics with accompanying fuel savings? At least two companies, Edscha Trailer Systems and Sturdy-Lite, market a rolling tarp system that allows any conventional flatbed or lowboy trailer to be converted into a van-style unit and back to a flatbed in minutes. The system works by rolling the fabric tarp on a track that is attached to the side of the trailer. No aerodynamic performance data are available at this time, and a great deal of testing would be required to determine the benefits of this approach compared to normal loading of freight on a flatbed, since so many freight configurations are possible. Notice that the Sturdy-Lite solution incorporates a full-height vertical nose plate, where many flatbed trailers have none, and a rear curtain frame (see Figure 5-23).

Combination of Solutions

In 2005, Walmart initiated a program with a stated goal of doubling the fuel economy of its (then) 7,200-tractor private carrier fleet from 6.5 mpg to 13 mpg by 2015 (see Figure 5-24).

Collaborating with suppliers Peterbilt and Great Dane, evaluated components included an *12-ft 6-in. trailer height* (down from industry standard 13-ft 6-in.), 12-in. trailer drop belly (to recover internal volume except over the tractor axles), an *aerodynamic tractor with matching lower roof fairing and cab extenders, trailer skirts (as well as the drop box), a rigid 2-ft boat tail*, and an auxiliary power unit (for idle management) (italics indicate aerodynamic features).

Walmart reports a 12 percent improvement in fuel economy with this first-generation package of components and design alterations, using the SAE J1321 test procedure. Interestingly, the trailer height reduction of 12 in. reduced the frontal area by 7 percent in this combination; correspondingly, a 3 to 4 percent fuel consumption reduction would be expected from this change alone. Note that some loss of cargo capacity results from Walmart's changes, and the new trailer height is not compatible with existing loading docks. Components still being evaluated in subsequent generations include a hybridized diesel-electric system and *auto-deploying trailer skirts*.

FIGURE 5-24 Walmart's 2008 low fuel consumption tractor trailer. SOURCE: Courtesy of Walmart.

As part of the TMA-DOE collaborative study on tractor-trailer aerodynamics, Mack Trucks studied the separate and combined effects of the three aerodynamic technologies shown in Figure 5-25. The three features were a 24-ft long trailer skirt with 9-in. of road clearance, a pair of boat tail options (2-ft and 4-ft), and a unique flexible gap-closing system. The performance of this combined set of features was evaluated using the SAE J1321 on-road procedure at 65 mph. A fuel consumption reduction of 7.75 percent was measured, which suggests a 15 percent reduction in C_d. While the test parameters met all the precision criteria of the procedure, the prevailing winds and yaw conditions were not reported.

A recent DOE initiative includes a project valued up to \$2 million for the development, evaluation, and deployment of advanced aerodynamic trailer technology. The DOE has selected Navistar International Corporation (Fort Wayne, Indiana) for the 30-month project with a goal to design, demonstrate, and bring to market a tractor-trailer combination and tire package that can reduce the fuel consumption of a heavy vehicle by at least 15 percent. Following development, a commercial fleet will evaluate the benefits of the new technology package through real-world use. After the term of the project, the team members will make this fuel-efficient technology package available for sale. Team members on the program include Frito Lay, Kentucky Trailer, Freight Wing, Michelin, and DOE's Lawrence Livermore National Laboratory.

FIGURE 5-25 Mack truck with aerodynamic device combination. SOURCE: TMA (2007), pp. 57, 58, 60.

Straight Trucks with Aerodynamic Treatment

Trucks with box van bodies are most amenable to the solutions found on van trailers. Nose cones have been widely applied, but little or no SAE J1321 data are available (see Figure 5-26). Many straight trucks operate in urban and suburban settings, where aerodynamic treatments are unlikely to offer a significant fuel savings, but those few straight trucks used routinely in highway operations could see significant benefits.

One trailer aerodynamics manufacturer, Freight Wing, has fitted skirts to a van body straight truck, but no performance data are available (see Figure 5-27).

Aerodynamics of Motor Coach

Because of their frequent high-speed operation, some motor coaches may benefit significantly from aerodynamic drag reduction. However, there is limited information available in the literature, and little additional data were provided during the committee's site visits that speak directly to motor coach aerodynamics. As such, the estimates of motor coach aerodynamic drag reduction potential in this report are based on extending results of the line-haul analysis to the motor coach segment (see Table 5-8).

The following approaches were considered for optimizing the aerodynamics of the tractor trailer: cab streamlining, boat tailing, underbody treatments, gap treatments, wheel fairings, removal of the mirrors, and active flow control. Of these, gap treatments are clearly not applicable to motor coaches; and given the already-low ride height, skirts and underbody treatments likely do not offer a significant benefit. Active flow control has not yet been demonstrated; moreover, it is most beneficial on streamlined, as distinguished from bluff, bodies (Salari, personal communication, 2009).

Of the remaining options, cab streamlining and boat tails would appear to offer the best prospects for reducing aero-

FIGURE 5-26 Nose Cone fairing on face of straight truck. SOURCE: Photo provided with permission by FitzGerald Corporation, national marketer for Nose Cone Mfg. Co. Nose Cone is a registered trademark.

FIGURE 5-27 Laydon skirt on straight truck. SOURCE: Laydon Composites.

dynamic drag. Boat tails on a tractor trailer are assumed to reduce drag by 6 to 9 percent and to offer fuel consumption benefits of 3 to 5 percent. Additional cab streamlining in the line-haul segment is estimated to reduce drag by 6 to 8 percent and to reduce fuel consumption by 3 to 4 percent. In combination, these approaches would reduce drag by 12 to 18 percent and could offer a 6 to 9 percent reduction in fuel consumption (if average speeds are >60 mph). It is assumed that these benefits come at a similar cost to those assumed for the line-haul segment: $2,750 for streamlining and $1,500 to $2,000 for boat tails. Given the very low manufacturing volumes for motor coaches, the committee expects that actual costs are likely to be higher.

Aerodynamics of Class 2b Pickups and Vans

The potential for aerodynamic drag reduction in the Class 2b segment is estimated based on analysis conducted during the National Highway Traffic Safety Administration (NHTSA) rulemaking for the light-duty Corporate Average Fuel Economy (CAFE) standard published in 2009. Although the NHTSA analysis extends only to Class 2a vehicles (<8,500 lb GVW), it is assumed that similar aerodynamic improvements can be implemented in Class 2b. In general, Class 2b vehicles have form factors similar to that of their smaller counterparts considered in the NHTSA rulemaking, which suggests that this is a reasonable approximation.

The NHTSA rulemaking considered underbody treatments; streamlined hood, windshield, fenders, and grill; reduced ride height (analogous to trailer side skirts); low-drag side mirrors; optimized airflow pathways; and wheels or wheel wells. All together, the NHTSA rulemaking assumes that a 10 percent reduction in C_d is feasible. As in heavy trucks, this level of drag reduction is assumed to result in a 4 to 5 percent reduction in fuel consumption (see Table 5-9). Caution is necessary in the use of these fuel consumption estimates since they apply to a 60 to 65 mph average speed. If these trucks are used principally in a pickup/delivery duty where average speed is about 40 mph, the fuel consumption benefit of the aerodynamic component will shrink by 70 percent. At speeds below 40 mph, the benefit becomes insignificant.

These benefits are estimated to range from $60 to $116 in added cost. These cost estimates were applied independent of vehicle class, so no adjustment was applied to translate these to Class 2b vehicles.

Aerodynamic Summary by Sector

TIAX, in its report for the committee, summarized the aerodynamic fuel consumption potential reduction by time frame and application (see Table 5-10).

The committee believes the potential for aerodynamic improvement for the tractor-trailer application is under-

TABLE 5-8 Motor Coach—Applicable Aerodynamic Technologies

Technology	Fuel Consumption Reduction (%)	C_d Improvement (%)	Cost ($)
Boat tail	3-5	6-9	1,500-2,000
Vehicle streamlining	3-4	6-8	2,750

TABLE 5-9 Class 2b Van and Pickup—Applicable Aerodynamic Technologies

Technology	Fuel Consumption Reduction (%)	C_d Improvement (%)	Cost ($)
Vehicle streamlining	4-5	10	60-116

SOURCE: TIAX (2009), p. 4-50.

TABLE 5-10 Aerodynamic-Related Fuel Consumption Reduction Packages by Sector and by Time Frame

Sector	2013-2015 (%)	2015-2020 (%)
Tractor trailer	5.5	11.5
Class 6 box truck	3	6
Class 6 bucket truck	0	0
Motor coach	0	8
Class 2b pickup/van	3	3

SOURCE: TIAX (2009).

stated by TIAX. In 2013 to 2015 a 9 percent reduction in fuel consumption is achievable versus 5.5 percent based on the "full-package" average in Figure 5-15. In 2015 to 2020 a 15 percent reduction in fuel consumption is achievable versus 11.5 percent at 65 mph. This estimate is based on improved full-package trailer performance (as described in Figure 5-15), plus a next-generation tractor performance of 3 to 4 percent (see Table 5-5). These two performance values are combined and confirmed by the method of multiplication of consumptions.

Also, the committee has recognized the Class 6 Box Truck aerodynamic design features in Table 6-6 that are currently available for implementation and has adopted them in Table 5-10 to report a 3 percent fuel consumption reduction for the 2013 to 2015 period. The 3 percent fuel consumption reduction is achievable provided that such box trucks have an average speed significantly higher than the 30 mph typical for pick-up and delivery duty. At a 30 mph average, the fuel consumption would be about 1 percent.

AUXILIARY LOADS

In addition to driving the wheels, power from the engine cylinders is used for many requisite auxiliary loads. Compressed air is needed for the braking systems; air conditioners are used for driver and passenger comfort; and power-steering systems require power to drive the associated pumps, compressors, and fans; and finally an alternator is used to charge the vehicle's battery. The power to operate these systems comes from the engine cylinder and represents a use of fuel energy for functions other than putting power to the wheels. (The power required for running the coolant pump, fuel pump, and lubricating pumps is classified under the category of accessories, as shown in Figure 5-1.) The impact that driving these auxiliaries has on an engine's fuel consumption will be highly dependent on the engine's speed and the duty cycle on which the vehicle operates. Estimates that approximately 1.7 to 4.5 percent of fuel energy could be saved through the use of dedicated auxiliary power units (APUs) have been reported to the committee.[6]

[6]K.G. Duleep, Energy and Environmental Analysis. "Heavy Duty Trucks Fuel Economy Technology," presentation to the committee, December 5, 2008, Washington, D.C., Slide 23.

Tables 5-1, 5-2, and 5-3 show the power consumption for auxiliaries for different vehicles operating under representative operating conditions. The power consumed by auxiliaries can be significant, approximately 25 percent of the total power in the case of a transit bus with the air conditioner operating. The extent to which these auxiliaries are used is very dependent on the vehicle and its duty cycle, as shown in Tables 5-11 and 5-12.

TABLE 5-11 Examples of Power Requirement for Selected Auxiliary Loads

Parasitic Devices	Horsepower at 600 rpm	Horsepower at 1200 rpm	Horsepower at 1800 rpm
Air conditioning (R-12) (not including blower)			
90° Day/drawdown (hot cab)	7.5	7.5	7.5
90° Day/maintain (cool cab)	3.5	3.5	3.5
Air compressors			
Loaded	1.1	2.5	4.0
Unloaded	0.2	0.5	1.0
Blower motors (1/3 hp electric)	0.4	0.4	0.4
Fans			
On/off (engaged)	1.5	5.0	20.0
On/off (disengaged)	0.0	0.0	0.0
Viscous (engaged)	1.5	5.0	20.0
Viscous (disengaged)	1.1	2.3	6.0
Power takeoff (varies by application)	[a]	[a]	[a]
Power steering/unloaded			
Hydraulic	0.4	1.1	3.0
Wipers, operating			
Electric	0.3	0.3	0.3
Air	0.4	0.4	0.4

[a]Check service manual for HP.

SOURCE: TMC (n.d.)

TABLE 5-12 Auxiliary Use for Line-Haul Duty Cycles

Component Name	Line-haul Duty Cycle (% of time on)	Local-haul Duty Cycle (% of time on)
A/C compressor	50	50
Power steering	10	60
Air brake compressor	5	30
Engine fan	5	10
Alternator	100	100
Oil pump	100	100
Coolant pump	100	100

SOURCE: Society of Automotive Engineers (SAE) J1343, "Information Relating to Duty Cycles and Average Power Requirements of Truck and Bus Engine Accessories," August 2000.

Improvement in these auxiliary systems and the use of electric drives instead of direct mechanical drives from the engine offer potential to reduce fuel consumption. However, the fuel consumption reduction will be very application specific. Hendricks and O'Keefe (2002) suggest that reduction of fuel consumption by optimizing the handling of auxiliary loads is indeed highly dependent on the vehicle and its driving cycle. Also, it is important to realize that reducing the mechanical power requirement from the engine to drive auxiliaries will not necessarily be a one-to-one reduction in fuel consumption. In general, reducing the load on the engine will also result in a decrease in the engine's thermal efficiency. To gain the maximum benefit from reducing auxiliary power requirements and optimizing their operation, the engine will need to be re-optimized for the new duty cycle. This will need to be done for each vehicle application under consideration. However, it is technically feasible to save a portion of this 2.5 percent of the fuel's energy.

ROLLING RESISTANCE

Technologies for Reducing Rolling Resistance

Tire rolling resistance accounts for roughly one-third of the power required to propel a line-haul truck at highway speeds on level roads (Bradley and Nelson, 2009; Kenworth, 2008). The force resisting a rolling tire is primarily due to the inelastic cyclic tire deformation when rolling and the shear and compressive forces at the contact patch. The resistive force has been found to be nearly linearly proportional to the load on the tire, and hence it is convenient to define a coefficient of rolling resistance, C_{rr}, as follows:

$$C_{rr} = \text{resistive axial force} / \text{normal force}$$

For a given tire, C_{rr} is mildly dependent on temperature, tread wear, and velocity (LaClair, 2005), which would need to be accounted for in vehicle models and must be considered in on-road comparative testing. Inflation pressure has a pronounced impact on C_{rr}, as does wheel alignment (slip angle). Opportunities for managing these items are discussed below, along with tire technology.

C_{rr} is highly dependent on tire technology, and large reductions in C_{rr} have been seen in the past few decades. C_{rr} is dimensionless, with typical values for modern truck tires of 0.004 to 0.008. For convenience the C_{rr} is often expressed in values of kilograms of force of rolling resistance per metric ton of load (kg/T), which converts the values to 4.0 to 8.0 kg/T.

Improved Tires

The historical trend in C_{rr} reduction as presented by Melson (2007) is shown in Figure 5-28. A significant recent advancement in tire technology is the New Generation Wide-Base Single (NGWBS) truck tire. With this technology, one NGWBS tire (e.g., 455/55R22.5 and 445/50R22.5 dimensions) replaces two conventional dual tires, as shown in Figure 5-29. In today's tire usage, conventional dual tires have C_{rr} ranging typically from 5 to 8 kg/T, depending on the design. One of the significant benefits of the current NGWBS tire is its lower rolling resistance coefficient, with a range of 4 to 5 kg/T. As seen in Figure 5-30, dual tires span a range of C_{rr} and can also qualify as low rolling resistance per the SmartWay definition.[7]

Replacement of typical dual tires with NGWBS tires can give fuel economy improvements approaching 10 percent for combination trucks, based on modeling as well as real-world studies (LaClair, 2005; Capps et al., 2008). Furthermore, U.S. EPA studies (Bachman et al., 2005, 2006) have demonstrated reductions in oxides of nitrogen emissions. The range of fuel economy improvements from tests is 5 to 10 percent, influenced by the vehicle type, driving cycle, and baseline tire (see Figure 5-30). Tire rolling resistance is further impacted noticeably by wear (tread depth), an additional variable in test results (Goodyear, 2003, p. 73). An additional confounding factor to test would be the amount of circumference mismatch in dual tires to minimize scrubbing losses. Typical results from a modeling study of a Class 8 truck show how the impact of tire C_{rr} varies depending on drive cycle and load (see Table 5-13).

Modest further reductions in C_{rr} are projected. Melson (2007) shows the C_{rr} of trailer tires in particular is expected to drop slightly below 0.004. The Aerospace Corporation, in a 1982 study for the DOE, projected that C_{rr} would reach 0.0045 in the following decade, which has proven accurate. Consistent with these sources, Reinhart[8] reported to the committee that the NESCAFF/ICCT study team had determined that a C_{rr} of 0.0045 was an appropriate estimate for future technology.

Despite recent advances in tire technology, it should be noted that NGWBS tires, which provide the lowest levels of C_{rr}, are presently not available in tire dimensions used on many Class 3 to 6 vehicles, and tires with the very lowest rolling resistance levels may not be practical for all applications. Tires, like most products, must satisfy a range of performance criteria (e.g., rolling resistance, wear, noise, traction, durability, cost), and several inherent design tradeoffs exist when balancing the tire performance for a particular use. For example, tires designed for optimal mud or

[7]Through collaboration with industry, EPA determined a typical C_{rr} for Class 8 truck tires and determined a level of improvement that was technically feasible, then worked with industry to establish a voluntary performance threshold, approximately 15 percent above typical, which is expected to provide a 3 percent or greater reduction in fuel consumption relative to the best-selling tires. Every major original equipment manufacturer of new commercial truck tires now offers at least one model of tire that meets this voluntary performance requirement.

[8]T. Reinhart, "Heavy Duty Vehicle Fuel Consumption and Emissions Improvement: Final Simulation Results," presentation by Southwest Research Institute to the committee, March 9, 2009.

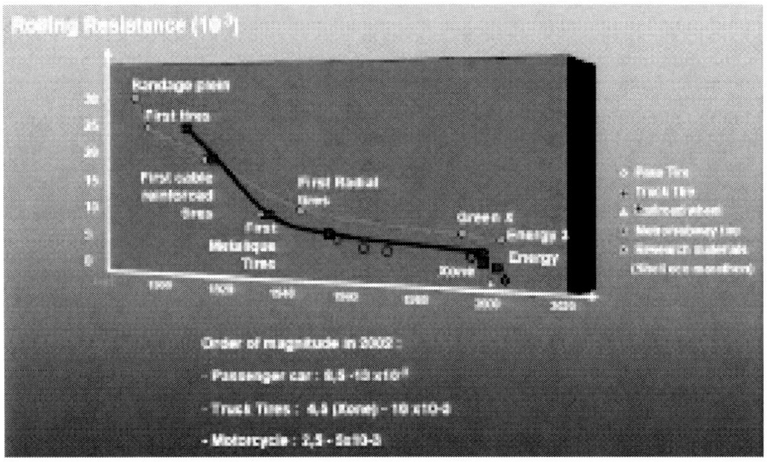

FIGURE 5-28 Rolling resistance technology, 1910-2002. SOURCE: MNA (2007). Courtesy of Michelin North America.

snow traction typically have more void in the tread pattern. This generally provides a tire with a higher C_{rr} than for tires with lower void content, since the reduced rigidity results in increased tread rubber deformation, in both compression and sheer, thereby increasing the hysteretic losses responsible for rolling resistance. Particular needs for individual consumers and tire uses will make it very challenging to have uniformly low rolling resistance for all vehicle applications, and it is likely that some vehicle applications will always use tires with C_{rr} values greater than the lowest levels available on the market. Also, fleet operating practices of running out partially worn tires from steer and drive axles on the trailer will often result in less than fully optimized rolling resistance at different stages of tire life. And even in long-haul vehicles, the practice of retreading steer tires (which legally cannot be operated on the steer axle after retreading) will assure that nearly 10 percent or more of tires in use will always be dual tires. Finally, retreaded tires generally have a somewhat higher level of rolling resistance than new tires. These factors must be taken into consideration when establishing any regulation regarding tire rolling resistance, and it should be well understood that fleet practices will have a very direct impact on the average level of rolling resistance for tires in actual use.

FIGURE 5-29 New-generation wide-base single tire (right) to reduce the rolling resistance of conventional dual tires (left). SOURCE: Presentation to the committee by C. Bradley and S. Nelson, Michelin Tire North America, "Truck Tires and Rolling Resistance," February 4, 2009. Courtesy of Michelin Tire North America.

Recent proposals have been made to create a tire fuel efficiency rating system for replacement tires and to allow point of sale information to be displayed to inform consumers of the role that tires have on fuel efficiency. The approach is similar to the Energy Star program and EnergyGuide labeling used for household appliances and other products. Pursuant to the Energy Independence and Security Act of 2007, a national tire fuel efficiency consumer information program was developed for passenger car tires by NHTSA (NHTSA, 2009b). The NHTSA proposed rulemaking proposes to require tire manufacturers to rate the fuel efficiency of their tires using the International Organization for Standardization (ISO) 28580 standardized test method for tire rolling resistance measurement. The tire industry has been working recently to develop this test standard, which is applicable to new passenger car, truck, and bus tires. EPA is considering this same standardized test method for future SmartWay definitions.

NGWBS tires also give a weight savings of roughly 340 kg in a typical combination tractor-trailer rig, allowing for an increase in payload capacity, which can further improve freight efficiency. Barriers to the adoption of NGWBS tires have included concerns relative to the following:

- Perceptions of the difficulty/downtime associated with flats

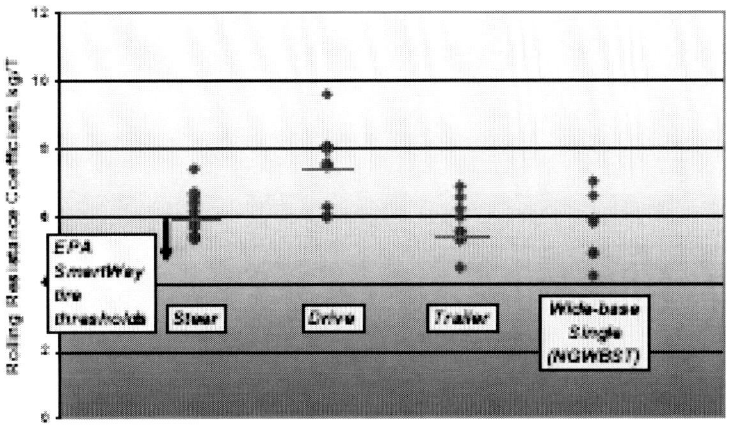

FIGURE 5-30 Example rolling resistance coefficients for heavy-duty truck tires. SOURCE: Presentation to the committee by C. Bradley and S. Nelson, Michelin Tire North America, "Truck Tires and Rolling Resistance," February 4, 2009. Courtesy of Michelin Tire North America.

- Safety/stability in the event of a tire failure
- Availability of replacement tires
- Increased road damage

Various studies have been conducted recently to address these above concerns for NGWBS tires, however, and results have demonstrated performance similar to that of dual tires. Furthermore, tire manufacturers have made efforts to minimize or mitigate any inconveniences that could be experienced on the road when using NGWBS tires.

Several of the concerns stated above can be attributed to past experiences and studies of the original 65-series on/off-road and regional single tires (e.g., 385/65R22.5 and 425/65R22.5). These tires are normally operated at elevated service inflation pressures, with consequently reduced tire footprint size and peak tire-road contact stresses that surpass those of traditional tires. Studies of the original "65-series singles" concluded that increased road damage could be attributed to these tires, and this naturally generated concern for state and federal departments of transportation when NGWBS tires were developed and placed into use. Although differences have been observed in various metrics of roadway stresses for dual and NGWBS tires, pavement damage is expected to be similar for these two types of tires for primary roads (Al-Qadi, 2007a,b).

TABLE 5-13 Results of Truck Model Showing Effect of Coefficient of Rolling Resistance, C_{rr}, on Fuel Economy for Several Drive Cycles

Heavy Truck (40 ton) 12.01 Diesel Engine	Fuel Consumption (gal/100 miles) C_{rr} = 5.5 kg/t	C_{rr} = 4.5 kg/t	Fuel Savings (gal/100 miles)	Percent
Suburban use, half loaded	16.53	15.98	0.55	3.3
Regional use, half loaded	15.13	14.53	0.59	3.9
Long-haul use, half loaded	12.62	11.94	0.68	5.4
Suburban use, fully loaded	21.29	20.57	0.72	3.4
Regional use, fully loaded	19.76	19.04	0.72	3.7
Long-haul use, fully loaded	14.92	14.02	0.89	6.0

SOURCE: C. Bradley and S. Nelson, Michelin Tire North America, "Truck Tires and Rolling Resistance," presentation to the committee, February 4, 2009. Courtesy of Michelin Tire North America.

For safety concerns, testing has shown that rapid air loss events on NGWBS tires do not compromise the stability and behavior of the vehicle (Bradley and Nelson, 2009). Additionally, rollover performance with NGWBS tires has been shown to be similar to or improved relative to dual-tire rollover performance (Knee et al., 2008).

Regarding flat tires, a vehicle cannot continue driving with a flat NGWBS tire without risking serious damage to the tire and wheel. Although it may be possible to continue to drive a vehicle for tire replacement if one dual tire goes flat, this practice is not recommended and may not be legal. (Federal Motor Carrier Safety Administration Regulation Title 49 Code of Federal Regulations Part 393.75 (a), states, "No motor vehicle shall be operated on any tire that . . . is flat or has an audible leak.")

In the United States, the use of NGWBS tires is legal in all 50 states on nonpermit-required vehicles (e.g., for loads up to 80,000 lb). There currently exist some state restrictions on NGWBS tire use where a state oversize or overweight permit is required for vehicle operation. These restrictions are premised on the studies of the original 65-series single tires. Furthermore, the use of NGWBS tires in Canada may require a special permit to operate at the Canadian dual-tire maximum load of 9,000 kg per axle in some provinces. Since many vehicles operate in both the United States and Canada, an amendment was passed in May 2008 to the Canadian Ministers of Transportation and Highway Safety memorandum of understanding on vehicle weights and dimensions approving the U.S dual-tire load parity of 7,700 kg axle (15,400 kg per tandem axles) for the NGWBS single tires.

Tire Pressure Maintenance and Effects

Rolling resistance is strongly affected by the pressure in a tire, increasing steadily as tire pressure decreases below the recommended inflation pressure. For truck tires, rolling resistance can be expected to increase by about 5 to 8 percent for a 20 percent reduction in pressure. This will typically yield a 2 to 3 percent loss in fuel economy in a Class 8 truck if all tires are underinflated (SmartWay; Goodyear, 2003, p.72). While trucking fleets generally monitor tire pressure more frequently than do personal car owners, substantially underinflated tires are likely responsible for a relatively large level of unnecessary fuel consumption. For passenger cars and light trucks, the NHTSA conducted a study following the passage of the Transportation Recall Enhancement, Accountability, and Documentation Act of 2000 in which tire pressure was measured on 11,530 cars and light trucks. The study found that 27 percent of passenger cars and 33 percent of light trucks have at least one tire significantly underinflated (by 20 percent or more). For truck tires a separate NHTSA study (Woodrooffe et al., 2008) found that tire underinflation is one of the three main causes of truck tire casing removals. The level of underinflation on truck tires across the U.S. fleet is not precisely known; however, the impact on fuel economy can be rather significant.

Given these facts, it is clear that tire pressure monitoring systems (TPMSs) can help improve the fuel consumption and safety of heavy trucks by ensuring that adequate pressure is maintained and, consequently that the rolling resistance is at the design level. There is already a moderate usage of TPMS in fleets, although a detailed quantification of the percentages of trucks equipped with TPMS is not available at this time. Automatic tire inflation (ATI) systems are also commercially available that maintain tire pressure at the desired level.

Use of nitrogen instead of air for tire inflation has been found to reduce pressure loss rate (NHTSA, 2009a). In static laboratory tests the inflation pressure loss for new tires inflated with nitrogen was approximately two-thirds of the loss rate of new tires inflated with air. Similar differences between nitrogen and air permeation rates in new tires were found under dynamic, loaded laboratory testing. This is attributed primarily to the higher diffusion rate of oxygen through rubber compared to nitrogen, which was observed and characterized decades ago. In certain fleet operations the use of nitrogen fill would be effective in place of or in addition to TPMS and would lessen the need for ATI systems. The same NHTSA study reported that inflating tires with nitrogen in place of air had no direct effect on laboratory rolling resistance performance. Walmart reported to the committee that it found a 1.0 to 1.5 percent reduction in fuel use after instituting nitrogen inflation protocols.

Tire/Wheel Alignment

The alignment of each axle has a direct impact on rolling resistance. Misalignment effects that result in a slip angle on a truck's tires (both toe settings and out-of-perpendicular (thrust) or out-of-parallel (scrub) tandem-axle alignment—see Figure 5-31) can strongly impact rolling resistance, in addition to negatively affecting tire wear and possibly vehicle handling. Camber effects, by comparison, are relatively small on rolling resistance, and the slip angle impact is thus the only alignment effect considered here. For a given slip

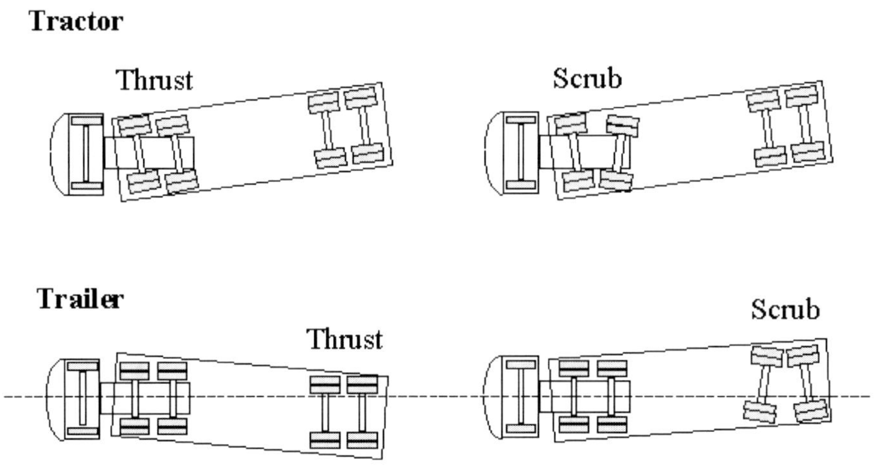

FIGURE 5-31 Tractor-trailer tandem-axle misalignment conditions. SOURCE: Kreeb and Brady (2006).

angle, α, on the tire, the added rolling resistance is given by the following relationship (Schuring, 1977):

$$F_R = F_{R_0} + C_\alpha \alpha^2$$

where C_α is the cornering stiffness of the tire.

An estimation of the impact of misalignment on fuel consumption can be made by assuming that the average misalignment on each wheel of the truck is approximately 0.1°. (This level is consistent with standard practices and recommendations for tandem axle alignment in the industry and typical toe settings.) Based on a cornering stiffness of 3 kilonewtons (kN) per degree, the increase in rolling resistance per tire is about 0.5 N, which for an average rolling resistance coefficient of 7 kg/T with a fully loaded vehicle represents about 0.4 percent of each tire's rolling resistance. Again assuming that rolling resistance accounts for about one-third of the mechanical energy needed to propel a long-haul truck, the average fuel economy impact is about 0.1 percent. Note that the quadratic dependence of rolling resistance on slip angle means that if the average slip angle is 0.2°, the fuel-savings potential increases by a factor of 4. In any case this analysis indicates that the loss in fuel economy on heavy trucks resulting from wheel alignment is on the order of 0.1 percent. It should be noted once again that these estimates in fuel consumption are intended to provide only order-of-magnitude assessment and that the assumptions used to make this evaluation are generalizations and large variations can exist among different vehicles.

Impact of Reduced Rolling Resistance for Different Truck Classes

Given the large fuel use in Class 8 trucks, coupled with the large impact of rolling resistance, most information about tire effects has been quantified only for this class. Lower vehicle loads result in a reduction in the total rolling resistance, and the relative impact that tires have on fuel economy therefore decreases for vehicles with lower mass (see estimates of relative energy losses among different vehicle applications in previous sections of this chapter). Also, reductions in rolling resistance do not generate the same level of fuel savings for drive cycles that include frequent stops and starts as for drive cycles with minimal amounts of braking, at least for traditional vehicles that do not employ regenerative braking. As a result of these factors, the contribution of rolling resistance to truck fuel use is less in delivery trucks and still less in refuse haulers or buses, roughly 10 percent of total energy consumed (the energy balances shown in Figure 5-1 and in Chapter 4). Hence, the potential improvements in truck fuel consumption are less for truck classes other than Class 8 line haul. As reported in site visits by the committee, the choice of low rolling resistance tire design for urban operation must consider the tire scrub caused by smaller turning radii. Increased probability of damage from curb impacts also was noted. Accordingly, it was reported to the committee that wide single tires were less desirable for this type of use (based particularly on the committee's site visits to Walmart). Tires for refuse haulers are often designed for resistance to cuts and rough or soft terrain encountered at dump sites and would not likely have the lowest levels of C_{rr}.

NHTSA has recently proposed new fuel-economy labeling for passenger car tires as well as recommended standard test procedures (NHTSA, 2009b). While the regulation would only apply to passenger car tires, this labeling can be voluntarily applied to tires for light trucks and Class 2b vehicles. The proposal for labeling stemmed from the 2006 Transportation Research Board (NRC, 2006) study of passenger car tires, which concluded that a 10 percent reduction in rolling resistance was technically and economically feasible, and that this would improve fuel economy by 1 to 2 percent. This is consistent with analysis available in the existing literature (LaClair, 2005). The committee believes these findings are applicable to Class 2b pickups and vans.

Rolling Resistance Summary

TIAX, in its report for the committee, summarized the rolling resistance fuel consumption potential reduction by range of years and by application in Table 5-14. The committee believes that a lower C_{rr} is achievable (i.e., 0.0045)

TABLE 5-14 Rolling Resistance Fuel Consumption Reduction Potential by Class

Truck Class	2013-2015 Fuel Consumption Benefit Relative to Baseline (%)	2016-2020 Fuel Consumption Benefit Relative to Baseline ($)	Assumptions
Class 8 tractor trailer	4.5	11	Base C_{rr} 0.0068 2013 = 0.0055 2016 = 0.0045 on more axles
Class 6 box truck	1.8	3	Same
Class 6 bucket truck	1.4	2.4	Same
Refuse truck	1.5	2.5	WBS not used
Urban bus	1.0	1.5	WBS not used
Motor coach	1.8	3	WBS not used
Class 2b pickup and van	1.0	2	10% C_{rr} reduction

NOTE: WBS, wide-base single.

on more axles in this time frame (2016), which would result in an 11 percent reduction in fuel consumption.

VEHICLE MASS (WEIGHT)

Truck weight (empty weight plus payload weight) affects fuel consumption (gallons per mile) primarily by reducing tire rolling resistance and unrecovered energy used when accelerating or grade climbing. The energy needed to overcome these resistances is essentially linearly dependent on truck weight. Truck weight effects are more conspicuous and dominate fuel consumption over duty cycles with frequent stops and accelerations, or on roads with notable grade. As an illustration of the significance of powering a vehicle of 80,000 lb GVW, maintaining 50 mph on a mere 3 percent grade triples the engine power demand compared to a level road (from 150 hp on a level road to 450 hp on the 3 percent grade; Caterpillar, 2006).

Reducing the unloaded (empty) weight of the truck and trailer, at a fixed payload, always benefits the metric of gallons per payload ton-mile and directly reduces fuel consumption in gallons per mile. This depends on whether the carrier adds payload to take advantage of the tare weight reduction while staying within the legal weight limit (80,000 lb for Class 8), or whether the carrier is at maximum cargo volume ("cubed-out"). For example, for a 35,000-lb truck/trailer that carries 45,000 lb of cargo, reducing weight by 1,000 lb increases capacity, and hence reduces the payload-specific fuel consumption, by 2.2 percent.

Vehicles with maximum GVW loads may still operate at reduced loads on the same trip after off-loading at stops, to the extreme example of tankers running empty on a return trip. During reduced-load vehicle miles traveled, the fuel consumption (gallons per mile) will conspicuously improve with reduced weight. For vehicles that are at maximum volume capacity, vehicle weight savings will always provide the more conspicuous reduction in fuel use.

Truck Weight and Payload Characteristics

In considering the fuel-savings opportunity through lightweighting technology, it is necessary to review the empty truck weight, maximum payload, and then the typically used payload weight in commerce. The range of truck weights and payload capacities is shown in Table 5-15. Focusing on Class 8 combination trucks for now, more specifically, a representative Class 8 sleeper cab tractor will weigh 16,000 to 19,000 lb, and a van trailer (nonrefrigerated) will typically weigh about 14,000 lb. The distributions of weight among the major components of the tractor are shown in Figures 5-32 and 5-33 (Smith and Eberle, 2003).

TABLE 5-15 Typical Weights of Trucks, Empty Versus Gross Weight

Truck Class	Gross Weight Range (lb)	Empty Weight Range (lb)	Payload Capacity Maximum (lb)	Payload Capacity Maximum (% of empty)
1c	0(3,200)-6,000	2,400-5,000	1,000	20
1t	0(4,000)-6,000	3,200-4,500	1,500	33
2a	6,001-8,500	4,500-6,000	2,500	40
2b	8,501-10,000	5,000-6,300	3,700	60
3	10,001-14,000	7,650-8,750	5,250	60
4	14,001-16,000	7,650-8,750	7,250	80
5	16,001-19,500	9,500-10,800	8,700	80
6	19,501-26,000	11,500-14,500	11,500	80
7	26,001-33,000	11,500-14,500	18,500	125
8a	33,001-80,000	20,000-26,000	54,000	200
8b	33,001-80,000	20,000-26,000	54,000	200

Truck Weights on the Road

How many trucks on the road are running at the maximum GVW and thus would benefit from lightweighting by increasing payload? With the demise of the VIUS report,[9] the most recent data are derived from truck scale reporting and weigh-in-motion devices and reporting from surveyed fleets. Figures 5-34 through 5-36 show the distribution of truck weights on the road, again focused mostly on five-axle Class 8 combination trucks. The data in Figures 5-34 through 5-36 were corroborated by direct input from DOT, which reported the following average weights from WIM stations (M. Onder, personal communication, U.S. Department of Energy, May 2009):

- Single-unit truck, 18,728 lb GVW
- Single tractor trailer truck, 54,145 lb GVW
- Multi tractor trailer truck, 59,091 lb GVW

These sources are consistent in concluding that a relatively modest fraction of trucks on the road are near the maximum GVW. Hence, with lightweighting technology, the reduction in fuel consumption will be very apparent as reduced gallons per mile (and per ton-mile).

Impact of Vehicle Weight on Fuel Consumption

As described elsewhere in this study, the truck weight impacts the power needed to propel the truck through rolling resistance, grade climbing, and accelerations. Clearly the duty cycle will influence the degree to which weight savings will reduce fuel consumption. It is also expected that the impact of weight reduction as a percentage of fuel consumption depends on the base weight of the truck.

For the heavier weight classes, a commonly used metric is the percentage change in fuel consumption per 1,000 lb of weight reduction. Data from a recent on-road study were analyzed to illustrate the impact of weight on real-world fuel consumption, shown in Figure 5-37.

[9]The Vehicle Inventory and Use Survey (VIUS), formerly known as TIUS (truck survey), was conducted and reported by the Census Bureau every 5 years starting in 1963. It was discontinued in 2002.

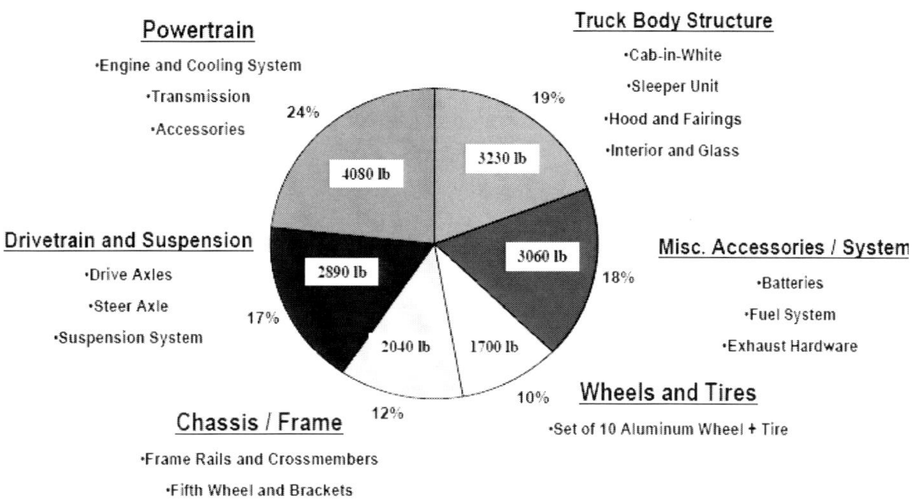

FIGURE 5-32 Weight distribution of major component categories in Class 8 tractors. SOURCE: Smith and Eberle (2003).

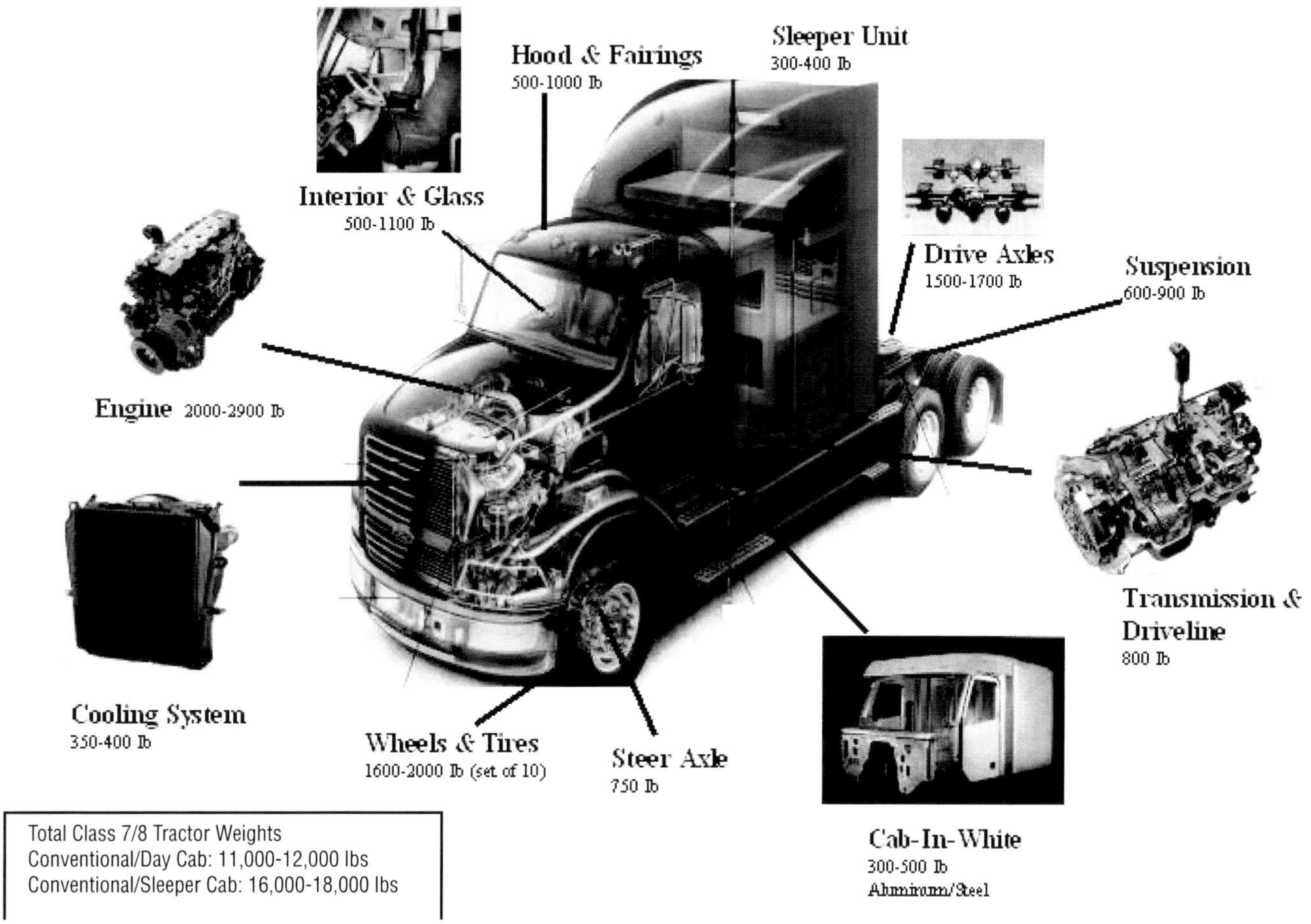

FIGURE 5-33 Typical weights of specific components in Class 8 sleeper tractors. SOURCE: Smith and Eberle (2003).

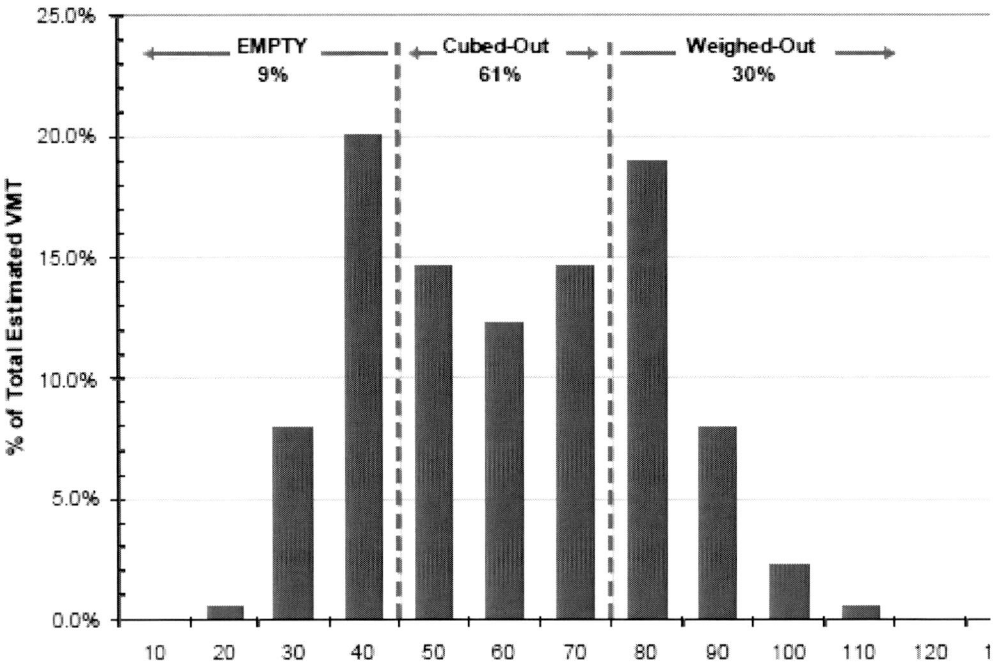

FIGURE 5-34 Truck weight distribution. SOURCE: M.J. Bradley and Associates (2009).

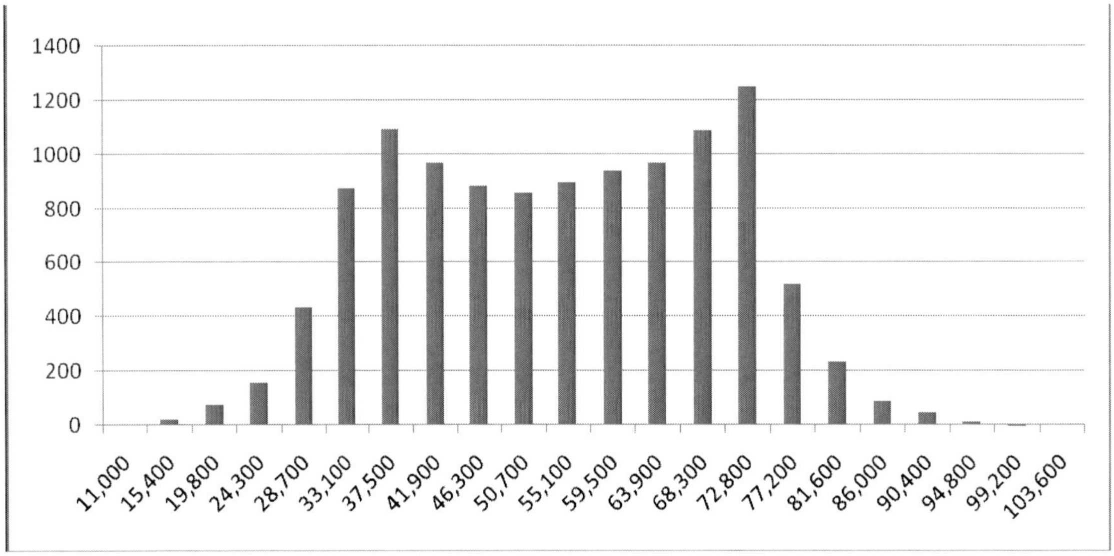

FIGURE 5-35 Truck weight distribution from 2008 weigh-in-motion. SOURCE: New West Technologies.

These data and similar information from numerous other sources are summarized in Table 5-16 for a wide range of vehicle classifications. Published tests on weight effects merely vary the payload for the most part and do not convey the potential fuel savings that could be achieved if engines were downsized or gear ratios changed. Typically, models are very effective and would be needed in examining those effects.

Another advantage of weight reduction technology is the ability afforded the truck manufacturer to maintain the axleload limits imposed by regulation. There is a legal load limit of 34,000 lb for a tandem axle, and 20,000 lb for a single axle on interstate roads, without a special permit (or some higher levels in specific states). With the 80,000-lb limit for maximum gross weight on most U.S. roadways, trucks usually will target about 12,000 lb on the steer axle and 34,000 lb

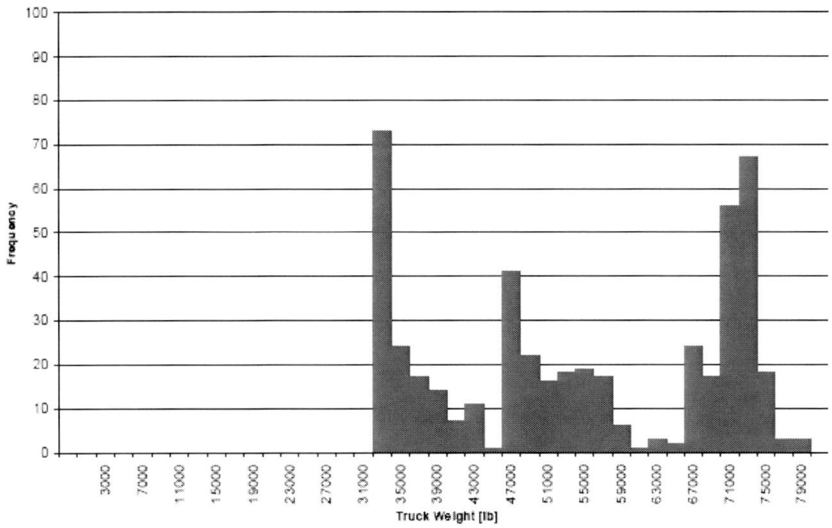

FIGURE 5-36 Truck weight versus trip frequency for six trucks of a single fleet operator. SOURCE: Capps et al. (2008).

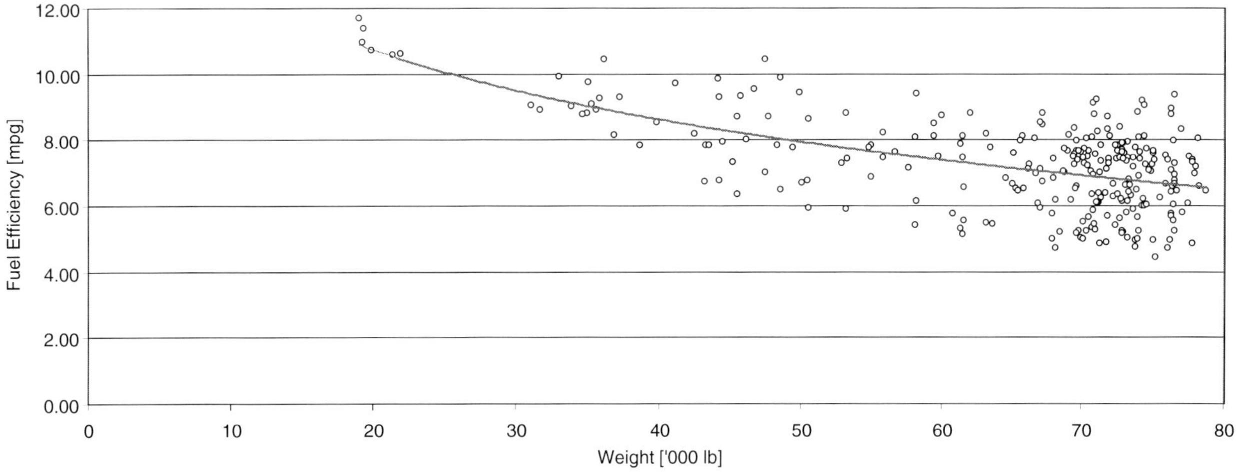

FIGURE 5-37. Effect of weight on truck fuel economy for a monitored fleet of six trucks with combination of dual and wide single tires for a variety of drive routes. SOURCE: Capps et al. (2008).

combined on each of the twin axle pairs for drive and trailer axles.

Potential for Lightweighting Trucks

Trucks, trailers, and buses are benefiting from greater use of lightweight materials and structures. Components already making use of aluminum include the cab structure, wheels, fifth wheel, bellhousing, and more (see Table 5-17).

Aluminum composite panels have been introduced on trailers, and the use of wood in trailers is diminishing. The barrier to additional use of aluminum or carbon composites is primarily cost effectiveness, with carbon fiber composites, for example, costing several times more per unit mass than aluminum. Some technical and cost-effectiveness issues with carbon composites and have been studied in DOE programs with industry (Rini, 2005). An illustration of the possible weight savings with the use of aluminum is shown in Figure 5-38. Some of the measures shown in the illustration are already entering commercial use.

While progress is being made in weight reduction through materials and design, certain weight-adding components have been necessary. Emissions control components are adding roughly 400 lb, and aerodynamic devices another 200 lb, but are deemed a positive tradeoff with aerodynamic drag reduction. Similarly, the weight addition from efficiency technologies such as waste heat recovery are projected to provide net benefits. In hybrid applications, batteries and

TABLE 5-16 Summary of Impacts of Weight on Fuel Consumption of Trucks by Class

Truck Class	Weight Range Studies	Reported Fuel Efficiency Impact	Comments	Source
Combination Class 8	65-80k	0.5% per 1000 lb reduction	5.96 mpg at 65k 5.4 mpg at 80k Highway drive cycle with some grade change	NESCCAF/ICCT (2009)
Combination Class 8	20-80k	0.7-1.0% per 1000 lb level terrain 1.2-1.5% on upward grade route <0.1% on downward grade	Derived from monitored operation of six similar trucks, combination of dual and single tire data	Capps et al. (2008)
Class 8	Not specified	0.4-1.0% fuel savings per 1000 lb weight reduction		EPA SmartWay presentations and fact sheet
Class 8	21-80k	2.0-2.4% per 1000 lb over hill climb and rolling terrain 1.0% on level terrain 1.6% in stop/go	Single truck driven over routes of different terrain at different weights	Strimer et al. (2005)
Class 6 hybrid	Not reported	Fuel economy increase 2.4% per 1000 lb	Model result, using CILCC drive cycle; mostly city and suburb, favors hybridization	NREL (2004)
Bus, 40 ft	35k baseline, conventional drivetrain	~2.0% fuel consumption reduction per 1000 lb	Central business district cycle	NREL (2002, 2004)
		1.66 percent fuel economy gain per 10 percent mass reduction	Hybrid	IFEU (2003)
	Conventional bus	3.75-7.5 percent per 10 percent wt reduction	Urban operation. Wt effects less inter-city	
Class 2b pickup or van	~8000-9000 lb	3-6% fuel economy improvement per 10% weight reduction;[b] higher values cited in urban delivery, 7.5% per 10% wt reduced	Adjusting rear axle ratio also done to give higher impact	Scheps, 2009[a] TIAX (2009) IFEU (2003)

[a]Randall Scheps, Aluminum Association, "The Aluminum Advantage: Exploring Commercial Vehicles Applications," presentation to the committee, Ann Arbor, Michigan, June 18, 2009.

[b]Larger impact assumes engine is resized.

other hybrid components add 300 to 1000 lb for trucks and even more in bus applications. Hybrid components are described elsewhere in Chapter 4.

Weight Reduction Summary

TIAX in its report for the committee summarized weight reduction fuel consumption reduction potential by ranges of years and application, as shown in Table 5-18.

IDLE REDUCTION

Idle reduction[10] technologies use a portion of a vehicle's engine output to power auxiliary systems, which are better suited to the functions they are designed for. Idle reduction technologies that reduce in-use idling in traffic or at work sites using electrification of auxiliaries and engine off at idle are discussed in the Diesel Engine and Hybrid Vehicle paragraphs of this chapter. Creep devices are discussed in the Transmission and Driveline Technology paragraph. This paragraph focuses on approaches to reduce the overnight hotel loads in Class 8 long-haul tractor trailers. The assessment of the technologies comes from the study TIAX conducted for the committee (TIAX, 2009).

Truck idle reduction technologies fall into five categories:

- Automatic shut-down/start-up systems
- Battery-powered
- Fuel perated heaters
- Auxiliary power units or generator sets
- Truck stop electrification

Each idle reduction technology targets a specific set of operational requirements. Benefits will depend on the amount of idling, climate, time of year, and types of auxiliary loads. There is not one solution that fits the northern (high heating loads) and southern (high cooling loads) climates. There is a great deal of information on idle reduction strategies, fuel economy, and costs on EPA's SmartWay Web site.

Automatic Shutdown/Startup Systems

Automatic engine idle management systems monitor cabin and engine temperatures and turn the engine on/off as

[10]"Idle reduction" in this report refers to technologies and practices that reduce the amount of time a vehicle idles the internal combustion engine.

TABLE 5-17 Summary of Weight-Reduction Estimates and Weight-Increase Offsets

Truck Class	Weight Reduction Potential	Weight Increase Potential	Comments	Source
Class 8 combination trucks	Wide singles with aluminum wheels afford up to 340 kg total weight savings over duals with steel; 100 lb savings, aluminum fifth wheel	300 lb, urea; 450 lb, APU; + 60-80 lb, DPF; EGR cooler systems, 40 lb	Some of these changes already on trucks in 2007	Michelin Rhein report
	3000 lb weight savings said achievable by tag axle, engine downsize, wide single tires, lighter trailer			Ogburn et al. (2008)
		Bottoming cycles and other waste heat recovery, 240 lb and up		NESCAFF/ICCT (2009)
	Hybridization can eliminate transmission and mechanical pumps and alternator	Batteries offset weight savings, 200-2000 lb		
	21st Century Truck Partnership agreed on goal of 20% weight reduction			Sullivan, 2007[a]
	Recent Volvo press release says 20% reduced weight feasible in 10 years			FleetOwner (2009)
	25% weight reduction estimate by American Iron and Steel Institute			*Transport Topics* (2009)
Trailers for Class 8	Latest trailers with new materials and composite structures are ~1,000 lb lighter than previous generation	Aerodynamic devices for trailers weigh about 50-200 lb	Lightweighting already under way	Etrucker.com, June 2006 utility trailer press release; Great Dane
Class 3-6 delivery			Already in place except urea; not much change 2007 and beyond	
	Engine; Navistar reports its engine up to 800 lb lighter than competitors' engines		Navistar does not use SCR for 2010	Navistar (2009)
Bus	Use of lightweight materials and designs has shown path to reduce bus weight by ~10,000 lb, an ~50 percent reduction		Fisher Coachworks, used advanced stainless steel	Gibbs, 2009[b] Wall et al. (2006)
	Use of aluminum demonstrated 3,000 lb reduction		China	Scheps, 2009[c]
Class 2b	Similar to passenger cars			See NAS reports on autos

[a]R. Sullivan, "Parasitic Energy Loss Reduction," presentation to the Committee on Review of the 21st Century Truck Partnership, Washington, D.C., February 8, 2007.
[b]J. Gibbs, Presentation to Calstart by the U.S. Department of Energy, Vehicle Technologies Program. March 2009.
[c]Randall Scheps, Aluminum Association, "The Aluminum Advantage: Exploring Commercial Vehicles Applications," presentation to the committee, Ann Arbor, Michigan, June 18, 2009.

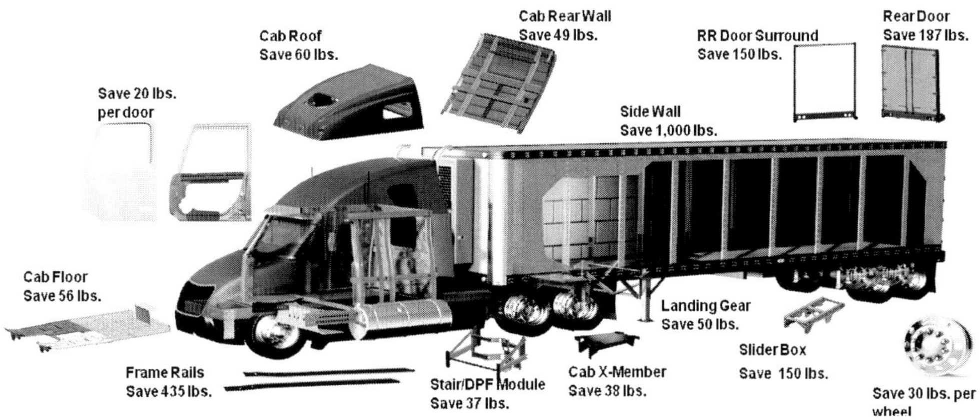

FIGURE 5-38 Weight reduction opportunities with aluminum.
SOURCE: Randall Scheps, Aluminum Association, "The Aluminum Advantage: Exploring Commercial Vehicle Applications," presentation to the committee, Ann Arbor, Mich., June 18, 2009.

TABLE 5-18 Weight-Reduction-Related Fuel Consumption Reduction Potential (percentage) by Class

	2013-2015	2015-2020
Tractor trailer	0.8	1.25
Class 6 box truck	2.0	4
Class 6 bucket truck	1.6	3.2
Refuse truck	—	1
Urban bus	3.0	6.25
Motor coach	0.7	1.05
Class 2b pickup and van	0.4	0.75

needed to maintain the desired temperatures. Several engine manufacturers offer electronically controlled optimized idling control devices. Higher end integrated systems include thermal storage. More advanced systems run the engine at higher load to maximize engine efficiency when it is on. Table 5-19 characterizes several such systems. Duleep[11] estimates that these systems reduce fuel consumption by 3 percent.

TABLE 5-19 Comparison of Automatic Shutdown/Startup Systems

Feature/Supplier	BBW IdleSmart	Cummins ICON™ Idle Control	TAS Temp-A-Start
Monitors temperature inside and out	✓	✓	✓
Auto starts engine	✓	✓	✓
Increases engine speed for maximum efficiency	✓	✓	
Monitors system (i.e., oil and cab temperatures, battery voltage)	✓	✓	✓
Heats and cools cab	✓	✓	✓
Idles down	✓	✓	
Shuts down engine	✓	✓	✓
Transferable to successive trucks	✓	✓	✓
Weight (lb)	30	na	14
Retail price	$3,750	$1,325	$2,500

SOURCE: TIAX (2009), p. 4-90.

Battery-Powered

Battery-powered idle reduction systems incorporate the use of either the vehicle's battery or a separate battery pack to power the heating and/or cooling of the cab and powering of onboard appliances. A few systems also offer shore power connections. Each of these systems provides 8 to 12 hours of runtime. Because of this limited duration, they may not be ideal for applications that require longer term (weekend) heating or cooling. Table 5-20 compares a variety of systems currently available as identified by the U.S. EPA's SmartWay Web site.[12] As shown, each system offers various features and each retails for $575 and $7,500. Both Kenworth and Peterbilt have introduced idle reduction systems on their products. The Kenworth system is called Clean Power, and operators could see as high as 8 percent reduction in fuel consumption. The system, however, depends greatly on the application. Operators with substantial heating requirements could see the benefit reduced to 1 to 2 percent. Both systems depend on the current battery capacity. Current systems use lithium-ion batteries, and manufacturers are looking at various chemistries for future generations. There is an expectation that the passenger car market will drive economies of scale and lower battery costs.

Fuel-Operated Heaters

Also known as direct-fired heaters, fuel-operated heaters are available to heat the cab, engine, or both. These heaters use significantly less fuel than the primary engine by supplying heat from either a combustion flame or a small heat exchanger, which is more energy efficient than using an engine cycle to achieve a heating affect. In model year (MY) 2007, there was about 10 percent market penetration for the direct-fired heater option. Table 5-21 shows a comparison of several fuel-operated heaters.

Auxiliary Power Units

Auxiliary power units (APUs) provide electricity and heat with the help of a small internal combustion engine equipped with a generator and heat recovery device. Cooling can be provided with the installation of an electric air conditioner in the cab. Many APU devices are available. In addition to those APUs compared in Table 5-22, Navistar announced in a press release:[13]

> Operating at 4.2 kW, the MaxxPower APU equates to best-in-class fuel consumption of 0.18 gallons per hour. It features EPA Tier IV certification and is in the process of acquiring CARB '08 emissions compliance. Several states currently offer incentives for trucks with idle reduction systems. Production is slated for later this year.

(Note that APUs need to be emissions compliant to operate in California, which can be accomplished with the installation of a $3,000 diesel particulate filter.)

APUs can also be fuel cell powered. This technology requires hydrogen for fuel or a reformer system for carbon-based fuels and the system can achieve the same fuel consumption improvement as conventional APUs.

[11] K.G. Duleep, Energy and Environmental Analysis, "Heavy Duty Trucks Fuel Economy Technology," presentation to the committee, Washington, D.C., December 5, 2008.

[12] See http:eps.gov/smartway/transport/what-smartway/idling-reduction-available-tech.htm.

[13] See http://www.rueters.com/article/pressRelease/idUS213209+07-May-2008+BW20080507.

TABLE 5-20 Idling Reduction Technologies

Feature/Supplier	Autotherm T-2500 Energy Recovery System	Bergstrom NITE	Dometic	Driver Comfort System	Glacier Bay ClimaCab	Idle Free Systems Reefer Link System	Kenworth Clean Power	Peterbilt ComfortClass	Safer Viesa	Sun Power Technologies
Heats cab	✓	✓	✓	✓	✓	✓	✓	✓		✓
Cools cab		✓	✓	✓	✓	✓	✓	✓	✓	✓
Powers on-board appliances				✓			✓	✓		
Circulates heated coolant from engine to heater coils	✓									
Runs off of vehicle battery	✓					✓			✓	
Runs off separate battery pack		✓	✓	✓	✓	✓	✓	✓		✓
Shore power connection				✓		✓	✓			
Automatic start	✓ at engine off	✓			✓					
Automatic shut down	✓	✓			✓					
Monitors system (coolant temperature, battery voltage)	✓	✓ (cab only)		✓	✓			✓		
Run time (hours)	na	10	10-15		10 (heating) 12 (cooling)		10 (cooling)	10	8	8-12
Thermal storage cooler							✓	✓		
Direct fueled heater						✓	✓			✓
Weight (lb)	5	210	na	520	161 (not including batteries)	200	550		126	440
Retail price	$575 to $710	$3,495	$3,500 to $7,500	$6,895		$7,995			$1,600	$6,900

SOURCE: TIAX (2009), p. 4-91.

TABLE 5-21 Comparison of Fuel-Operated Heaters

Feature/Supplier	Automotive Climate Control	Espar Heater System		Webasto Product North America	
		Cab	Engine	Cab	Engine
Heats cab	✓	4 models		2 models	
Manual or automatic control	Yes	Auto	Auto	Auto	
Battery-powered fuel-fired heaters	✓	✓	✓	✓	✓
Heats engine			2 models		3 models
Fuel use	1 gal/24 hr	1 gal/20 hr	1 gal/4-6 hr	1 gal/20 hr	0.03-0.24 gal/hr
Retail price	$920-$1,200	$1,000-$3,000		$1,000-$3,000	

SOURCE: TIAX (2009), p. 4-92.

Truck Stop Electrification

There are currently 138 truck stops equipped with truck stop electrification identified in the Alternative Fuel Data Center Truck Stop Electrification database.[14] Located in 34 states, these sites provide truckers the opportunity to "plug-in" to power heaters, air conditioners, lights, and other accessories. There are generally two types of electrified parking spaces: dual system and single system. A dual system requires equipment on both the truck and the ground. A single system requires equipment only at the truck stop. Table 5-23 shows a comparison of several truck stop electrification systems. As discussed in the National Research Council's (NRC's) review of the DOE 21st Century Truck Partnership (NRC, 2008), continuing efforts to standardize the electrical systems on trucks and at truck stops are needed.

[14]See http://www.afdc.energy.gov/afdc/progs/tse_listings.php.

TABLE 5-22 Comparison of Auxiliary Power Units

Supplier	Heats Cab	Cools Cab	Heats Engine	FC (gal/hr)	110V AC	Auto-on	Shorepower	Cost
Auxiliary Power Dynamics	Yes	Yes	Yes	0.25	Yes	Yes		
Black Rock Systems	26,000 Btu	26,000 Btu		0.20 or 0.30				$7,499 to $8,100
Carrier Transicold	Yes	Yes	Yes	0.2	Yes	Yes	Yes	
Comfort Master	31,000 Btu	31,000 Btu		0.25	Yes	Yes	Yes	$7,200 to $8,100
Craufurd Manufacturing	22,000 Btu	22,000 Btu						
Diamond Power Systems	14,500 Btu	14,500 Btu		0.26	Yes	Yes	Yes	$6,500
Double Eagle Industries	Yes	Yes	Yes	0.3	DC			$7,000 to $9,000
Flying J	Yes	Yes	Yes		Yes	Yes	Yes	$6,999
Frigette Truck Climate System	Yes	Yes			Yes			$6,000 to $7,500
Idlebuster	Yes	Yes	Yes		Yes	Yes	Yes	$6,900 to $7,750
Kohler	10,000 Btu	12,000 Btu			Yes	Yes		
Kool-Gen	Optional	Yes			Yes			$6,925
Mechron Power Systems	5,000-10,000 Btu	10,000-14,000 Btu		0.21	Yes			
Pony Pack	Yes	Yes	Yes	0.2	DC			$7,500
Rig Master Power	Yes	Yes	Yes	0.2	Yes			$6,300
Star Class								$5,995 to $6,500
Thermo King	13,000 Btu	Yes	Yes	0.04 to 0.14	Yes	Yes		$8,000 to $10,000 (+$3,000 DPF for California)
TRIDAKO Energy Systems	30,000 Btu	24,000 Btu	Yes	0.4				$8,499
Truck Gen	Yes	Yes		0.2	Yes		Yes	$6,000 to $7,000

SOURCE: TIAX (2009) p. 4-94.

Idle Reduction System Comparison

Idle reduction systems differ in a number of respects. Some of the key discriminators are fuel use per time; functionality (heating ability, cooling ability, and electric loads); infrastructure requirements; and cost. These features are summarized in Table 5-24 for each of the five idle reduction strategies examined. Cells that are shaded light green correspond to favorable attributes; cells that are shaded yellow correspond to mild drawbacks; and cells that are shaded dark orange correspond to major drawbacks. The fuel-savings benefits in Table 5-24 are estimated using the following assumptions:

- An idling engine consumes 0.8 gallons of fuel per hour. This is based on a SmartWay estimate for idle engine fuel consumption, and is consistent with a 4-kW accessory load with an engine operating at 15 percent efficiency. The 21st Century Truck Partnership estimates a typical accessory load of 3 to 5 kW for a line-haul truck (NRC, 2008). It is possible that current engines have lower idle fuel consumption rates.
- A line-haul truck is assumed to operate under hotel loads for between 1,500 to 2,400 hours per year. This range is meant to bracket the range between a medium- and high-mileage vehicle.
- A line-haul truck is assumed to use 20,000 gallons of fuel per year.
- It was assumed that a direct-fire heater is in use for 600 to 800 hours per year. SmartWay estimates 800 hours per year of heater fuel use.[15]

TABLE 5-23 Comparison of Truck Stop Electrification Systems

Supplier	Cost ($)	Service Fee ($)
Dual Systems		
Phillips and Temro	125	
Shurepower	200-2,000	0.50 per hour
Teleflex (Proheat)	2,500	
Xantrex	1,500, inverter/charger 1,500, electric HVAC	2,500 per space
Single Systems		
Cabaire		
Craufurd Manufacturing		8,550 per space
IdleAire Technologies	10, adapter	2.18 per hour retail 1.85 per hour fleet

SOURCE: TIAX (2009), p. 4-93.

INTELLIGENT VEHICLE TECHNOLOGIES

IVT combines information about the state of the vehicle, the environment around the vehicle, and Telematics to provide assistance to the driver. Telematics refers to the integration of Global Positioning System (GPS) technology with computers and mobile communications technologies. Although IVT is commonly applied to active safety systems (i.e., crash avoidance and crash mitigation), for purposes of this report the definition is broadened to specifically address fuel consumption reduction.

[15]See http:www.epa.gov/smartway/transport/calculators/calculator explanation.htm.

TABLE 5-24 Comparison of Idle Reduction Systems

	Heating	Cooling	Electric	Requires Recharge Infrastructure	Service Fee	Emissions Control Needed?	Idle time avoided per year	Fuel Use (gal/hr)	% Benefit	Cost
Engine Ctrl	Yes	Yes	Yes	No	No	No	1,500 to 2,400	~0.5	3%	$1,000 to $4,000
Heater	Yes	No	No	No	No	No	500 to 800	0.2 to 0.3	1.3 to 2.3%	$1,000 to $3,000
APU	Yes	Yes	Yes	No	No	In CA	1,500 to 2,400	0.2 to 0.3	4 to 7%	$6,000 to $8,000[43]
Battery	Yes	Yes	Some	Yes	No	No	1,500 to 2,400	-	5 to 9%	$3,000 to $8,000
Shore Power	Yes	Yes	Yes	Yes	Yes	Yes	1,500 to 2,400	-	5 to 9%	~$100

SOURCE: TIAX (2009), p. 4-95.

IVT-Enabling Technologies

Although computing technologies have advanced greatly over the past decade, the most significant enablers for IVT are the advances in GPS and mobile communications. The first step in the GPS modernization program occurred in May 2000 when the U.S. Department of Defense ended the use of Selective Availability (SA). SA was an intentional degradation of civilian GPS accuracy implemented on a global basis from the GPS satellites. Prior to its deactivation, civil GPS readings could be off by up to 100 m. After SA was turned off, civil GPS accuracy instantly improved by an order of magnitude, thereby benefiting civil and commercial users worldwide. In 2004 about 1 million fleet vehicles in the United States were equipped with GPS devices (Murphy, 2004).

Mobile communications, particularly cell phone usage, has grown immensely in recent years. Cell phone users in the United States have increased from 50 million a little over a decade ago to more than 250 million in 2007 (CIA, 2007). Many cell towers are located along the interstate system, which allows mobile travelers to have cellular service even in the more remote rural regions of the United States.

Dedicated Short Range Communications (DSRC) is a block of spectrum in the 5.850 to 5.925 GHz band allocated by the Federal Communications Commission in 2003 to enhance the safety and productivity of the transportation system. The DSRC service involves vehicle-to-vehicle and vehicle-to-infrastructure communications, helping to protect the safety of the traveling public. The band is also eligible for use by nonpublic safety entities for commercial or private DSRC operations.

Adaptive Cruise Control

Description. Adaptive cruise control (ACC) augments conventional cruise control by sensing the traffic ahead with a radar or laser sensor mounted on the front of the vehicle. When there is no vehicle ahead, ACC operates the same as conventional cruise control. However, when the forward-looking sensor detects a vehicle ahead that is traveling at a slower speed, the vehicle speed is slowed to the speed of the preceding vehicle through actuation of the throttle or mild brake action. The appropriate separation distance between vehicles is then maintained automatically. The desired separation distance can be set by the driver within limits and is a function of the vehicle speed. Typically, this is the distance the vehicle would travel in the range of 2 to 3 seconds. Anything significantly larger than that would likely result in "cut-ins" by a third vehicle.

Applications. All vehicles that regularly travel on urban and rural interstate roads. Adaptive Cruise Control systems are available on current vehicles (primarily as a safety feature) but are not widely adopted.

Cost. $1,100 to $3,000, depending on application and included features.

Benefit. Cruise control is a driver aid that relieves driver workload and provides smooth acceleration and deceleration for navigating grades. In traffic, conventional cruise control is little used because a constant speed cannot be maintained. The driver must therefore revert to manual control. ACC has fuel economy benefits because, even in traffic, all accelerations and decelerations are smaller and have a smooth profile and there is little accelerator pedal "dither" compared to manual driving. A field operational test using 108 nonprofessional drivers and 10 passenger cars equipped with ACC in southeast Michigan (Koziel et al., 1999) indicated a 10 percent fuel consumption reduction compared with manual driving. For professional drivers, the benefit is more likely to be around 1 percent (TIAX, 2009, p. 4-99).

Predictive Cruise Control

Description. Predictive cruise control is an enhancement of conventional cruise control whereby the current vehicle location and topography of the upcoming road provided by a GPS receiver are used to calculate the target cruise speed within an upper and a lower limit. When a truck approaches a hill, the truck accelerates prior to beginning the climb. During the climb, the target cruise speed is continuously calculated and the truck is allowed to slow to the lower limit. When approaching a downhill, the truck slows down prior to beginning the descent but is allowed to increase to the upper limit during the descent. Fuel savings accrue because there is less need to accelerate while on the uphill climb and less time spent in the lower gears. In addition, less fuel is consumed compared to conventional cruise control during the downhill phase because the truck is allowed to slow to a lower limit prior to the descent of a hill rather than maintaining a fixed cruise speed.

Applications. All vehicles that regularly travel on rural interstate roads. Daimler Trucks of North America will be offering this feature in 2009.

Cost. $861 to $1,561, depending on type of vehicle.[16]

Benefit. The reduction in fuel consumption will vary depending on the topography the truck is traveling. For example, traveling on relatively flat terrain would yield little benefit. Traveling in the hills of Tennessee, on the other hand, could yield significant fuel savings. Experimental results reported by Hellstrom et al. (2007) show a 3.5 percent fuel use reduction without an increase in trip time for a Class 8 truck traveling 150 km between Norrkoping and Sodertalje, Sweden, compared to conventional cruise control. Simulation results reported by Lattermann et al. (2004) of a 75,000 lb class 8 truck traveling on a 25 km stretch of I5 around Portland, Oregon, showed a 4 to 5 percent reduction in fuel consumption with an increase in trip time of 0.3 to 1.4 percent compared to conventional cruise control. Other estimates would suggest an improvement of 1 to 3 percent (TIAX, 2009, p. 4-99).

Adaptive Cruise Control with Real-Time Traffic Information

Description. This concept utilizes real-time traffic information to anticipate changes in traffic speed and then adjusts the set cruise speed accordingly to reduce large accelerations and decelerations. The real-time traffic data can come from imbedded loop detectors in the highway such as the Performance Measurement System (PeMS) used in California[17] or traffic probe vehicles that carry special cell phones able to communicate their position and velocity in real time. PeMS provides average traffic speed and density every 5 minutes at a resolution of 0.3 to 3 miles. A communications link, such as cell phones or SRDC, is required for the vehicle to acquire the traffic information from nearby vehicles or the infrastructure.

Applications. All vehicles that regularly travel rural and urban interstate roads. There is some usage in California, but its low. There are currently no other commercial applications in existence.

Cost. No cost data are available.

Benefit. In research reported by Kohut et al. (2009), simulations were run using real traffic speed data for a trip from Palo Alto to San Jose, California, and validated vehicle and engine models of a passenger car. The simulations calculated the reduction in fuel used when varying the look-ahead distance of traffic flow. The simulations showed a 5 to 7 percent reduction of fuel for a trip time that changes 3 percent or less when the look-ahead distance was varied from 1,200 to 2,000 m.

Predictive Control of Hybrid Electric Vehicles

Description. Real-world optimal fuel consumption for hybrid electric vehicles (HEVs) is possible using predictive control based on GPS with a topographical database of the road ahead and, possibly, real-time traffic information, and then controlling the power split ratio (PSR) of the internal combustion engine and the electric motor. The main controller uses current operating information such as battery state of charge, engine efficiency, and emission maps, to establish the PSR at each instant. A navigation controller uses traffic and GPS information to predict the future driving state of the vehicle and modifies the PSR to charge the battery (if, for example, it is predicted that the vehicle will change from highway to city driving, where the electric motor will be required) or to deplete the battery for improved fuel economy in anticipation of a downgrade, where regeneration may be expected (Kessels, 2007).

[16]David Kayes, Daimler Trucks, personal communication, June 23, 2009.

[17]California Performance Measurement System, http://pems.eecs.berkeley.edu.

Applications. All HEV applications. Current commercial usage is unknown.

Cost. Similar to that cost of predictive cruise control.

Benefit. Simulation results reported by Rajagopalan and Washington (2002), Johannesson et al. (2007), and Kim et al. (2008) show a fuel economy improvement of 3 to 9 percent versus nonpredictive control. The SENTIENCE program described by Walker (2008), which is underway, is intended to provide much-needed validation of simulation results.

Electronic Tow Bar

Description. The electronic tow bar is a concept whereby a lead vehicle is driven manually and a following vehicle is driven automatically by a vehicle controller that maintains a set distance between the two vehicles. In a sense, it is similar to ACC except that the gap between the lead and trailing vehicles is much smaller to take advantage of the aerodynamic drag reduction from the slipstream effect. Because of this small separation distance, typically one-half to one truck length, precise control must be maintained by the trailing vehicle to prevent the vehicles from contacting if the lead vehicle suddenly brakes. Although this concept is not yet commercially available, the experimental hardware configurations described by Fritz (1999), Fritz et al. (2004), and Lu et al. (2004) generally use radar, laser, or optical sensors on the trailing vehicle to measure the separation distance. In addition, vehicle-to-vehicle communications are used to provide information on the state of lead vehicle (e.g., vehicle speed, acceleration, pedal position) to the vehicle controller in the trailing vehicle. This gives additional lead time to the trailing vehicle to respond to sudden accelerations or braking of the lead vehicle.

Applications. The electronic tow bar would be most applicable to line-haul trucks using the interstate system. Although it is possible to extend the concept to more than two vehicles, it is unlikely that this would be allowed, with the possible exception of truck-only lanes.

Cost. Anticipated costs of around $500 to $2,600 for additional sensors and active safety features (Baker et al., 2009).

Benefit. Bonnet and Fritz (2000) conducted experiments on two heavy-duty semitrailer Mercedes-Benz trucks of type ACTROS 1853 LS, both having cab-over-engine design. The lead truck was 32,000 lb and the trail truck was 62,000 lb. For the trail truck with the separation distance varied between 8 and 16 m, the fuel consumption reduction ranged from 15 to 21 percent at 80 km/hour and from 10 to 17 percent at 60 km/hour when compared with the truck driving in isolation. For the lead truck, the fuel consumption reduction was between 5 and 10 percent at 80 km/h and between 3 and 7 percent at 60 km/hour.

Browand et al. (2004) report on experiments conducted on two Freightliner 2001 Century Class trucks with 53-foot van trailers. The tractors were engine-forward design. One vehicle was 32,000 lb and the other was 64,000 lb. At a constant speed of 55 mph, the measured fuel savings at a spacing of 10 m were 10 percent and 6 percent, respectively, for the trail and lead truck. In the spacing range of 3 to 10 m, fuel consumption savings were in the range 10 to 12 percent for the trail truck and 5 to 10 percent for the lead truck, with the larger values of savings occurring at the shorter spacing.

Navigation and Route Optimization

Description. In its simplest form, navigation and route optimization consists of an in-vehicle device that contains a GPS receiver, a database that includes map information and points of interest, and a display that allows the driver to enter a desired destination and view the map. The device then calculates a route based on one of several criteria, such as fastest time or shortest distance. While en route, the driver receives visual and verbal turn-by-turn instructions to the destination. Navigation devices specially designed for the trucking industry may have features that will inform the driver about tolls, road restrictions, hazmat routes and preferred truck routes (e.g., parkways that do not allow trucks, directions that do not require right-hand turns). In addition, it may include points of interest such as truck stops, rest stops, weigh stations, and other services useful to the driver.

For the fleet operator, the route is usually planned in the back office using route optimization software and downloaded to the vehicle either before the vehicle leaves the terminal or while the vehicle is on the road. The route optimization software may reside on a single personal computer, the fleet's central computer, or a Web-based application and has sophisticated algorithms that take into account historical traffic data as a function of time of day and up-to-date speed limits on the planned route. The planned route will yield a path with minimum fuel consumed. It is also possible for the fleet operator to provide immediate changes to its routes in case of incidents such as road construction or weather-related road closures.

Dynamic vehicle routing differs from the static routing previously described in that it includes real-time traffic information in addition to historical traffic data for computing the optimized route. Real-time traffic information from imbedded loop detectors or traffic probe vehicles using smartphones allow the route planning software to do en route rerouting to avoid traffic congestion. This concept, relatively new, has potential drawbacks in that the driver may not receive the new route in a timely manner to avoid the congestion or the alternate route may itself become congested due to the additional traffic.

Applications. Pickup and delivery applications, regional and long-haul operations, fleet operators.

Cost. Navigation device, between $400 and $800 plus monthly service fees that range from $20 to $40 (TIAX, 2009, p. 4-98). Route optimization software starts at $10,000 (Bennett, 2008).

Benefit. Bennett (2008) reports the case of a fleet operator where route optimization software reduced fleet mileage between 5 percent and 10 percent annually. For line-haul operations, which spend a comparatively small amount of time in congested driving conditions, the estimated fuel-savings potential is up to 1 percent (TIAX, 2009, p. 4-98).

FINDINGS AND RECOMMENDATIONS

Aerodynamics

Finding 5-1. At highway speeds, aerodynamic loads consume more power than any other load on current tractor-trailer vehicles. Aerodynamic features can significantly reduce these loads, but their value diminishes rapidly as average vehicle speed goes down. In low-speed operation, aerodynamic features have little value.

Finding 5-2. Four areas of the tractor-trailer combination have been identified as critical for achieving aerodynamic improvements:

- Tractor streamlining
- Management of airflow around the tractor-to-trailer gap
- Management of airflow under the trailer
- Management of airflow at the rear of the trailer

Finding 5-3. By the 2015 to 2020 time frame, the use of aerodynamic features can provide fuel consumption reductions of about 15 percent for tractor-van trailer vehicles operating at 65 mph. The potential benefits for other classes of vehicles are significantly less.

Finding 5-4. Many tractor and trailer aerodynamic features are damage prone in low-speed operation. The cost of repairing these features may be a significant barrier to implementation for some applications, and broken aerodynamic components could become road hazards.

Recommendation 5-1. Regulators should require that aerodynamic features be evaluated on a wind-averaged basis that takes into account the effects of yaw. Tractor and trailer manufacturers should be required to certify their drag coefficient results using a common industry standard.

Auxiliary Loads

Finding 5-5. Auxiliary loads can consume up to 2.5 percent of fuel, so fuel consumption reductions of 1 percent to 2.5 percent are feasible. Electrification of these auxiliaries, mostly in hybrid vehicles, will reduce some of this loss.

Rolling Resistance

Finding 5-6. Technological advances have lowered the coefficient of rolling resistance of tires by roughly 50 percent since 1990, but further reductions are expected to be less dramatic. The use of low rolling resistance tires, such as wide-based singles, show 4 percent to 11 percent reductions in fuel consumption with models and on-road tests, depending on terrain, weight, and choice of baseline tire.

Finding 5-7. Tire pressure monitoring, automatic inflation systems, and nitrogen inflation are all effective in avoiding wasting fuel due to underinflation and improve vehicle safety.

Recommendation 5-2. There are numerous variables that contribute to the range of results of test programs. An industry standard (SAE) protocol for measuring and reporting the coefficient of rolling resistance is recommended to aid consumer selection, similar to that proposed for passenger car tires.

Vehicle Mass (Weight)

Finding 5-8. Results from tests and computer models summarized in this chapter show that the impact of weight on truck fuel consumption will range from 0.5 to 1.0 percent per 1000 lb on level roads to over 2 percent per 1,000 lb on hilly terrain and for driving cycles with frequent accelerations. These results are primarily for Class 8 combination trucks. For Class 8 trucks at full weight capacity, the payload-specific fuel consumption is reduced by about 2 percent per 1,000 lb.

Finding 5-9. Design progress and the use of lightweight materials for major components, such as the engine, drivetrain, wheels and tires, and chassis, have been estimated to save weight up to 20 percent beyond current technology by the 21st Century Truck Partnership and separately by one manufacturer. This could amount to as much as 5,000 lb over the next decade. A fuel consumption reduction of about 5 percent could be achieved.

Idle Reduction

Finding 5-10. There are a number of technologies and products available for reducing idle fuel use in class 8 heavy-duty vehicles. It is reported that up to 9 percent fuel consumption reduction is available, but it is dependent on the hotel power load factor. The committee has used 5 percent to 9 percent and TIAX used an average of 6 percent fuel consumption reduction potential.

Intelligent Vehicle Technologies

Finding 5-11. In general, intelligent vehicle technologies provide fuel consumption reductions by taking advantage of knowledge of the vehicle's location, terrain in the vicinity of the vehicle, congestion, location of leading vehicles, historical traffic data, and so forth, and altering the speed of the vehicle, the route the vehicle travels, or, in the case of hybrid electric vehicles, altering the power split ratio. This fuel savings may not show up in any fuel consumption test.

Finding 5-12. A number of the technologies—adaptive cruise control, predictive cruise control, and navigation and route optimization—are being applied by the trucking industry without any regulation because the owners and operators view the reduction in fuel costs as good business.

Finding 5-13. Based on experiments to date, the electronic tow bar concept of trucks traveling closely spaced in tandem can provide significantly lower fuel consumption, 8 percent to 15 percent, compared with the same vehicles traveling separately.

BIBLIOGRAPHY

Al-Qadi, I. 2007a. New generation of wide-base tire and its impact on trucking operations, environment, and pavements. Transportation Research Board 86th Annual Meeting, Paper No. 07-2432, January 21-25, Washington, D.C.

Al-Qadi, I. 2007b. Impact of wide-based tires on pavement and trucking operation. Presented at the International Workshop on the Use of Wide-Base Tires, Federal Highway Administration, Turner-Fairbank Highway Research Center, October 25-26, McLean, Virginia.

Bachman, L., A. Erb, and C. Bynum. 2005. Effect of single wide tires and trailer aerodynamics on fuel economy and NO_x emissions of Class 8 line-haul tractor-trailers. SAE Paper 2005-01-3551. Warrendale, Pa.: SAE International.

Bachman, L., A. Erb, B. Anthony, C. Bynum, B. Shoffner, H. De La Fuente, and C. Ensfield. 2006. Fuel economy improvements and NO_x reduction by reduction of parasitic losses: Effect of engine design. SAE Paper 2006-01-3474. Warrendale, Pa.: SAE International.

Baker, H., R. Cornwell, E. Koehler, and J. Patterson. 2009. Review of Low Carbon Technologies for Heavy Goods Vehicles. Prepared by Ricardo for the U.S. Department of Transportation. RD.09/18261.6. June.

Bennett, S. 2008. Smart routing to save fuel. Light & Medium Truck Magazine. August.

Bonnet, C., and H. Fritz. 2000. Fuel consumption reduction in a platoon: Experimental results with two electronically coupled trucks at close spacing. SAE Paper 2000-01-3056. Warrendale, Pa.: SAE International.

Bradley, C., and S. Nelson. 2009. Truck Tires and Rolling Resistance: Modeling and Impact on Fuel Consumption. Presentation by Michelin Tire North America to the committee, February 4.

Bradley, M.J., and Associates LLC. 2009. Setting the Stage for Regulation of Heavy-Duty Vehicle Fuel Economy and GHG Emissions: Issues and Opportunities. Washington, D.C.: International Council on Clean Transportation. February.

Browand, F., J. McArthur, and C. Radovich. 2004. Fuel Savings Achieved in the Field Test of Two Tandem Trucks. California PATH Research Report UCB-ITS-PRR-2004-20.

Capps, G., O. Franzese, B. Knee, M.B. Lascurain, and P. Otaduy. 2008. Class-8 Heavy Truck Duty Cycle Project Final Report. Report No. ORNL/TM-2008/122. Oak Ridge, Tenn.: Oak Ridge National Laboratory. December.

CARB (California Air Resources Board). 2008. Final Regulation Order to Reduce Greenhouse Gas Emissions from Heavy-Duty Vehicles. Available at http://www.arb.ca.gov/regact/2008/ghghdv08/ghgfro.pdf.

Caterpillar. 2006. Understanding Tractor-Trailer Performance. Brochure. Available at http://ohe.cat.com/cda/files/867819/7/LEGT6380.pdf. Peoria, Ill.: Caterpillar.

CIA (Central Intelligence Agency). 2007. CIA World Fact Book.

Cooper, K. 2004. Commercial vehicle aerodynamic drag reduction: Historical perspective as a guide. Pp. 9-28 in The Aerodynamics of Heavy Vehicles: Trucks, Buses, and Trains (R. McCallen, F. Browand, and J. Ross, eds). New York: Springer.

DOE (U.S. Department of Energy). 2006. 21st Century Truck Partnership Roadmap and Technical White Papers. Doc. No. 21CTP-003. Washington, D.C.: DOE. December.

DOE. 2008. Areas of Loss in Typical On-Highway Vehicles, and/or Review. Washington, D.C.: DOE.

ECD (European Council Directive 89/297/EEC). 1989 Underride Protection for Unprotected Road Users. April13. Available at http://eur-lex.europa.eu/LexUriServ/LexUriServ.do?uri=CELEX:31989L0297:EN:NOT.

Englar, R. 2005. The application of pneumatic aerodynamic technology to improve drag reduction, performance, safety, and control of advanced automotive vehicles. Proceedings of the 2004 NASA/ONR Circulation Control Workshop, Part 2, pp. 957-995. June. Available from http://ntrs.nasa.gov/search.jsp?R=421681&id=4&as=false&or=false&qs=Ne%3D25%26Ns%3DHarvestDate%257c0%26N%3D255%2B32

EPA (U.S. Environmental Protection Agency). 2009. Websites for the EPA SmartWay Program, including http://www.epa.gov/smartway/transport/what-smartway/tractor-trailer-tech.htm (accessed 13/09/09) and http://www.epa.gov/smartway/transport/what-smartway/tractor-trailer-markuse.htm (accessed September 13, 2009).

FleetOwner. 2009. Superlight trucks are coming. FleetOwner Web Magazine. June 25. Available at http://fleetowner.com/trucking_around_world/building-lighter-trucks-0625/.

Fritz, H. 1999. Longitudinal and lateral control of heavy duty trucks for automated vehicle following in mixed traffic: Experimental results from the CHAUFFEUR Project. Proceedings of the 1999 IEEE International Conference on Control Applications. Kohala Coast-Island of Hawaii, State of Hawaii, August.

Fritz, H., A. Gern, H. Schiemenz, and C. Bonnet. 2004. CHAUFFEUR Assistant—A Driver Assistance System for Commercial Vehicles Based on Fusion of Advanced ACC and Lane Keeping. IEEE 2004 Intelligent Vehicles Symposium. Parma, Italy. June.

Goodyear (Goodyear Tire and Rubber Company). 2003. Radial Truck Tire and Retread Manual. Manual No. 700-862-932-505.

Hellstrom, E., M. Ivarsson, J. Aslund, and L. Nielson. 2007. Look Ahead Control for Heavy Trucks to Minimize Trip Time and Fuel Consumption. Presented at the Fifth IFAC (International Federation of Automatic Control) Symposium on Advances in Automotive Control, Aptos, Calif., August 20-22.

Hendricks, T., and M. O'Keefe. 2002. Heavy Vehicle Auxiliary Load Electrification for the Essential Power System Program: Benefits, Tradeoffs, and Remaining Challenges. Golden, Colo.: National Renewable Energy Laboratory.

IFEU (Institut für Energieund Umweltforschung, Heidelberg GmbH). 2003. Energy Savings by Lightweighting. Heidelberg, Germany: IFEU.

Johannesson, L., M. Asbogard, and B. Egardt. 2007. Assessing the potential of predictive control for hybrid vehicle powertrains using stochastic dynamic programming. IEEE Transactions on Intelligent Transportation Systems, Vol. 8, No. 1. March.

Kenworth. 2008. White paper on fuel economy. Paper No. WP-FE08. Kenworth Truck Company. August. Available at http://www.kenworth.com/FuelEconomyWhitePaper.pdf.

Kessels, J. 2007. Electronic Horizon: Energy Management Using Telematics Information. Presented at the IEEE 2007 Vehicle Power and Propulsion Conference, September.

Kim, T., C. Manzie, and H. Watson. 2008. Fuel economy benefits of look-ahead capability in a mild hybrid configuration. Proceedings of the 17th World Congress of IFAC, Seoul, Korea, July.

Knee, B., O. Franzese, N. Wood, M. Arant, D. Hall, S. Nelson, R. Hathaway, M. Keil, P. Pollock, D. Pape, S. Yeakel, and J. Petrolino. 2008. Heavy Truck Rollover Characterization (Phase A) Final Report. National Transportation Research Center, Inc., Knoxville, Tenn.

Kohut, N., K. Hedrick, and F. Borrelli. 2009. Integrating Traffic Data and Model Predictive Control to Improve Fuel Economy. Presented at the 12th 2009 IFAC Symposium on Transportation Systems, Redondo Beach, Calif., September.

Kreeb, R., and S. Brady. 2006. Tractor-Trailer Dynamic Alignment and Research. Report No. FMCSA-PSV-06-002. February.

LaClair, T. 2005. Rolling Resistance. Pp. 475-532 in The Pneumatic Tire (J.D. Walter and A.N. Gent, eds.). Washington, D.C.: U.S. Department of Transportation, National Highway Traffic Safety Administration.

Lattermann, F., K. Neiss, S. Terwen, and T. Connolly. 2004. The predictive cruise control: A system to reduce fuel consumption of heavy duty trucks. SAE Paper 2004-01-2616. Warrendale, Pa.: SAE International.

Lu, X., S. Shladover, and J. Hedrick. 2004. Heavy-duty truck control: Short inter-vehicle distance following. Proceedings of the 2004 American Control Conference, Boston, Mass., June.

Melson, J. 2007. Progress in Tires. Presentation by Michelin at the International Workshop on the Use of Wide-Base Tires, Federal Highway Administration, Turner-Fairbank Highway Research Center, McLean, Virginia, October 25-26.

MNA (Michelin Tires North America). 2007. International Workshop on the Use of Wide-Base Tires, Federal Highway Administration, Turner-Fairbank Highway Research Center, McLean, Virginia. October.

Murphy, S. 2004. How to choose a GPS fleet management system: What features are right for you? Management Quarterly, Vol. 45.

Navistar. 2009. MaxxForce 9 & 10. Brochure No. CGE 634 3/2009. Available at http://www.navistar.com/InternationalTrucks/PageContent/2010Emissions/pdf/MaxxForce9.pdf.

NESCCAF/ICCT, Southwest Research Institute, and TIAX. 2009. Reducing Heavy-Duty Long Haul Combination Truck Fuel Consumption and CO_2 Emissions. Washington, D.C.: International Council on Clean Transportation (ICCT). October.

NHTSA (U.S. Department of Transportation, National Highway Traffic Safety Administration). 2009a. The Effects of Varying the Levels of Nitrogen in the Inflation Gas of Tires on Laboratory Test Performance. Report No. DOT HS 811 094. Washington, D.C.: U.S. Department of Transportation. March.

NHTSA. 2009b. Notice of Proposed Rulemaking, Tire Fuel Efficiency Consumer Information Program, 49 CFR Part 575 [Docket No. NHTSA-2008-0121]. RIN 2127-AK45. June.

Nose Cone. 2009. Nose Cone Manufacturing Co. See http://www.nosecone.com/aptruck.htm. Accessed September 16, 2009.

NRC (National Research Council). 2006. Transportation Research Board Special Report 286: Tires and Passenger Car Fuel Economy. Washington, D.C.: The National Academies Press.

NRC. 2008. Review of the 21st Century Truck Partnership. Washington, D.C.: The National Academies Press.

NREL (National Renewable Energy Laboratory). 2002. Milestone Report MP-540. Prepared for the U.S. Department of Energy by NREL. Golden, Colo.: NREL.

NREL. 2004. Milestone Report 20040901. Prepared for the U.S. Department of Energy by NREL. Golden, Colo.: NREL.

Ogburn, M., L. Ramroth, and A. Lovins. 2008. Transformational Trucks: Determining the Energy Efficiency Limits of a Class-8 Tractor-Trailer. Snowmass, Colo.: Rocky Mountain Institute. July. Available at http://www.rmi.org/rmi/Library/T08-08_TransformationalTrucksEnergyEfficiency.

O'Keefe, M., and K. Vertin. 2002. Analysis of Hybrid Electric Propulsion Systems for Transit Buses. Report No. NREL/MP 540-32858. Golden, Colo.: National Renewable Energy Laboratory. October.

Peterbilt, Inc. 2009. Images from "Peterbilt Model 389" (http://www.peterbilt.com/trad389.2.aspx) and "Peterbilt Model 387" (http://www.peterbilt.com/aero387.2.aspx). Accessed October 10, 2009.

Rajagopalan, A., and G. Washington. 2002. Intelligent control of hybrid electric vehicles using GPS information. SAE Paper 2002-01-1936. Warrendale, Pa.: SAE International.

Rini, N. 2005. Carbon Fiber SMC for Class 8 Vehicle Hoods, FY 2005 Progress Report for High-Strength Weight Reduction Materials. Prepared by Volvo for the U.S. Department of Energy, Washington, D.C.

SAE (Society of Automotive Engineers). 2000. Recommended Practice J1343, Information Relating to Duty Cycles and Average Power Requirements of Truck and Bus Engine Accessories. Warrendale, Pa.: SAE International. August.

SAE Recommended practices SAE J1252, SAE J1263, and SAE J1321.

Schuring, D. 1977. A new look at the definition of tire rolling loss. Presentation at the Tire Rolling Losses and Fuel Economy—An R&D Planning Workshop, SAE Conf. Proc. P-74, 31-37. Warrendale, Pa.: SAE International.

Smith, M., and C. Eberle. 2003. Heavy Vehicle Mass Reduction Goals and Manufacturing Challenges. Presentation at the U.S. Department of Energy Workshop on Tooling Technology for Low Volume Vehicle Production, October 28, 2003, Seattle, Wash.

Strimer C., D. Carder, M. Gautam, and G. Thompson. 2005. Impact of vehicle weight on truck behavior and emissions, using on-board measurement. SAE Paper 2005-01-3788. Warrendale, Pa.: SAE International.

Sturdy-Light. 2009. Vango Rolling Tarp Systems. Bristol, Tenn.: Sturdy-Light. Available at http://www.sturdy-lite.com/files/vangospg.php. Accessed August 25, 2009.

TIAX. 2009. Assessment of Fuel Economy Technologies for Medium- and Heavy-Duty Vehicles. Report prepared for the National Academy of Sciences by TIAX LLC, Cupertino, Calif., July 31.

TMA (Truck Manufacturers Association). 2007. Test, Evaluation, and Demonstration of Practical Devices/Systems to Reduce Aerodynamic Drag of Tractor/Semitrailer Combination Unit Trucks. Prepared for the National Energy Technology Laboratory (NTEL) under contract number DE-FC26-04NT42117. Morgantown, W. Va.: NTEL. April.

TMC (Technology and Maintenance Council). No date. A guide to improving commercial fleet fuel efficiency. TMC Fuel Economy Digest. Available at http://tmc.truckline.com.

Transport Topics. 2009. Financial and Operating Information for America's 100 Largest For-Hire and Private Fleets. Available at http://www.ttnews.com/tt100/TT100_web_FH09.pdf. Accessed September 14, 2009.

Volvo. 2009. Volvo Trucks Website: http://www.volvo.com/trucks/na/en-s/products/vn/vn.htm. Accessed September 13, 2009.

Walker, J. 2008. SENTIENCE: Using Electronic Horizon Data to Improve Hybrid Vehicle Fuel Economy. Road Transport Information and Control, RTIC 2008 and ITS United Kingdom Members' Conference, May.

Wall, E., et al. 2006. Progress Report for High Strength Weight Reduction Materials, FY 2006. Washington, D.C.: U.S. Department of Energy. March.

Wood, R. 2006. A Discussion of a Heavy Truck Advanced Aerodynamic Trailer System. Presented at the International Forum for Road Transport Technology (IFRTT) 9th International Symposium on Heavy Vehicle Weights and Dimensions, June 18-22, 2006. Pennsylvania State University, State College, Penn. Available at http://www.mne.psu.edu/ifrtt/ConferenceProceedings/ISHVWD_9_2006/docs/pdfs/session%209/s9-4%20140.pdf.

Wood, R. 2009. Aerodynamic Yaw, Drag and Fuel Economy. SOLUS-Solutions and Technologies LLC and Old Dominion University. August.

Woodrooffe, J., O. Page, D. Blower, and P. Green. 2008. Commercial Medium Tire Debris Study. Report prepared by Virginia Polytechnic Institute and State University for the U.S. Department of Transportation, National Highway Traffic Safety Administration (NHTSA). Report number DOT HS 811 060. Washington, D.C.: NHTSA.

6

Costs and Benefits of Integrating Fuel Consumption Reduction Technologies into Medium- and Heavy-Duty Vehicles

The costs and benefits of particular measures to reduce fuel consumption can be estimated with some degree of precision. Before manufacturers or users invest in a particular technology, they must have an idea of the likely payoff, in terms of cost, reliability, performance, and fuel consumption. In addition to the direct costs and benefits discussed here, there are a number of indirect costs, indirect benefits, and unintended consequences related to the implementation of energy-efficient technologies in commercial vehicles.

In this chapter the committee further discusses the fuel consumption technologies identified in Chapters 4 and 5 with regard to their performance in different vehicle classes, which will largely determine what technologies will be used in various vehicles. The committee also discusses the costs of these technologies and their cost-benefit ratio. Operating and maintenance (O&M) costs are discussed as well as indirect effects.

The committee evaluated a wide range of fuel-saving technologies for medium- and heavy-duty vehicles. Some technologies, such as certain aerodynamic features, automated manual transmissions, and wide-base single low-rolling-resistance tires, are already available in production. Some of the technologies are in varying stages of development, while others exist only in the form of simulation models. Reliable, peer-reviewed data on fuel-saving performance are available only for a few technologies in a few applications. As a result, the committee had to rely on information from a wide range of sources, (e.g., information gathered from vehicle manufacturers, component suppliers, research labs, and major fleets during site visits by the committee), including many results that have not been duplicated by other researchers or verified over a range of duty cycles.

There is a tendency among researchers to evaluate technologies under conditions that are best suited to that specific technology. This can be a serious issue in situations where performance is strongly dependent on duty cycle, as is the case for many of the technologies evaluated in this report. One result is that the reported performance of a specific technology may be better than what would be achieved by the overall vehicle fleet in actual operation. Another issue with technologies that are not fully developed is a tendency to underestimate the problems that could emerge as the technology matures to commercial application. These issues often result in implementation delays as well as a loss of performance compared to initial projections. As a result of these issues, some of the technologies evaluated in this report may be available later than expected, or at a lower level of performance than expected. Extensive additional research would be needed to quantify these issues, and regulators will need to allow for the fact that some technologies may not mature as expected.

The fuel-saving technologies that are already available on the market generally result in increased vehicle cost, and purchasers must weigh the additional cost against the fuel savings that will accrue. In most cases, market penetration is low at this time. Most fuel-saving technologies that are under development will also result in increased vehicle cost, and in some cases, the cost increases will be substantial. As a result, many technologies may struggle to achieve market acceptance, despite the sometimes substantial fuel savings, unless driven by regulation or by higher fuel prices. Power train technologies (for diesel engines, gasoline engines, transmissions, and hybrids) as well as vehicle technologies (for aerodynamics, rolling resistance, mass/weight reduction, idle reduction, and intelligent vehicles) are analyzed in Chapters 4 and 5. Figure 6-1 provides estimates for potential fuel consumption reductions for typical new vehicles in the 2015 to 2020 time frame, compared to a 2008 baseline.

The technologies were grouped into time periods based on the committee's estimate of when the technologies would be proven and available. In practice, timing of introduction will vary by manufacturer, based in large part on individual company product development cycles. In order to manage product development costs, manufacturers must consider the overall product life cycle and the timing of new product introductions. As a result, widespread availability of some technologies may not occur in the time frames shown.

The percent fuel consumption reduction (% FCR) num-

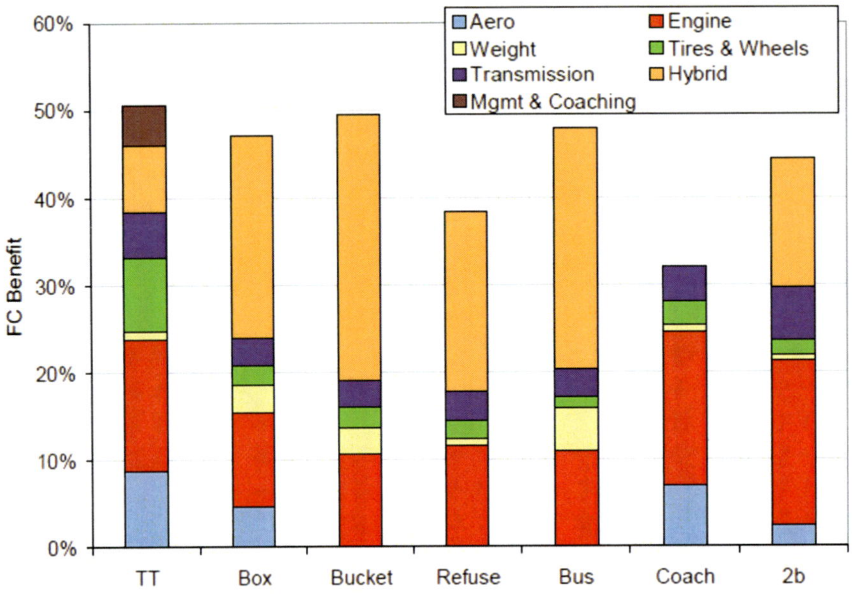

FIGURE 6-1 Comparison of 2015-2020 new-vehicle potential fuel-saving technologies for seven vehicle types: tractor trailer (TT), Class 3-6 box (box), Class 3-6 bucket (bucket), Class 8 refuse (refuse), transit bus (bus), motor coach (coach), and Class 2b pickups and vans (2b). NOTE: TIAX (2009) only evaluated the potential benefits of driver management and coaching for the tractor-trailer class of vehicles. It is clear to the committee that other vehicle classes would also benefit from driver management and coaching, but studies showing the benefits for specific vehicle classes are not available. For more information, see the subsection "Driver Training and Behavior" in Chapter 7. Also, potential fuel reductions are not additive. For each vehicle class, the fuel consumption benefit of the combined technology packages is calculated as follows: [% $FCR_{package}$ = 100 [1 – (1 – {% FCR_{tech1}/100}) (1 – {% FCR_{tech2}/100)} … (1 – {% FCR_{techN}/100})]. Values shown are for one set of input assumptions. Results will vary depending on these assumptions. SOURCE: TIAX (2009).

bers shown for individual technologies and other options are not additive. For each vehicle class, the % FCR associated with combined options is as follows:

$$\% \, FCR_{package} = 100 \, [1 - (1 - \{\% \, FCR_{tech1} /100\}) \, (1 - \{\% \, FCR_{tech2}/100\}) \ldots \{(1 - \{\% \, FCR_{techN} /100\})]$$

where % FCR_{techx} is the percent benefit of an individual technology.

DIRECT COSTS AND BENEFITS

Technology Applications to Specific Vehicle Classes

The technologies discussed in Chapters 4 and 5 can be consolidated into 12 categories and then a judgment made regarding the benefits by broad vehicle class. Table 6-1 shows the applications in which these technologies will be effective. Some technologies are broad enough to be applied to all classes: tires and wheels, weight reduction, transmission and driveline, engine efficiency, and hybridization. Others are more specific to the class of vehicle, such as replacing gasoline engines with diesel engines (dieselization), which is applicable only to Class 2 to 7 vehicles, where gasoline engines are offered today.

Fuel Consumption Reductions

The fuel consumption reductions identified in Chapters 4 and 5 can be summarized in a broad matrix of vehicle applications versus technology, for the years 2015 to 2020, as shown in Table 6-2 and Figure 6-1. The most effective

TABLE 6-1 Technologies and Vehicle Classes Likely to See Benefits

Technologies	Tractor Trailer	Urban Bus	Motor Coach	Class 3-7	Class 2b	Refuse Truck
Trailer aerodynamics	X					
Cab aerodynamics	X		X	X	X	
Tires and wheels	X	X	X	X	X	X
Weight reduction	X	X	X	X	X	X
Transmission and driveline	X	X	X	X	X	X
Accessory electrification	X	X	X	X	X	
Overnight idle reduction	X					
Idle reduction	X	X	X	X		
Engine efficiency	X	X	X	X	X	X
Waste heat recapture	X		X			
Hybridization	X	X	X	X	X	X
Dieselization				X	X	

TABLE 6-2 Fuel Consumption Reduction (percentage) by Application and Vehicle Type

Application	Engine	Aerodynamics	Rolling Resistance	Transmission and Driveline	Hybrids	Weight
Tractor trailer	20	11.5	11	7	10	1.25
Straight truck box	14	6	3	4	30	4
Straight truck bucket	11.2	0	2.4	3.2	40	3.2
Pickup truck (gasoline)	20[a]	3	2	7.5	18	1.75
Pickup truck (diesel)	23[b]	3	2	7.5	18	1.75
Refuse truck	14	0	1.5	4	35	1
Transit bus	14	0	1.5	4	35	1.25
Motor coach	20	8	3	4.5	NA	1.05

[a]Compared to a baseline gasoline engine.
[b]Compared to baseline diesel engine.
SOURCE: Adapted from TIAX (2009).

technologies in terms of fuel consumption reduction are as follows:

- Hybridization
- Replacement of gasoline engines with diesel engines
- Improvement in diesel engine thermal efficiency
- Improvement in gasoline engine thermal efficiency
- Aerodynamics, especially on tractor-trailer applications
- Reduced rolling resistance
- Weight reduction

Costs, Cost-Benefit, and Implementation of These Technologies

The committee determined the direct costs of the technologies in several ways:

- Estimates of presenters and manufacturers for the retail price equivalent (RPE), for the components and/or package.
- Dealer's data book list prices multiplied by 0.6 to estimate RPE.[1]
- Complete vehicle cost premiums from various publications, such as for a hybrid bus.

In the following discussion and tables, the committee presents fuel consumption, in percent reduction, as a range or as one number representing its best estimate.[2] Similarly, the capital costs are presented as a range or as one number representing the committee's best estimate. The number for costs is then divided by the number for fuel consumption reduction (dollar cost/% fuel consumption reduction, $/%), and the result is called capital cost per percent reduction (CCPPR). Most of the focus is on the 2015 to 2020 time frame, but data are also presented for the 2013 to 2015 time frame. The applications discussed are tractor trailer, straight truck, pickup truck and van, refuse truck, transit bus, and motor coach. After considering these vehicle applications, alternative metrics for cost-benefit ratio are presented.

Tractor Trailer

This category of vehicles includes Class 8 tractors equipped with so-called fifth wheels for hitching to one or more trailers. The baseline vehicle for fuel consumption estimates is an older-generation aerodynamic tractor (drag coefficient, C_d = 0.63 to 0.64) with a sloped hood, roof fairings, aero bumpers, standard dual tires (coefficient of rolling resistance, C_{rr} = 0.0068), standard 53-ft box van trailer, a diesel engine with peak thermal efficiency of 41 to 42 percent, and cycle thermal efficiency of 37 to 39 percent on a long-haul cycle with long periods of constant speed operation, camshaft-driven unit injection at 2,000 bar, variable geometry turbocharger, cooled exhaust gas recirculation (EGR), a diesel particulate filter (DPF) for particulate matter (PM) control, a cylinder pressure limit of 200 bar, and a 10-speed manual transmission with overdrive.

Engine (2015 to 2020). Improved thermal efficiency from 42 percent peak to 52.9 percent peak efficiency compared to the 2008 baseline, which represents a 20 percent reduction in fuel consumption. The baseline 2010 technology includes a DPF at a cost of $7,000 and a selective catalytic reduction (SCR) catalyst at a cost of $9,600. The 2015-2020 technology includes SCR with improved nitrogen oxides (NO_x) conversion efficiency, a U.S. Environmental Protection Agency (EPA) required onboard diagnostic (OBD) system with closed loop controls, a high-pressure common rail fuel system with higher injection pressure of about 3,000 bar, piezo-electronic fuel injectors, increased cylinder pressure capability to 250 bar, and a bottoming cycle. The incremental cost is in the range of $23,000 in addition to the cost of 2010 emissions aftertreatment hardware. Besides the increase in

[1]Personal communication from Dave Merrion to the committee, May 27, 2009.
[2]The cost data are primarily from TIAX (2009) unless noted otherwise.

up-front capital cost, O&M costs will increase due to the increased complexity of the engine system. The combined total cost must be taken into account in any analysis, and the cost may make the advanced technologies unattractive for many applications, particularly low-mileage applications.

Aerodynamics. In the 2015-2020 time frame, the committee projects that a tractor-trailer combination C_d of about 0.45 should be feasible.[3] This requires developments going beyond the existing EPA SmartWay specification, which results in drag coefficients in the 0.50 to 0.55 range. To achieve this improvement, changes in vehicle operations and infrastructure (such as loading docks) are likely to be necessary. Costs are estimated for one tractor and three trailers, since this is the typical ratio of tractors to trailers in the field. An improvement in C_d from 0.63 to 0.45 will result in a fuel savings of 11.5 percent on a high-speed, long-haul duty cycle and 15 percent at 65 mph. This savings is very dependent on the actual vehicle duty cycle. High average speeds will result in achieving the projected savings, while applications in more congested areas with lower average speeds will achieve a smaller benefit. The cost of this package for the tractor is in the range of $2,700 to $3,500 (TIAX, 2009, Table 5-1). The trailer aerodynamics cost $3,000 per trailer. The overall cost of aerodynamic features for a tractor and three trailers is estimated at $12,000. Aerodynamic features also bring with them O&M costs that need to be considered. The primary cost is repair or replacement due to damage encountered in operation. For example, many fleets today avoid the use of fuel-tank skirts on tractors, because these can have a very short life in the field in some applications. Trailer skirts are also known to be damage prone. Future aerodynamic improvements may impose additional costs due to changes in vehicle operation, infrastructure, or capability, and these costs must be taken into account.

Rolling Resistance. Widespread implementation of wide-base single tires with low rolling resistance is expected to be feasible in 2015 to 2020 for both tractors and trailers. Using rolling resistance values projected by Michelin and the EPA, the NESCCAF/ICCT (2009, p. 51) report projects a benefit of 11 percent compared to standard dual tires. The cost of applying wide-base singles to one tractor and the accompanying three trailers is projected by NESCCAF to be $4,480 (NESCCAF/ICCT, 2009, p. 95) and by TIAX (2009, p. 5-2) to be $3,600.

Transmission and Driveline. The manual 10-speed transmission, with overdrive, is considered the baseline transmission, along with a tandem rear axle with various ratios that can be specified when the vehicle is purchased. An alternate technology that is already available on the market is the automated manual transmission (AMT). With an AMT the clutch is only used to launch the vehicle and then the transmission shifts automatically. The AMT does not have the powershift capability of a fully automatic transmission and thus does not have the productivity improvement of an automatic. The power interruption during an AMT shift is very similar to that of a manual transmission, but use of the controller to determine shift points tends to remove some of the driver effect on fuel consumption variability. In addition, low-viscosity synthetic lubricants can be used in the transmission (manual or AMT) and throughout the driveline. This combination of AMT and synthetic lubricants can reduce fuel consumption by 2 to 8 percent and has a projected cost of $5,800. Fully automatic transmissions do not offer significant productivity (trip time) or fuel savings in long-haul operations. Therefore, fully automatic technology is not expected to gain significant share in the long-haul market. However, in urban and suburban driving, a switch to fully automatic transmissions can result in significant fuel savings (up to 5 percent) and significant productivity improvements at a cost of about $15,000. Another driveline option that is feasible for 2015 and beyond is the 6 × 2 tractor layout. Most tractors have two drive axles in a configuration referred to as 6 × 4 (six wheels, with four of them driven). A 6 × 2 tractor has only one drive axle, and the second rear axle only carries the weight of the truck. The use of a single drive axle saves about 1 percent of fuel consumption, at the expense of lost traction. This option may not be feasible for applications in areas with significant snow or for vehicles that must also operate off highway. Truck purchasers may be worried about loss of resale value if they purchase 6 × 2 tractors, since the resale market places a premium on tractors that can be used in a wide variety of applications.

Hybrid Tractor Power Trains. The use of hybrid power trains in Class 8 tractor trailers has been assigned a low priority in the long-haul market, by manufacturers, due to the typical duty cycle of mainly constant speed long-haul operation, which provides little opportunity for battery or hydraulics to store and release energy. As it turns out, the highway tractor trailer spends a significant amount of time on arterial highways, on grades that provide regenerative braking opportunities, and idling. NESCCAF/ICCT (2009, p. 54) showed fuel consumption reduction of 5.5 to 6 percent on one sample long-haul duty cycle including some suburban segments and some grades. For those trucks that idle overnight to support the "hotel load" of the sleeper, NESCCAF/ICCT showed an additional benefit of about 4 percent, for a total benefit of 10 percent. TIAX estimates the cost of the parallel hybrid modeled in the NESCCAF report at $25,000 in the 2015 to 2020 time frame, which assumes that significant volumes will be reached by 2015. Note that there is little actual field data available on the application of hybrids to Class 8 tractor trailers, so only simulation data were used to estimate fuel

[3]Note that the estimates provided here apply only to standard 53-ft van trailers. Other trailer types will allow some degree of aerodynamic improvement, but the baseline C_d, the degree of improvement available, and the cost of the improvement will all vary substantially by trailer type.

savings. O&M costs will also be a critical factor in hybrid applications. Insufficient data are available today to project the life of batteries, which will be a major cost borne in most cases by the second or third owner of the vehicle. In many cases, later owners may elect not to repair a failed hybrid system (because of the high cost) as long as the vehicle is able to operate without using the hybrid system.

Idle Reduction. Several idle reduction technologies were discussed in Chapter 5 and are summarized in Table 6-3. Note that the benefit of an idle timer is not additive with APU or other hotel load systems, since the benefits claimed by these systems already include idle elimination. The idle timer cannot be effective if the engine is used to support hotel loads, but it can reduce idling at loading docks and other times where the hotel load is not required. If a hybrid system is used, idle reduction comes at no extra cost.

The CCPPR ratios of technologies, expressed in dollars per percent fuel consumption reduction, are summarized in Table 6-4 for Class 8 tractor trailers in the 2015 to 2020 time frame. Table 6-5 presents the same information as Table 6-4 in a slightly modified format, and includes the total potential benefit that can be achieved for Class 8 tractor trailers in the 2015 to 2020 time frame. Note that the total percentage benefit is not simply a sum of the individual benefits. As incremental improvements are added, each percentage benefit applies to an already reduced fuel consumption, not to the baseline fuel consumption. As a result, the total benefit $P_{overall}$ is calculated as follows:

$$1 - P_{overall} = (1 - p_1)(1 - p_2)\ldots(1 - p_n).$$

Using this formula, the total overall benefit available is 50.5 percent, whereas a simple summation of the benefits would suggest a 65 percent overall benefit.

Tractor-Trailer Summary

All the technologies listed in Table 6-5 may be implemented in production in the 2015 to 2010 time frame, taking into account the comments made in the following subsections.

TABLE 6-3 Idle-Reduction Packages

Technology	Fuel Consumption Savings (%)	Cost ($), Retail Price Equivalent
Engine electronic control unit acts as an idle timer (programmable)	2-3	0
Direct-fired heater	1-3	1,000-3,000
Small auxiliary power unit or battery system	4-6.5	5,000-8,000
Large auxiliary power unit	5-8	8,000-12,000
Hybrid system used for idle elimination	4-6.5	(cost covered by hybrid)

TABLE 6-4 Technology for Class 8 Tractor Trailers in the 2015-2020 Time Frame

Technology	Capital Cost ($)	Mean Improvement in Fuel Consumption (%)	CCPPR ($/%)
Diesel engine	23,000[a]	20	1,150
Aerodynamics (tractor trailer)	12,000	11.5	1,043
Rolling resistance	3,600	11	327
Transmission and driveline	5,800	4	1,450
Hybrid (includes idle reduction)	25,000	10	2,500
Weight reduction	13,500	1.25	10,800

NOTE: CCPPR, capital cost per percent reduction.
[a]Not including $9,000 selective catalytic reduction.

TABLE 6-5 Tractor Trailers Benefit from Advances in Every Technology Category

Category	Description	% Fuel Consumption Benefit	Capital Cost ($)	CCPPR ($/%)	Weight (lb)
Aerodynamics	Improved SmartWay tractor + three aerodynamic trailers	11.5	12,000	1,043	750
Engine	Advanced 11-15L diesel with bottoming cycle	20	23,000	1,150	800
Tire	Improved WBS on tractor + three trailers	11	3,600	327	−400
Transmission and driveline	AMT, reduced driveline friction	7	5,800	829	80
Hybrid	Mild parallel hybrid with idle reduction	10	25,000	2,500	400
Management and coaching	60 mph speed limit; predictive cruise control with telematics; driver training	6	1,700	283	—
Idle reduction	Included with hybrid system	—	—	—	—
Total added weight	Added components	−1	—	—	+2,030
Weight reduction	Material substitution—2,500 lb.	1.25	13,500	10,800	−2,500
TOTAL	2015-2020 Package	50.5	84,600	1,674	−470

NOTE: For each vehicle class, the fuel consumption benefit of the combined technology packages is calculated as follows: [% $FCR_{package}$ = 100 [1 – (1 – {% FCR_{tech1}/100 }) (1 – {% FCR $_{tech2}$/100}) … (1 – {% FCR_{techN}/100})] where % $FCR_{tech\ x}$ is the percent benefit of an individual technology. CCPPR, capital cost per percent reduction.
SOURCE: TIAX (2009), p. 5-2.

Diesel Engines. Diesel engines used in Class 8 tractors are expected to begin improving their thermal efficiency as emissions regulations stabilize after 2010. A peak thermal efficiency of 53 percent may be achievable in the 2015-2020 time frame (a 20 percent improvement over current engines). Given the cost of some of the technologies required to achieve this improvement, it is likely that this level of improvement will only be achieved with the use of incentives or regulation. Because of the high cost and difficult packaging of some advanced technologies, some of these technologies will probably be restricted to certain high-mileage applications.

Aerodynamics. The current average drag coefficient is about 0.63, and the best currently available is 0.50 to 0.55 (SmartWay-certified tractors and trailers). In the 2015 to 2020 time frame, drag coefficients around 0.45 should be feasible. Some of these changes may require changes in the way vehicles operate, including infrastructure changes, such as loading dock height change. Changing a fundamental parameter such as loading dock height would require a huge investment that will not occur without regulatory stimulus. The cost and fragility of drag reduction features need to be dealt with in order to make them more attractive from a cost-benefit standpoint. Aerodynamic improvements for other trailer configurations (bulk hauler, flatbed, tanker, etc.) are not well understood and require further research.

Rolling Resistance. Widespread implementation of wide-base single tires should be possible in the 2015 to 2020 time frame. These tires have an attractive CCPPR ratio and thus may achieve broad application without incentives or regulation.

Transmission and Driveline. For tractor-trailer combination vehicles, automated manual transmissions combined with low-friction driveline components and lubricants should be widely implemented in the 2015 to 2020 time frame. Other transmission concepts may become feasible in that time frame, but the technology is not far enough along to make any reliable forecasts. The widespread implementation of the 6 × 2 tractor configuration is very uncertain, given concerns about limited usefulness of the vehicle to subsequent owners.

Hybrid Tractor Power Trains. Hybrid power trains for tractor-trailer vehicles are technically feasible for the 2015 to 2020 time frame. The question will be whether the high cost and added weight are justified by the potential fuel savings of around 6 percent for tractors that do not include sleeper operation and 10 percent for sleepers. Widespread application of hybrids in tractor-trailer applications is very unlikely without substantial incentives or regulation.

Idle Reduction. Several forms of idle reduction technology are already available, including both onboard systems (APUs, diesel-fired heaters, etc.) and remote systems such as truck stop heating, ventilation, and air conditioning (HVAC) systems. If hybrid systems are adopted for tractor-trailer trucks, they will provide a built-in idle reduction capability. It is not clear which of the existing technologies will come to dominate the idle reduction market or if some future technology will dominate. It is possible that no one idle reduction technology will become dominant.

Straight Truck

This category of vehicle is very broad, including Class 3 to 8 straight (i.e., nontrailer) trucks, so it includes cut-away vans, parcel delivery vehicles, beverage delivery vehicles, shuttle buses, cab over engine cabs, conventional cabs, both gasoline and diesel engines, and various forms of work trucks described in Chapters 1 and 3. The baseline vehicle, for fuel consumption estimates, is a pickup and delivery Class 6 regional haul, traveling about 150 miles per day at an average speed of 30 mph. The 2007 certification, 6-9L diesel engine, has a cycle thermal efficiency of 31 to 35 percent and a peak thermal efficiency of 40 to 41 percent (TIAX, 2009, Table 4-5). The engine has high-pressure common rail fuel injection (1,800 bar), is equipped with a turbocharger, cooled EGR, a DPF, and about 175 bar cylinder pressure. The vehicle has no aerodynamic treatment and standard tires with steel wheels. The transmission is a six-speed automatic with no anti-idle technologies.

Engine. The diesel engine is considered baseline in this application, but the gasoline engine penetration was 42 percent in 2008 and has been increasing in the past few years. The increase in gasoline engine penetration is due to the recent cost increase of diesel engines, especially from addition of the DPF in 2007. Furthermore, the prospect of SCR or advanced EGR in 2010 is expected to serve as an additional sales deterrent for those classes whose duty cycles are more urban and have relatively low annual vehicle miles traveled (VMT). Dieselization of this class provides a large fuel consumption reduction (about 30 percent), but the incremental cost of the diesel engine beyond 2015 will be high due to further emissions compliance modifications and implementation of EPA-required OBD systems with closed loop controls.

Also problematic to diesel sales is that the diesel fuel consumption cost benefit was nearly wiped out in mid-2008 when the diesel fuel price per gallon increment was about \$0.80 (20 percent) above gasoline (itself at \$4.00). While in October 2009 this differential was only 5 percent,[4] it may shrink when product demand rises with an improving economy. In addition to the 2010 emissions, the advancing diesel technology beyond 2015 will likely include 42 percent cycle

[4]See http://tonto.eia.doe.gov/oog/info/gdu/gasdiesel.asp; accessed October 6, 2009.

thermal efficiency and 45 percent peak thermal efficiency, facilitated by increased injection pressure (2,000 to 2,400 bar), improved SCR conversion efficiency, and some engines equipped with multistage turbochargers and accompanying increased cylinder pressure (about 200 bar), plus reduced parasitic losses. These improvements result in a 13 to 16 percent reduction in fuel consumption, but principally only for those applications where heavier loaded duty cycles are the norm. The cost premium is $11,000 to $14,000, largely for SCR or advanced EGR and OBD (TIAX, 2009, Table 3-8).

Selection of a diesel or gasoline engine from the developing engine technologies will consider cost benefits associated with widely varying duty cycle and durability needs.

Diesel engine technology continues to be strongly developed mainly for Class 8 tractor-trailer duty, and the smaller Class 3 through 7 diesel engines benefit from this development. Gasoline engine technology is also developing (as discussed in Chapter 4 and the Class 2b pickup section of Chapter 5) and may capture an increasing proportion of medium-duty trucks in the sector (i.e., Classes 3 to 7 and perhaps even some of the lighter duty vehicles of Class 8). These are straight trucks with relatively low VMT or with both lighter average loads and lower average speeds. Bucket trucks and most service trucks are among the best examples. Diesel engines in this sector will best serve those operations that require higher loads, higher annual VMTs, higher average speed applications, and longer durability.

Aerodynamics. The aerodynamic packages for medium-duty trucks are not part of the OEM/dealer standard options like they are on tractor trailers. Aerodynamic features (roof deflectors, fuel tank fairings, box skirts, mirrors, etc.) are available (four of the six identified features) as options and are dealer installed (except cab streamlining). Table 6-6 identifies the aero opportunities for this style of truck.

A 1.5 percent fuel savings reflects the base truck's 30 mph average duty-cycle speed. Larger benefits can be achieved for those trucks that operate at higher average speeds, but these higher speeds are not typical of a pickup and delivery duty cycle described. The package cost is more than $3,000, which provides an extremely poor value for a truck averaging 30 mph.

Rolling Resistance. Currently available low-rolling-resistance SmartWay dual tires can give a fuel consumption reduction of 1 to 2 percent at a cost of $120 per rear axle. The reduced C_{rr} duals provide a simple payback in 4 months for the assumed baseline truck. Current wide base single (WBS) tires can give a fuel consumption reduction of 2 to 4 percent at a cost of $450 per rear axle (TIAX, 2009). The WBS tires require about 8 months for payback, but early evaluations have shown poor tread wear in tighter turning conditions of urban operation. Yet the WBS tires could be used in this sector where the duty cycle is mainly arterial and interstate. As SmartWay has helped provide the greatest incentive for low-C_{rr} tires, nearly all of that program has been tractor trailer focused, and the Class 3 to 7 sector has not had much attention toward this feature.

It is expected that the next-generation low-rolling-resistance duals in circa 2013 will reduce fuel consumption from the 2007 base by 2.8 to 3.5 percent, also at a cost of $120 per axle (TIAX, 2009, Table 3-8). Applications are likely that cannot use low-rolling-resistance tires because of traction issues. Examples include dump trucks, cement mixers, and service trucks that must occasionally operate off road.

Transmission and Driveline. The six-speed fully automatic transmission (AT) is considered baseline in this sector, especially for pickup and delivery operation in urban and suburban areas. In those cases fuel consumption favors the AT, as does improved productivity (in terms of trip speed; Allison Transmission). Further, the AT is used by many operators for safety reasons (two hands on the wheel) and for reduced driver training needs. Manual transmissions and, increasingly, automated manual transmissions (AMTs) will be favored for higher annual VMTs in arterial and interstate routes, as they will result in fuel consumption improvements.

TABLE 6-6 Straight Box Truck Aerodynamic Technologies

Technology	FC Benefit[a] (%)	C_D Improvement (%)	Capital Cost ($)	Status
Aerodynamic Devices				
Roof deflector	2-3	7-7.5	500-800	Available
Chassis fairings	0.5-1	2.5-3	400-500	Not available
Box skirts	2-3	4.5-5	500-1,000	Demos
Box fairing	0.5-1	2.4-2.7	500-650	Available
Cab streamlining	1-2	5-6	750	Available
Aft box taper	1.5-3	7.6-8	1,000	Not available
Aerodynamic Packages				
Combination of straight truck aerodynamics	5-8[b]	20	3,000-3,500	Not available

[a]Fuel consumption (FC) benefit is critically dependent on duty-cycle average speed.
[b]About 1.5 percent for the baseline average speed of 30 mph.
SOURCE: TIAX (2009), Table 3-8.

For these conditions a reduction of 4 percent may be achieved (TIAX, 2009, p. 4-70). These sorts of evolutions will occur slowly as particular operations experiment with different transmission solutions for small fuel consumption reductions or productivity improvements. Altogether, it is not likely a major shift will occur.

Hybrid Power Trains. This vehicle sector has received the most attention for hybridization, after passenger cars and urban buses. Electric systems are in early production, and both advanced electric and hydraulic systems are in development and/or demonstration. The fuel consumption reduction for an Eaton electric system in a Navistar or Kenworth truck currently is 20 to 30 percent with a cost in 2014 of $20,000. The cost will be driven by incorporation of higher volume Li-ion batteries. This cost is nearly half that of the low-volume (batches of hundreds at a time) units offered in fall 2009, which use nickel metal-hydride batteries. The same system with an electric power takeoff, appropriate for a bucket truck, will reduce fuel consumption by 30 to 40 percent with a 2014 cost of $30,000.

Many companies, original equipment manufacturers and suppliers, are focused on this segment for electric and hydraulic hybrids. Some companies are in production, such as Freightliner M2 Business Class, Navistar Durastar, ISE, Azure, Kenworth, Peterbilt, and Workhorse Custom Chassis. A Duke University study[5] estimates that 4,850 hybrids will be produced by 2010; 29,000 by 2015; and 60,000 by 2020. This early penetration will be helped by EPA grants[6] of $50 million for deploying hybrids, which provide federal tax credits for incremental costs as follows: 20 percent tax credit for 30 to 40 percent fuel reduction, 30 percent tax credit for 40 to 50 percent fuel reduction, and 40 percent tax credit for more than 50 percent fuel reduction. (Note that these credits have caps of $7,500, $15,000 and $30,000 for vehicles weighing 8,501 to 14,000 lb, 14,001 to 26,000 lb, and more than 26,000 lb, respectively.)

Cost-Benefit. The CCPPR of technology expressed in dollars per percent fuel consumption reduction is summarized in Table 6-7 for Class 6 straight trucks in the 2015 to 2020 time frame. A 1 percent fuel savings equals about $225 per year (at $3.00 per gallon fuel price).

Class 3 to 6 bucket trucks are also prevalent in this sector, and by virtue of the hybrid electrical system will substantially benefit from electrification of the vehicle's power takeoff (PTO) unit (see Table 6-8). One troublesome factor for bucket trucks in this sector is their average annual mileage of 13,300 miles as found by one study of 31 utilities.[7] Such mileage will greatly limit the potential payback of a hybrid system, and is likely to lead utilities to consider converting to gasoline to reduce cost.

Summary: Box and Bucket Class 6 Straight Trucks

Diesel engines may continue to dominate this sector, especially for the medium- and heavy-duty applications with higher annual vehicle miles traveled. Gasoline engines should capture increasing market share because of both technology improvements and the increasing diesel price differential due to emission control systems. Hybrid solutions may further incentivize gasoline engines, since with a hybrid the total power demand on the engine shrinks by 25 to 40 percent.

Hybridization in this sector will be the strongest contributor to reduced fuel consumption due to both percentage of power supplied by the hybrid and the expected cost benefit in the period beyond 2015 with the takeover by Li-ion battery solutions as forecast by most observers (Research and Markets, 2009). Simple payback for the baseline straight box truck in this section is about three years. However, the substantially low VMTs of the bucket truck application put simple payback at eight years, even considering that idle fuel consumption (20 percent of the total) will be replaced by the battery (TIAX, 2009, p. 2-5). These estimates are with $3.00 per gallon fuel.

Both hybrid system and engine efficiency developments have been government incentivized as these have the highest development costs and incur the greatest price increases and with lack of success would impose the greatest loss of fuel consumption reduction opportunity for this sector. The EPA hybrid grant program described previously is an example of such an incentive. Nevertheless, the industry today is remarkably immature, and low-volume high incremental price hybrid offerings could use some incentives to "get ready" for the 2015 marketplace.

Pickup Truck and Van (Class 2b)

Class 2b includes vehicles of 8,500 lb to 10,000 lb gross vehicle weight (GVW). For EPA and California Air Resources Board emission certification purposes, the classes are as follows:

- 8,500 to 19,500 lb light to heavy duty EPA[8]
- 8,500 to 10,000 lb California MDV4
- 10,000 to 14,000 lb California MDV5

The baseline vehicle is a pickup truck or van with a 6- to 8-liter gasoline engine, naturally aspirated, port fuel injected, and a four-speed automatic transmission.

[5]See Lowe et al. (2009).

[6]See http://www.epa.gov/otaq/diesel/projects.htm.

[7]T. Reinhart, personal communication to committee members, October 9, 2009, and phone interview with S. Bibono of Chatham Consulting, October 2009. Chatham does benchmarking surveys for the utility industry.

[8]Vehicles up to 10,000 lb GVW used for personal transportation are classified as medium-duty passenger vehicles and are subject to light-duty vehicle legislation. Also complete heavy-duty diesel vehicles under 14,000 lb can be chassis emission certified rather than engine-dyno certified.

TABLE 6-7 Class 3 to Class 6 Straight Box Truck with 2015-2020 Technology Package

Category	Description	Fuel Consumption Benefit (%)	Capital Cost ($)	CCPPR ($/%)	Weight (lb)
Aerodynamics	Aero cab, skirts, round corners	6	3,250	542	300
Engine	Advanced 6-9L Engine	14	13,000	929	400
Tire	Improved low rolling resistance duals	3	300	100	—
Transmission and driveline	8-speed automatic transmission, reduced driveline friction, aggressive shift logic	4	1,800	450	—
Hybrid	Parallel hybrid	30	20,000	667	400
Management and coaching	—	—	—	—	—
Idle reduction	—	—	—	—	—
Total added weight	Added components	−4.4	—	—	+1,100
Weight reduction	Material substitution—1,000 lb	4	4,770	1,193	−1,000
Total	2015-2020 package	47.1	43,120	915	+100

NOTE: For each vehicle class, the fuel consumption benefit of the combined technology packages is calculated as follows: [% $FCR_{package}$ = 100 [1 – (1 – {% FCR_{tech1}/100 }) (1 – {% FCR $_{tech2}$/100}) … {(1 – {% FCR_{techN}/100})] where % $FCR_{tech\,x}$ is the percent benefit of an individual technology. CCPPR, capital cost per percent reduction.
SOURCE: TIAX (2009).

Fuel consumption can be reduced through vehicle modifications and systematic incorporation of advanced technologies into the power train. Engine fuel consumption can be reduced by either applying advanced technologies to the spark ignition engine or substituting a diesel engine for the spark ignition engine. Typical applications for a Class 2b vehicle would be as an urban delivery vehicle or a work vehicle for a small contractor. For this application the average speed would not be high and there would be frequent stops. As such, the highest potential for reducing fuel consumption will reside with engine improvements, hybridization, and transmission improvement. Average vehicle miles traveled are also typically fairly low, which limits the payback of fuel-saving technologies.

Aerodynamics. The extent to which aerodynamic treatment of the vehicle will be a cost-effective approach to reducing fuel consumption will be highly dependent on the application of the vehicle. If this class of vehicle is used primarily by contractors or local delivery services, the benefit of adding aerodynamic treatments would be minimal. The report by TIAX estimates a potential fuel consumption reduction of 3 percent with an incremental cost of $100 from aerodynamic treatments (TIAX, 2009).

Rolling Resistance. Like aerodynamic improvement, the potential for fuel consumption reduction through low rolling resistance tires will be heavily dependent on the application. Again, assuming that the likely use for these vehicles will be local contractor-type work or urban delivery, the benefit from low rolling resistance tires will most likely be small: 2 percent at an incremental cost of approximately $10 (TIAX, 2009).

TABLE 6-8 Class 3 to 6 Bucket Truck with 2015-2020 Technology Package

Category	Description	% Benefit	Capital Cost ($)	CCPPR ($/%)	Weight (lb)
Aerodynamics	—	—	—	—	—
Engine	Adv. 6-9 L engine	11.2	13,000	1,161	400
Tires	Improved low-rolling-resistance duals	2.4	300	125	—
Transmission and driveline	Reduced driveline friction	3.2	1,800	450	—
Hybrid	Parallel hybrid with electric power takeoff	40.0	30,000	667	650
Management and coaching	—	—	—	—	—
Idle reduction	—	—	—	—	—
Total added weight	Added components	3.4	—	—	1,050
Weight reduction	Materials substitution—1000 lb	3.2	4,770	1,193	−1,000
Total	2015-2020 package	49.6	49,870	1,005	50

NOTE: For each vehicle class, the fuel consumption benefit of the combined technology packages is calculated as follows: [% $FCR_{package}$ = 100 [1 – (1 – {% FCR_{tech1}/100 }) (1 – {% FCR $_{tech2}$/100}) … {(1 – {% FCR_{techN}/100})] where % $FCR_{tech\,x}$ is the percent benefit of an individual technology. CCPPR, capital cost per percent reduction.
SOURCE: TIAX (2009), p. 5-3.

Transmission. Incorporating a six- or eight-speed automatic transmission with reduced driveline friction that incorporates a shift logic aimed at minimizing fuel consumption could potentially reduce fuel consumption by approximately 7 percent with an incremental cost on the order of $1,000 (TIAX, 2009).

Power Train. Hybridizing the vehicle power train would have high potential for reducing fuel consumption for the application assumed to be typical for this vehicle class. The load-leveling, accessory electrification and electric launch capability of the hybrid power train could reduce fuel consumption on the order of 18 percent for this application. The incremental cost would be approximately $9,000 (TIAX, 2009).

Engine. Starting from a base engine that is gasoline fueled, spark ignited, naturally aspirated, and port fuel injected, many technologies could be introduced to reduce fuel consumption. Variable valve actuation (variable valve timing—VVT, or variable valve lift—VVL) could reduce fuel consumption by 1 to 3 percent at incremental costs of $120 to $750. Cylinder deactivation could reduce fuel consumption by 2 to 3 percent at an incremental cost of around $75.[9] Implementing direct injection while still operating at stoichiometric could reduce fuel consumption by 2 to 3 percent at an incremental cost in the range of $550 to $950. Turbocharging and downsizing the direct-injected stoichiometric engine could reduce fuel consumption an additional 2 percent at an incremental cost of approximately $1,200. Further improvements could be made by invoking lean burn operation to the stoichiometric direct injection turbocharged engine. A reduction in fuel consumption of 10 to 14 percent might be achieved by doing this. The incremental cost of lean burn over the direct-injected turbocharged stoichiometric engine is approximately $750. This incremental cost is the estimate for the exhaust aftertreatment system required for the lean burn engine. Finally, if homogeneous charge compression ignition (HCCI)-like combustion can be implemented, a reduction in fuel consumption of 10 to 12 percent could be achieved relative to a stoichiometric direct injection engine. The incremental cost for HCCI would be around $685.

It is important to realize that there can be redundancy as well as synergy in applying these technologies. To get a more reliable estimate of the reduction in fuel consumption expected, the application of these technologies as packages in a simulation applied to the application of the engine in question should be evaluated.

[9]VVT, VVL, and cylinder deactivation fall into the class where their cumulative effect would not be additive. Variable valve actuation and cylinder deactivation applications will achieve some, or all, of their benefit by reducing pumping losses. Consequently if all of these technologies were used on the same engine, the reduction in fuel consumption would *not* be the sum of the individual estimates given in this paragraph.

Diesel Engine. If a diesel engine were used instead of a spark ignition engine, fuel consumption could be reduced about 19 to 24 percent over the base engine at a cost of approximately $8,000 to $9,000.

As one of the tasks in the committee's contract with TIAX, a projection of fuel consumption reduction achievable for Class 2b pickups and vans was done for a selected technology package. For this package aerodynamic improvements were assumed. Low rolling resistance tires, an eight-speed transmission coupled to a parallel hybrid system, and minor light weighting were also included. The package also included an advanced turbocharged downsized stoichiometric direct-injected gasoline engine. The improvements were projected relative to the current baseline engine described above. This package results in projected fuel savings of 44.5 percent at a cost of $14,710, with no increase in vehicle weight. The time frame in which it is likely that these technologies would be incorporated in a vehicle at significant market penetration is 2015 to 2020. A breakdown of the impact and incremental cost of the technologies in the package is shown in Table 6-9. The committee also concluded that similar fuel consumption benefits can be achieved with an advanced diesel engine in place of a baseline gasoline engine.

Refuse Truck (Refuse Packer)

This vehicle class is distinguished by its unique duty cycle, its weight (Class 8 vehicle), and its excellent potential for hybridization. The packer cab design is often a low cab-over-engine (LCOE) to aid ingress and egress by the operators. Its diesel engine is typically 9 to 11 liters in size with 280 to 325 horsepower and drives through an automatic transmission.

For comparison of fuel consumption estimates, the baseline truck has an urban duty cycle of about 700 load stops over 25 pickup miles a day, plus two round trips to a landfill for 50 additional miles. The truck has no aerodynamic devices and has a standard tandem axle with dual tires on steel wheels and a six-speed automatic transmission. The engine is a diesel of 11 liters displacement with peak thermal efficiency of 41 to 42 percent and cycle thermal efficiencies of 34 to 37 percent, cam actuated electronic unit injection (2,300 bar), variable-geometry turbocharger, and cooled EGR, plus DPF and 200-bar cylinder pressure. The packer PTO hydraulic pump is engine driven.

Engine. The current engine is forecast to be substantially evolved to an improved thermal efficiency of 49 percent (peak) which will provide a 14 percent reduction in fuel consumption, from a 2008 baseline. The 2020 technology includes optimized SCR with improved NO_x conversion efficiency, OBD with closed loop controls, higher injection pressure, increased cylinder pressure, and turbocompounding. The incremental cost is in the range of $14,000 to $16,000, which is a 2010 engine with SCR plus $3,000 to

TABLE 6-9 Class 2b Pickups and Vans with 2015-2020 Technology Package

Category	Description	Benefit (%)	Capital Cost ($)	CCPPR ($/%)	Weight (lb)
Aerodynamics	—	3	100	33	—
Engine	5-8 L turbocharged downsized s-GDI gasoline engine	23	4,000	174	—
Tire	Improved low rolling resistance	2	10	5	—
Transmission and driveline	8-speed automatic transmission, reduced driveline friction, aggressive shift logic	7.50	1,000	133	—
Hybrid	Parallel hybrid	18	9,000	500	300
Management and coaching	—	—	—	—	—
Idle reduction	—	—	—	—	—
Total added weight	Added components	–0.75	—	—	+300
Weight reduction	3%—~300 lb	0.75	600	800	–300
Total	2015-2020 package	44.5	14,710	331	0

NOTE: For each vehicle class, the fuel consumption benefit of the combined technology packages is calculated as follows: [% $FCR_{package}$ = 100 [1 – (1 – {% FCR_{tech1} /100 }) (1 – {% FCR $_{tech2}$/100}) ... {(1 – {% FCR_{techN}/100})] where % $FCR_{tech\ x}$ is the percent benefit of an individual technology. CCPPR, capital cost per percent reduction.
SOURCE: TIAX (2009).

$4,000 (TIAX, 2009). The automatic transmission will then use eight gears.

Rolling Resistance. The use of low rolling resistance tires has some application in refuse packers, but their low vehicle speed profile and the need for good traction makes the application questionable.

Hybridization. Several hybrid concepts are being applied to refuse packers, including starter alternator motor, electric hybrid, and hydraulic hybrid. Demonstration systems are as follows:

- Crane Carrier Company (electric and hydraulic) for the City of New York Department of Sanitation (CNY-DOS; expected 30 to 50 percent reduction in fuel use; Calstart, 2008),
- Mack Trucks Inc. 120-kW integrated starter alternator motor, parallel hybrid electric operating with a 600-V Li-ion battery pack (expected 30 percent fuel consumption reduction), also for the CNY-DOS (Walsh, 2009, p. 1),
- Crane Carrier and Bosch Rexroth Corp. (parallel hydraulic hybrid) and Crane Carrier and ISE Corporation (series electric hybrid; Calstart, 2008).

It is too early to know which of the several hybrid variants offer the best combination of cost and performance in the refuse packer. But a hydraulic hybrid may perform well in such applications where hundreds of launch/stop cycles characterize the longest portion of the operation day. It is noted that high-density residential packers using an automated side loader arm will achieve up to 1,200 launch/stops in a 10-hour day. This orientation for a hydraulic hybrid is a consequence of both energy recovery efficiency during heavy braking and incremental cost.

The Eaton parallel electric system is being applied to the Class 6 and 7 Navistar Durastar, Kenworth T270/T370, and Perterbilt 300 Series, which claim a 20 to 30 percent fuel consumption advantage at a cost of $38,000 to $40,000.[10] The Mack system is integrated in Mack's own Class 8 LCOE Terra-Pro refuse chassis and is equipped with a Mack 11-liter engine certified for EPA 2010 with an SCR NO_x aftertreatment system. Interestingly, Mack has captured about 50 percent of the LCOE Class 8 refuse sales for decades.

Idle Reduction. Idle timers and stop-start systems are ideal for refuse packer cycles except when the engine power take-off is needed. The engine pump for packing demands high horsepower during the packing cycle, and the recovered braking energy stored by the hybrid system is reserved for vehicle launch assist.

Summary. Table 6-10 captures a likely solution package for the refuse packer that is expected to be offered beyond 2015.

Refuse Truck Summary

The above technologies for diesel engines and hybrids will be production implemented by 2020. Refuse truck fuel consumption reduction will most substantially result from improvements and innovations in the power train system, which then becomes increasingly complex. Other technologies will be applied as spinoffs from Class 8 tractor-trailer applications (e.g., reduced transmission and driveline friction, accessory electrification, weight reduction, lower rolling resistance tires).

This sector is one of relatively low annual mileage (about 20,000 miles), but the current average fuel consumption is quite high (33 to 40 gallons/100 miles), owing to the intensity

[10]N. Naser, "Oshkosh Truck Corporation–AHHPS," presentation to the committee, Washington, D.C., February 8, 2007, slide 16.

TABLE 6-10 Class 8 Refuse Packer with a Hydraulic Hybrid System, 2015-2020

Category	Description	% Benefit	Capital Cost ($)	CCPPR ($/%)	Weight (lb)
Aerodynamics	—	—	—	—	—
Engine	Advanced 11-15 L engine	14	14,800	1,057	500
Tire	Improved low-rolling-resistance duals	2.50	300	120	—
Transmission and driveline	Reduced driveline friction	4	2,700	675	—
Hybrid	Parallel hydraulic hybrid	25	30,000	1,200	1,000
Management and coaching	—	—	—	—	—
Idle reduction	—	—	—	—	—
Total added weight	Added components	–3	—	—	+1,500
Weight reduction	Material substitution—500 lb	1	3,000	3,000	–500
Total	2015-2020 package	38.4	50,800	1,323	+1,000

NOTE: For each vehicle class, the fuel consumption benefit of the combined technology packages is calculated as follows: [% $FCR_{package}$ = 100 [1 – (1 – {% FCR_{tech1} /100 }) (1 – {% FCR_{tech2}/100}) … {(1 – {% FCR_{techN}/100})] where % $FCR_{tech\ x}$ is the percent benefit of an individual technology. CCPPR, capital cost per percent reduction.
SOURCE: TIAX (2009).

of their launch/stop duty. The industry today is remarkably immature, and low-volume, high-incremental price hybrid offerings could use some incentives to "get ready" for the 2015 and later marketplace.

Transit Bus

This category can also be called urban or city bus. The baseline is a 40-ft bus weighing about 40,000 GVW with an 8- to 9-liter diesel engine. The 2008 baseline engine has a thermal efficiency of 40 to 41 percent (peak) with high-pressure common rail or unit fuel injection (1,600 to 2,000 bar), variable geometry turbocharging, cooled EGR, active DPF, and 170- to 180-bar cylinder pressure. It is assumed to have no aero shaping and standard dual tires, steel wheels, six-speed automatic transmission, low-speed idle, but no idle reduction technology. The driving cycle is urban, low speed (12 mph average), 150 to 250 miles per day, high air conditioning and alternator load on the engine (30 to 40 kW). Transit buses have an average VMT of 35,167 and an average fuel consumption of 31 gallons per 100 miles (TRB, 2009, Table 5.14).

Engine. Improved thermal efficiency of 49 percent (peak), which is a 16.4 percent reduction in fuel consumption, will be available in the 2015 to 2020 time frame (TIAX, 2009). This reduction includes OBD with closed loop controls, improved NO_x conversion in the SCR system, higher fuel injection pressure (3,000 bar using common rail injection and multiple injections per cycle), increased cylinder pressure (220 to 230 bar), advanced EGR, dual-stage turbocharging, and electrically powered accessories. The incremental cost is in the range of $12,200 to $13,700, which is a 2010 engine with SCR plus $2,000 to $4,000.

Aerodynamics. No improvement anticipated.

Rolling Resistance. Transit buses gain only minor benefit from low rolling resistance tires Table 6-11), and they are unlikely to adopt wide-based singles. Benefits are estimated on the basis of the Class 8 tractor trailer and are scaled using a factor of 0.4 to account for the differences in vehicle duty cycle.

Transmission and Driveline. Transit buses use fully automatic transmissions, allowing the driver to keep both hands on the wheel and have uninterrupted shifts (which gives increased vehicle drivability and productivity). Opportunities for reducing fuel consumption include friction reduction, reducing transmission parasitic losses, using more aggressive shift logic, and increasing the number of transmission gears (see Table 6-12).

Weight. Given their low speed, stop-and-go duty cycle, transit buses can benefit greatly from weight reduction. TIAX (2009) used data from various sources and developed Table 6-13 to represent the summary.

Hybridization (Hybrid Power Trains). Urban transit buses have been undergoing hybridization since the early 2000s, and many reports are available regarding fuel consumption, capital costs, operating costs, and life-cycle costs. It is estimated that hybrid power trains constituted 25 to 30 percent of the market in 2009 (TIAX, 2009). Most systems on the road use parallel diesel configuration, although diesel series hybrids and gasoline series hybrids are also offered. The fuel consumption improvement varies from city to city, manufacturer to manufacturer, and month to month; Tables 6-14 to 6-16 give examples from various sources.

It should be noted the transit bus market is small, on the order of 5,000 buses per year, and hybrids already have 25 percent market share, so production volumes will not increase much beyond present levels. At this point, the hybrid

TABLE 6-11 Transit Bus Tire and Wheel Technologies

System	Capital Cost ($)	Fuel Consumption Benefit (%)
Low-rolling-resistance dual	30 per tire	0.8-1.2
Next-generation dual	30 per tire	1.6-2
Tire monitor	900	0.25

SOURCE: TIAX (2009).

TABLE 6-12 Driveline and Transmission Strategies for Transit Buses

Strategy	Fuel Consumption Benefit (%)	Cost ($)
Appropriate specifications	1-3	—
Aggressive shift logic and early lockup	0.5-1	100
Reduced automatic transmission parasitics and friction reduction	1	0-500
8-speed automatic transmission	2-3	1,100-1,650

SOURCE: TIAX (2009).

TABLE 6-13 Weight Reduction Cost and Benefit for Transit Buses

Weight Reduction (lb)	Cost per Pound ($)	Fuel Savings per 1,000 lb (%)
0-800	2-4	2-3
800-1,600	4-8	2-3
1,600-2,800	8-10	2-3

SOURCE: TIAX (2009).

bus market depends on very large government subsidies. Unless costs drop substantially, this is unlikely to change.

Transit Bus Summary

The most effective fuel consumption reduction technologies are improved diesel engines and hybridization. The most cost-effective technologies are the engine technologies and the hybrid when taking into account the 80 percent Federal Transit Administration support and the Clean Fuel Grant Program incremental hybrid cost credit. Note that in the current situation, cost effective for operators and cost effective for taxpayers are two different concepts.

Motor Coach

The American Bus Association (2006) states that there are more than 33,000 motor coaches operating in the United States. With an average fuel economy of 5.7 miles per gallon, a corresponding fuel consumption (FC) of 17.5 gallons/100 miles, and an annual average mileage per motor coach of 56,000,[11] motor coaches in the United States consume nearly 330 million gallons of fuel annually. Motor coaches typically operate over the highway in high-mileage, high-speed duty cycles. Occasionally, they also operate in urban cycles. While motor coaches provide a high-efficiency means of travel on a passenger-mile/gallon basis, a significant amount of their duty cycle is spent idling to maintain a comfortable interior temperature and air brake pressure.

Motor coach results are shown in Table 6-17. Here, since these vehicles spend a good portion of their time on interstates, the committee has included improved aerodynamics, an advanced 11- to 15-liter diesel engine with bottoming cycle, low-rolling-resistance tires, eight-speed automatic transmission, and minor light-weighting. This package results in fuel savings of 32 percent at a cost of $36,350 and a reduction in vehicle weight of 400 lb.

Engine. Improved thermal efficiency from 41.5 to 53 percent peak efficiency from 2008 baseline, which is a 21.7 percent reduction in fuel consumption. The engine uses an exhaust energy recovery system with improved engine controls.

Aerodynamics. Because of their frequent high-speed operation, motor coaches may benefit significantly from aerodynamic drag reduction. However, there is limited information available in the literature and little additional data were provided during committee site visits that speak directly to motor coach aerodynamics. As such, our estimates of motor coach aerodynamic drag reduction potential are based on extending results of the line-haul analysis to the motor coach segment. Cab streamlining and boat tailing offer the best prospects for reducing aerodynamic drag.

Rolling Resistance. Motor coaches can benefit from low rolling resistance tires, wide-base singles, or automatic inflation systems. Wide-base singles are available for motor coaches, but they are not widely deployed. There is only a single axle on a motor coach that uses dual tires (and hence can take advantage of a WBS). Benefits are estimated from the Class 8 tractor trailer and are scaled using a factor of 0.9 to account for differences in duty cycle and weight.

Transmission and Driveline. Motor coaches are unique in that they all use automatic transmissions but spend much of their time on the highway. This is because automatic transmissions offer a smoother ride, which is important in a passenger-driven application. That said, European motor coaches tend to use manual or automated manual transmissions. The committee has assumed the following to charac-

[11] "Commercial Bus Emissions Characterization and Idle Reduction: Idle and Urban Cycle Test Results," June 14, 2006, prepared for the American Bus Association, U.S. Department of Energy, U.S. Environmental Protection Agency, and U.S. Federal Highway Administration. Available at http://www.buses.org/files/download/motorcoach_idling_study.pdf.

TABLE 6-14 Results for Urban Transit Buses—Selected Sources

New York City[a]	Series 50 diesel, non-EGR	Orion VII Hybrid, non-EGR
Average mpg	2.65	3.45
WMATA[b]	Automatic transmission diesel	Hybrid
Dyno mpg	3.85	4.8
Revenue mpg	3.18	3.51
Vehicle cost	$349,000	$522,000
Capital cost/mile	$0.81	$1.19
Operating cost/mile	$1.19	$1.13
Total cost/mile	$2.00	$2.32
New York City[c]	Series 50 diesel with diesel particulate filter	Orion VII Hybrid
Average mpg	2.33	3.18
Vehicle cost	$280,000-$300,000	$450,000-$550,000
Seattle/King County[c]	Diesel 60 ft	Hybrid 60 ft (New Flyer/Allison)
Fuel	2.50 mpg	3.17 mpg
Fuel cost/mile	$0.79	$0.62
Maintenance cost/mile	$0.46	$0.44
Total cost/mile	$1.25	$1.06
Chapel Hill[d]	Diesel	Hybrid
	$240,000	$530,000
New York City[d]	Hybrid	
	3.19 mpg, 34% better than diesel	

NOTE: Hybrid 26 to 52 percent (average 37 percent) improvement in fuel efficiency; hybrid reduced costs: extended brake life 50 to 100 percent. Clean Fuels Grant Program: 90 percent of incremental hybrid cost. Federal Transit Administration: 80 percent of cost of standard diesel. Operating cost of hybrid is 15 percent lower than diesel. EGR, exhaust gas recirculation.

[a]Barnitt (2008).
[b]Bus Engineering (2008).
[c]Ranganathan (2006).
[d]USA Today, January 22, 2008.

terize transmission and driveline opportunities in the motor coach segment:

- Motor coaches generally use technologies similar to those used by tractor trailers to improve driveline efficiency but adopt technologies similar to the transit bus to enhance transmission efficiency.
- Because the motor coach does not have two drive axles, unlike the tractor trailer, it cannot benefit from switching to a single drive axle.
- Because the motor coach spends more time at high speed, the efficiency benefit of using an eight-speed transmission is estimated to be lower.

The transmission assumed for 2015 to 2020 is an eight-speed AT along with reduced driveline friction and aggressive shift logic.

Weight Reduction. Estimates for the benefit of weight reduction in motor coaches are drawn from the IFEU (2003)

TABLE 6-15 Hybrid Technology Cost and Benefits for Transit Buses

Architecture	Incremental Capital Cost ($)	Fuel Consumption Benefit (%)	Introductory Year	Incremental Weight (lb)
Gasoline series	200,000	25-35[a]	Available—~150 on road	2,000
Diesel series	220,000	30-40	Available—~ 75 on road	2,600
Diesel parallel and dual mode or dual mode	200,000	22-35[b]	Available—2,000-3,000 on road	940-2,840

[a]SOURCE: TIAX (2009).

TABLE 6-16 Urban Transit Buses Can Benefit from Hybridization and from Weight Reduction

Category	Description	Benefit (%)	Capital Cost ($)	CCPPR ($/%)	Weight (lb)
Aero	—	—	—	—	—
Engine	Advanced 6-9 L engine	14	13,000	929	500
Tire	Improved low-rolling-resistance duals	1.50	300	200	—
Transmission and driveline	8-speed automatic transmission, reduced driveline friction, aggressive shift logic	4	1,800	450	—
Hybrid	Diesel series hybrid	35	22,000[a] 220,000	6,286	1,500
Management and coaching	—	—	—	—	—
Idle reduction	—	—	—	—	—
Total added weight	Added components	–5	—	—	+2,000
Weight reduction	Material substitution—2,500 lb	6.25	15,300	2,448	–2,500
Total	2015-2020 package	47.8	52,400[a] 250,400	1,096 5,238	–500

NOTE: For each vehicle class, the fuel consumption benefit of the combined technology packages is calculated as follows: [% $FCR_{package}$ = 100 [1 – (1 – {% FCR_{tech1}/100 }) (1 – {% FCR_{tech2}/100}) ... {(1 – {% FCR_{techN}/100})] where % $FCR_{tech\ x}$ is the percent benefit of an individual technology. CCPPR, capital cost per percent reduction.

[a]With 90% federal subsidy of the incremental cost.

SOURCE: TIAX (2009).

report, which estimates 2.4 percent fuel savings per 10 percent weight reduction. Data were estimated for an 18 metric ton vehicle traveling on a primarily highway duty cycle. For a 36,000-lb coach, this fuel-savings estimate equates to 0.7 percent per 1,000 lb saved. Costs are estimated using the same logic as that described for the transit bus. The committee assumed a 36,000-lb coach, which weighs approximately the same as a tractor trailer, and so no scaling is needed.

Motor Coach Summary

The most cost-effective technology was the advanced diesel engine with a bottoming cycle. Hybrids were not considered for motor coaches by TIAX because of the high cost of hybrid systems and relatively low benefit on a commuter motor coach duty cycle (less than 12 percent fuel consumption reduction). TIAX discusses two potential hybrid archi-

TABLE 6-17 Motor Coaches Benefit from Aerodynamics and from Engine Improvements, Including Waste-Heat Recovery

Category	Description	Benefit (%)	Capital Cost ($)	CCPPR ($/%)	Weight (lb)
Aero	—	8	4,500	563	300
Engine	Adv. 11-15 L engine with bottoming cycle	20	23,000	1,150	800
Tire	Improved low-rolling-resistance duals + wide-base single tires	3	450	150	—
Transmission and driveline	8-speed automatic transmission, reduced driveline friction, aggressive shift logic	4.50	2,400	533	—
Hybrid	—	—	—	—	—
Management and coaching	—	—	—	—	—
Idle reduction	—	—	—	—	—
Total added weight	Added components	–0.75	—	—	+1,100
Weight reduction	Materials substitution—1,500 lb	1.05	6,000	5,714	–1,500
Total	2015-2020 package	32.0	36,350	1,136	–400

NOTE: For each vehicle class, the fuel consumption benefit of the combined technology packages is calculated as follows: [% $FCR_{package}$ = 100 [1 – (1 – {% FCR_{tech1}/100 }) (1 – {% FCR_{tech2}/100}) ... {(1 – {% FCR_{techN}/100})] where % $FCR_{tech\ x}$ is the percent benefit of an individual technology. CCPPR, capital cost per percent reduction.

SOURCE: TIAX (2009).

tectures for the motor coach application: a parallel system similar to that used in transit buses and a parallel system similar to that prototyped for Class 8 tractor trailers.

SUMMARY OF FUEL CONSUMPTION AND COST DATA

As shown in Table 6-18, the 2015-2020 packages reduce fuel consumption by 32 to 51 percent; all of the vehicle classes, with the exception of the motor coach and refuse hauler, are over 40 percent. The costs of these packages range from \$14,000 (Class 2b) to \$250,000 (transit bus). However, the other applications lie in a narrow band from \$40,000 to \$85,000. Many of the technologies benefit from anticipated reductions in the cost of hybrid vehicles during this time frame. The CCPPR estimates range from \$331 to \$5,232 per percent improvement, depending on vehicle class. For five of the seven vehicle classes, the benefits range from \$915 to \$1,674 per percent fuel consumption reduction. The cost of saving a gallon of fuel per year ranged from a low of \$8.42 for tractor-trailer trucks to \$48.95 for transit buses. Saving a gallon of fuel in a transit bus costs almost six times more than saving a gallon in tractor-trailer operations, but transit buses now enjoy huge subsidies for the implementation of fuel-saving technologies. This huge range in cost shows that care must be taken if cost-effective fuel-savings are to be realized in practice.

TABLE 6-18 Fuel Consumption Improvement, Cost, and CCPPR, 2015 to 2020 Vehicle Technology

	TT	Box	Bucket[a]	Refuse	Bus	Coach	2b
Fuel Consumption Reduction							
Aerodynamics	11.5%	6%	—	—	—	8%	3%
Engine	20%	14%	11.2%	14%	14%	20%	23%
Weight	1.25%	4%	3.2%	1%	6.25%	1.05%	0.75%
Tire	11%	3%	2.4%	2.5%	1.5%	3%	2%
Transmission	7%	4%	3.2%	4%	4%	4.5%	7.5%
Hybrid	10%	30%	40%	25%	35%	—	18%
Management	6%	—	—	—	—	—	—
Idle reduction[b]	—	—	—	—	—	—	—
Subtotal[c]	51.0%	49.4%	51.3%	40.2%	50.4%	32.5%	44.9%
Added weight (lb)[d]	2,030	1,100	1,050	1,500	2,000	1,100	300
Adjusted fuel consumption total	50.5%	47.1%	49.6%	38.4%	47.8%	32.0%	44.5%
Cost (\$)							
Aerodynamics	12,000	3,250	—	—	—	4,500	100
Engine	23,000	13,000	13,000	14,800	13,000	23,000	4,000
Weight	13,500	4,770	4,770	3,000	15,300	6,000	600
Tire	3,600	300	300	300	300	450	10
Transmission	5,800	1,800	1,800	2,700	1,800	2,400	1,000
Hybrid	25,000	20,000	30,000	30,000	220,000	—	9,000
Management	1,700	—	—	—	—	—	—
Idle reduction[b]	—	—	—	—	—	—	—
Total	84,600	43,120	49,870	50,800	250,400	36,350	14,710
Cost Benefit (\$/Percent Fuel Consumption Benefit)							
Aerodynamics	1,043	542	—	—	—	563	33
Engine	1,150	929	929	1,057	929	1,150	174
Weight	10,800	1,193	1,193	3,000	2,448	5,714	800
Tire	327	100	100	120	200	150	5
Transmission	829	450	450	675	450	533	133
Hybrid	2,500	667	750	1,200	6,286	—	500
Management	283	—	—	—	—	—	—
Idle reduction[b]	—	—	—	—	—	—	—
All packages	1,674	915	1,006	1,323	5,232	1,135	331

NOTE: The baseline year for the analysis is 2008. For each vehicle class, the fuel consumption benefit of the combined technology packages is calculated as follows: [% $FCR_{package}$ = 100 [1 – (1 – {% FCR_{tech1}/100}) (1 – {% FCR_{tech2}/100}) ... (1 – {% FCR_{techN}/100 })] where % $FCR_{tech\ x}$ is the percent benefit of an individual technology.

[a]The 2015-2020 hybrid capital costs are forecasted to be reduced by up to 47 percent from the 2013-2015 costs. This is due to technology and volume changes but may not be fully realized.

[b]Overnight idle reduction is assumed to be implemented as part of the hybrid package.

[c]For each vehicle class, the fuel consumption benefit of the combined technology packages is calculated as noted above.

[d]The fuel consumption penalty resulting from weight increases due to the various technology packages has not been included in the results.

SOURCE: TIAX (2009), p. 5-8.

The near-term (2013-2015) analysis (Table 6-19) was done in a manner similar to the analysis for the 2015-2020 time frame. The technologies that are included vary by market segment but, broadly speaking, include a high-cost first-generation hybrid system for urban vehicles and aerodynamic improvements coupled with first-generation waste heat recovery on highway vehicles. All of the technologies include significant engine efficiency improvements as indicated. The near-term packages have projected fuel consumption benefits that range from 14 to 40 percent, while costs range from $4,435 (for the Class 2b pickup and van segment) to $215,600 (for the transit bus, which includes a high-cost hybrid system). However, most of the vehicle classes lie in a relatively narrow band that shows 30 to 40 percent fuel consumption reduction at costs that range from $44,000 to $62,000. The CCPPR of these packages is $1,270 to $1,860 per percent benefit for five of the vehicle classes, but ranges from $207 (for Class 2b vehicles) to $5,890 per percent (for transit buses).

Cost-Effectiveness Metrics

In this chapter, CCPPR has been presented in terms of dollars per percent reduction in fuel consumption. This allows comparison of fuel-saving technologies across a range of vehicle classes and applications. Unfortunately, as a true metric of cost-benefit ratio, dollars per percent fuel savings has many drawbacks. The committee would not have used this metric if enough information had been available in time to support the use of a better metric. The committee believes that a metric of dollars per gallon saved per year is a much more useful measure of cost-benefit ratio. This metric allows

TABLE 6-19 Fuel Consumption Improvement, Cost, and Cost Effectiveness, 2013-2015 Vehicle Technology

Technology	TT	Box	Bucket[a]	Refuse	Bus	Coach	2b
Fuel Consumption Reduction							
Aerodynamics	5.5%	—	—	—	—	—	3.0%
Engine	10.5%	9.0%	7.2%	10.5%	9.0%	10.5%	14.0%
Weight	0.8%	2.0%	1.6%	—	3.0%	0.7%	0.4%
Tire	4.5%	1.8%	1.4%	1.5%	1.0%	1.8%	1.0%
Transmission	5.0%	1.5%	1.2%	1.5%	1.5%	2.0%	4.5%
Hybrid	—	22.0%	35.0%	20.0%	30.0%	—	—
Management	3.0%	—	—	—	—	—	—
Idle reduction	6.0%	—	—	—	—	—	—
Subtotal[b]	30.6%	32.7%	42.2%	30.5%	39.7%	14.5%	21.4%
Added weight (lb)[c]	1,530	800	1,050	1,500	2,000	800	0
Adjusted fuel consumption total	30.0%	30.5%	40.2%	28.4%	36.6%	14.0%	21.4%
Cost ($)							
Aerodynamics	9,000	—	—	—	—	—	100
Engine	14,200	10,300	10,300	12,300	10,300	14,200	3,000
Weight	6,000	1,590	1,590	—	4,800	3,000	225
Tire	900	300	300	300	300	240	10
Transmission	5,200	200	200	200	200	300	1,100
Hybrid	—	38,000	50,000	40,000	200,000	—	—
Management	1,000	—	—	—	—	—	—
Idle reduction	8,000	—	—	—	—	—	—
Total	44,300	50,390	62,390	52,800	215,600	17,740	4,435
Cost Benefit ($/Percent Fuel Consumption Benefit)							
Aerodynamics	1,636	—	—	—	—	—	33
Engine	1,352	1,144	1,144	1,171	1,144	1,352	214
Weight	8,000	795	795	—	1,600	4,286	563
Tire	200	167	167	200	300	133	10
Transmission	1,040	133	133	133	133	150	244
Hybrid	—	1,727	1,429	2,000	6,667	—	—
Management	333	—	—	—	—	—	—
Idle reduction	1,333	—	—	—	—	—	—
All strategies	1,475	1,652	1,552	1,859	5,890	1,268	207

NOTE: For each vehicle class, the fuel consumption benefit of the combined technology packages is calculated as follows: $[\% \; FCR_{package} = 100 \; [1 - (1 - \{\% \; FCR_{tech1}/100\}) \; (1 - \{\% \; FCR_{tech2}/100\}) \ldots (1 - \{\% \; FCR_{techN}/100\})]$ where $\% \; FCR_{tech \; x}$ is the percent benefit of an individual technology.

[a]The 2015-2020 hybrid capital costs are forecasted to be reduced by up to 47 percent from the 2013-2015 costs. This is due to technology and volume changes but may not be fully realized.

[b]For each vehicle class, the fuel consumption benefit of the combined technology packages is calculated as noted above.

[c]The fuel consumption penalty resulting from weight increases due to the various technology packages has not been included in the results.

SOURCE: TIAX (2009), p. 5-7.

a direct comparison of the costs of various fuel-saving technologies in terms of how many dollars it costs to reduce a vehicle's fuel consumption by one gallon per year. Since the overall goal of energy policy is to reduce fuel consumption at the least cost to society, the committee believes that the metric of dollars invested per gallon of fuel saved is a very good metric to consider.

To calculate the proposed metric, information is need on the average VMT and fuel consumption of the target vehicle class. Given this, it is easy to determine the number of gallons consumed per year by vehicles in the target class, and from there to determine the cost to reduce that fuel consumption in units of dollars per gallon saved per year. Some examples:

- The fuel consumption reduction potential for the tractor-trailer application, in the 2015 to 2020 time frame, is 51 percent at a cost of $84,600. Assuming an average VMT of 120,000 miles per year and an average fuel consumption of 18 gallons/100 miles (data from VIUS, 2002), this results in a cost of $7.68 per gallon saved per year.
- The fuel consumption reduction potential for Class 6 box and bucket trucks, in the 2015 to 2020 time frame, is 47.1 percent for box trucks and 49.6 percent for bucket trucks. The resulting cost for box trucks is $43,120 and the cost for bucket trucks is $49,870. For Class 6 box trucks with an average VMT of 25,000 miles per year and an average fuel consumption of 12.5 gallons per 100 miles (committee estimates based on VIUS data), the cost is $29.29 per gallon saved per year. In the case of bucket trucks, where average VMT is 13,300 miles per year (survey result reported by Chatham Consulting) and average fuel consumption is 20 gallons per 100 miles (committee estimate, including fuel spent idling to support bucket operation), the cost is $37.80 per gallon saved per year.
- The fuel consumption reduction potential for the Class 2b pickup and van application, in the 2015 to 2020 time frame, is 44.5 percent at a cost of $14,710. With an average VMT of 14,000 miles per year and an average fuel consumption of 7 gallons per 100 miles (committee estimates), the cost is $33.73 per gallon saved per year.
- The fuel consumption reduction potential for refuse trucks, in the 2015 to 2020 time frame, is 38.4 percent at a cost of $50,800. At an average VMT of 20,000 miles per year and an average fuel consumption of 35 gallons per 100 miles (committee estimates), the cost is $18.90 per gallon saved per year.
- The fuel consumption reduction potential for transit bus applications, in the 2015 to 2020 time frame, is 47.8 percent at a cost of $250,400. With an average VMT of 35,167 miles per year and an average fuel consumption of 31 gallons per 100 miles (TRB, 2009), the cost is an extremely high $48.05 per gallon saved per year.
- The fuel consumption reduction potential for the motor coach application, in the 2015 to 2020 time frame, is 32 percent at a cost of $36,350, which results in a cost benefit of $1,117 per percent fuel consumption reduction. With a VMT of 56,000 miles per year and an average fuel consumption of 17.5 gallons per 100 miles (American Bus Association, 2006), the cost is $11.59 per gallon saved per year.

Table 6-20 summarizes a comparison of CCPPR measured by dollars per percent fuel saved, compared to a metric of dollars per gallon saved per year. A breakeven fuel price is also provided in the table. This breakeven price is the price of fuel such that the value of the discounted fuel savings over 10 years is just equal to the cost of the technology. This price does not necessarily reflect how truck buyers would evaluate technologies, since they often do not plan to own the truck for a full life, have a different discount rate, and would need to consider O&M costs. However, a lifetime breakeven price can be a useful metric for considering the social costs and benefits of regulation. Note that while the cost-effectiveness in terms of dollars per percent fuel saved is best for Class 2b, when measured in terms of dollars per gallon saved or breakeven fuel price, the Class 2b technologies prove to be very expensive. Tractor-trailer trucks, which require a relatively large investment per percent fuel saved, are actually the best bargain in terms of cost per gallon of fuel saved. The

TABLE 6-20 Fuel Consumption Reduction Potential for Typical Vehicles, 2015-2020, and Cost-Effectiveness Comparisons for Seven Vehicle Configurations

			Cost-Effectiveness Metric		
Vehicle Class	Fuel Consumption Reduction (%)	Capital Cost ($)	Dollars per Percent Fuel Saved	Dollars per Gallon Saved per Year	Breakeven Fuel Price[a] ($/gal)
Tractor-trailer	51	84,600	1,674	7.68	1.09
Class 6 box truck	47	43,120	915	29.30	4.17
Class 6 bucket truck	50	49,870	1,006	37.80	5.38
Class 2b pickup	45	14,710	331	33.73	4.81
Refuse truck	38	50,800	1,323	18.90	2.69
Transit bus	48	250,400	5,232	48.05	6.84
Motor coach	32	36,350	1,135	11.59	1.65

NOTE: Values shown are for one set of input assumptions. Results will vary depending on these assumptions.

[a]Calculated assuming a 7 percent discount rate and a 10-year life, excluding incremental operating and maintenance costs associated with the technologies.

SOURCE: Adapted from TIAX (2009).

transit bus stands out as a very expensive approach (without subsidies) to fuel saving by both metrics. Note that both of these metrics include only initial purchase cost. Increases in O&M costs are not included.

OPERATING AND MAINTENANCE COSTS

Driver pay has historically been the highest expense for all sectors of the trucking industry. However, due to recent increases in energy costs, diesel fuel cost per mile (CPM) equals or exceeds driver pay as the top cost for many motor carriers. Costs per mile are generally divided into two groups, vehicle and driver based. Significant operational costs include:

- Vehicle-Based
  - Fuel and engine oil
  - Truck/trailer lease or purchase payments
  - Repair and maintenance
  - Fuel taxes
  - Truck insurance premiums
  - Tires
  - Licensing and permits
  - Tolls
- Driver-Based
  - Driver wages
  - Driver benefits
  - Driver bonuses

The results of the American Transportation Research Institute (2008) survey of operating costs in terms of CPM and cost per operating hour (CPH) are shown in Table 6-21.

The technologies presented earlier in this chapter will significantly reduce fuel costs for fleets and customers. The addition of hardware such as turbocompounding, aerodynamic devices, hybrid power train components, and waste heat recovery systems (bottoming cycles) will increase complexity and potentially maintenance costs. Use of OBD, as is required on exhaust emission devices, should allow information to be made available to reduce repair and maintenance costs.

TABLE 6-21 Motor Carrier Marginal Expenses

Expenses	Cost per Mile ($)	Cost per Hour ($)
Vehicle-Based		
Fuel-oil costs	0.634	33.00
Truck/trailer lease or purchase payments	0.206	10.72
Repair and maintenance	0.092	4.79
Fuel taxes	0.062	3.23
Truck insurance premiums	0.060	3.12
Tires	0.030	1.56
Licensing and overweight-oversize permits	0.024	1.25
Tolls	0.019	0.99
Driver-Based		
Driver pay	0.441	16.59
Driver benefits	0.126	6.56
Driver bonus payments	0.036	1.87
Total Marginal Costs	1.73	83.68

SOURCE: Adapted from American Transportation Research Institute (2008).

TABLE 6-22 Incremental Operations and Maintenance Costs

Technology	Cost per Mile ($)
Aerodynamics	No increase
Single wide tires and advanced lubricants	0.0040
Hybrid power train	0.0060
Turbocompounding	0.0003-0.0007
Bottoming cycle	0.0030

According to the NESCCAF/ICCT (2009)report, O&M costs account for 5 percent of the total capital cost every 100,000 miles. An assessment was also made of incremental O&M costs for new technologies (Table 6-22).

The estimated O&M costs provided in Table 6-22 are projections, because many of the technologies are not in the field yet. Field experience may result in significant changes to these projections. For example, the committee has heard anecdotal evidence of substantial costs related to the repair and maintenance of aerodynamic devices that are damaged in service. The committee did not have the resources to undertake a complete study of operating and maintenance costs for all applications of future technologies. Additional study is required to refine estimates of O&M costs for fuel-saving technologies, since in some cases O&M costs can represent a substantial portion of the overall cost. Policymakers will need to consider fleet fuel savings (cost per gallon saved) as a parameter in their studies and evaluations.

INDIRECT EFFECTS AND EXTERNALITIES

Overview

The first part of this chapter presented the direct costs and benefits associated with various technology packages aimed at improving the efficiency of medium- and heavy-duty vehicles. Capital and O&M costs were identified and weighed against fuel-saving benefits. Understanding these direct costs and benefits is critical, as the economics of technology implementation are a primary decision attribute for manufacturers, carriers, and operators. However, direct costs and benefits are not the only costs and benefits associated with the application of new technologies. There are also *indirect* costs, benefits, effects, and externalities (impacts that are not put in market terms) that should be addressed. Some of these indirect effects represent unintended consequences associated with technologies or policies designed to spur greater fuel efficiency in medium- and heavy-duty vehicles.

This part of the chapter presents a number of indirect costs and benefits including:

- Fleet turnover effects
- Ton-miles traveled and the rebound effect
- Vehicle class shifting by consumers
- Environmental co-benefits and costs
- Congestion
- Safety impacts
- Incremental weight effects
- Manufacturability and product development.

The committee identified these indirect costs and benefits as important to discuss, although the committee recognizes this is not an exhaustive list. Assessment of possible indirect effects during policy development will help avoid or mitigate negative unintended consequences.

Fleet Turnover Effects

The implementation of regulations that increase the capital costs of new vehicles could have an effect on consumer purchase decisions, especially when access to capital is limited. In particular, instead of purchasing a new (more expensive) vehicle, consumers may likely choose to maintain their existing vehicle in order to extend its life. If these existing vehicles are less efficient than new ones, the overall effect of the regulation may be dampened or even counterproductive. The issue of how new technologies and regulations will affect new vehicle prices and operating costs—and the impact on fleet turnover from those cost effects—is an area that needs further analysis.

The purchase of new medium- and heavy-duty vehicles is a capital-intensive prospect. For example, the latest Class 8b tractor rigs commonly cost more than $100,000 (CARB, 2008). In addition, the adoption of efficiency improvement strategies such as aerodynamic retrofits or advanced engine designs, as shown in Tables 6-18 and 6-19, can add thousands of dollars to the cost of a truck. Accordingly, incremental vehicle cost increases associated with new fuel economy (or other) standards combined with truck owner budget constraints are likely to impact truck purchasing decisions at the fleet level.

The impact of increasing new vehicle purchase and operating costs will likely be twofold. First, some vehicle owners may decide to accelerate their purchase schedule, obtaining a new vehicle before the adoption of a new standard, thereby deferring the incremental cost of the standard until their next purchase cycle. Buyers may also be concerned about the reliability of unproven technology, further increasing their incentive for early purchases. This approach also allows truck buyers to observe the performance of the new technology secondhand, without incurring the associated risk.

Such early purchase behavior is referred to as "pre-buy." The associated dip in purchases in the time immediately following the introduction of the new, more costly vehicles meeting the standard is referred to as "low-buy." Such impacts have been observed recently in association with the adoption of the 2004 and 2007 heavy-duty emission standards and their associated price increases. Second, some fraction of vehicle owners may simply defer purchasing new trucks for some period of time, keeping their older vehicles in operation longer than they would have otherwise. The net impact of these altered purchasing patterns is to delay the full impact of any new standards that entail notable incremental costs. Figure 6-2 shows sales of Class 8 trucks from 1990 to 2007.

Truck sales are highly cyclical, responding to general economic conditions. As Figure 6-2 suggests, however, pre-buy behavior can alter somewhat the pattern of sales. In particular, there appears to be a general industry consensus that the sizable peak in 2006 were largely attributable to pre-buy behavior in advance of more stringent and costly NO_x and PM standards being introduced in the following years. The size of the peak appears to roughly correspond to the size of the incremental cost increase for the new standard—from about $1,000 for the 2004 standard[12] to between $7,000 and $10,000 for the 2007 standards.

An economic analysis of pre-buy and low-buy impacts for Class 8 trucks between 2005 and 2008 (NERA, 2008) utilized a vehicle scrap model along with a price elasticity model to predict increases and decreases in annual sales.[13] The model was relatively successful in predicting pre-buy increases although somewhat less successful in predicting low-buy decreases.

Both the NERA modeling exercise and subsequent data also showed that the low-buy "dip" was actually more substantial than the pre-buy "peak" by a ratio of about 3 to 2, meaning that there was a net decrease in sales over this period. A net downturn in sales also indicates that a portion of vehicle owners may be keeping their older units on the road longer (assuming freight demand levels do not decrease substantially). The aggregate impact of all of these factors was estimated to result in a net increase in national annual NO_x emissions in 2010 of more than 50,000 tons, relative to the case without pre-buy/low-buy and elasticity effects. This represents about 1 percent of expected NO_x emissions from all on-road sources.[14]

Although not identical to the situation presented by a potential fuel efficiency standard, where some or all of the incremental vehicle cost may eventually be recouped through future fuel savings, buyer responses to the cost increases associated with previous NO_x and PM standards could provide

[12]The 2004 heavy-duty engine standards actually began to penetrate the market at substantial levels in 2003 as a result of the consent decree pull-ahead.

[13]The elasticity model assumed a 1 percent increase in sales price translated to a 1.9 percent decrease in sales.

[14]See http://www.epa.gov/ttn/chief/trends/index.html for national emission inventory trends.

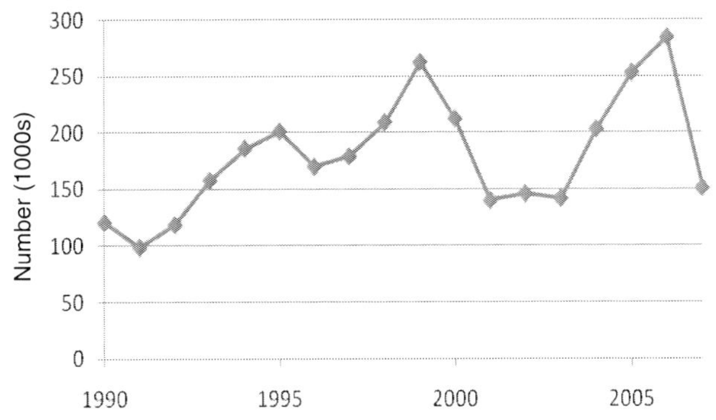

FIGURE 6-2 New retail Class 8 truck sales, 1990-2007. SOURCE: TRB (2009).

a rough sense of the possible pre-buy and low-buy impacts associated with future fuel efficiency regulations.

In addition to pre-buy influences, other factors can have a confounding effect on vehicle purchasing behavior, making it difficult to predict sales changes on a consistent basis. NERA notes other factors influencing sales prices, including fluctuations in steel prices. The likelihood of keeping an older vehicle on the road longer will also involve decisions about resale markets, variable O&M costs, and future demand levels. In fact, the California Air Resources Board has concluded that predicted economic activity in the transportation sector is by far the most important determinant impacting heavy-duty vehicle purchase decisions. From this perspective, much of the "low-buy" in heavy truck sales in 2007 may actually be due to the general economic downturn rather than a retreat from pre-buy expenditures. More research is needed to explore this effect.

In summary, during periods of stable or growing demand in the freight sector, pre-buy behavior may have a significant impact on purchase patterns, especially for larger fleets with better access to capital and financing. Under these same conditions, smaller operators may simply elect to keep their current equipment on the road longer, which is all the more likely given continued improvements in diesel engine durability over time. On the other hand, to the extent that fuel economy improvements can offset incremental purchase costs, these impacts will be lessened. Nevertheless, when it comes to efficiency investments, most heavy-duty fleet operators require relatively quick payback periods, on the order of 2 to 3 years. As such, pre-buy effects may still be a factor impacting the pace of new technology introduction.

Ton-Miles Traveled and the Rebound Effect

To understand the aggregate systemwide effect on fuel consumption, not only must the effect of fuel-saving measures on fuel consumption per ton-mile be considered, but also whether fuel consumption improvements reduce or raise total operating costs. Reductions in cost will lead to truck traffic increases, thereby partially offsetting the individual truck fuel savings. This effect is the so-called rebound effect and has been measured in many statistical studies pertaining to the light-duty vehicle fleet. Regulations that lead to increases in costs will have the opposite effect.

In light-duty vehicles the studies measure the increased driving that occurs in reaction to the reduction in the variable cost of driving a mile as a result of the application of fuel-saving technologies. In medium- and heavy-duty trucking, this "rebound" is a more complex phenomenon and has been studied less than the light-duty vehicle effect.

The problem can be conceptualized as follows. If investment in new technology is cost effective as seen by the private firm, and lowers the cost of truck transport, two types of reactions would be expected. First, lower costs in this very competitive industry are likely to be passed on to shippers as lower prices. The lower prices will reduce delivered prices of freight, which to some degree will lead to higher demand for the products being shipped and hence more shipments. Over a long term the cost of shipping will affect decisions on production and distribution center locations, which will also affect the demand for trucking services. In addition, the lower cost of shipping by truck relative to other modes, particularly rail, will lead to more freight being diverted from rail to being shipped by truck.

If the application of technology pushes beyond the private cost-effective level, and the costs of shipping increase, the response will be the reverse—higher costs will reduce shipments by truck. This case may be socially efficient if the higher cost is truly reflecting the additional social costs of climate change and oil security due to fuel consumption—costs that are difficult to determine. In this case the goal of the standard is to push technology adoption beyond the point a purely private decision would take it.

The aggregate effects of truck costs on the demand for shipping are summarized statistically by the own-price elasticity of trucking. The own-price elasticity measures the total percentage increase in the demand for shipments relative to the percentage change in the price of shipping. The modal-shift response, which is part of the total response, is summarized statistically by the cross-price elasticity. The cross-price elasticity measures the percentage change in rail shipments for a given percentage change in truck shipping rates.

Studies measuring own-price and cross-price elasticities have concentrated on long-haul freight movements since these movements are most subject to intermodal competition, represent a significant fraction of freight movement, and account for 78 percent of fuel consumed by medium- and heavy-duty trucks. The results of these studies range widely, depending on the type of product being shipped, the geography of the shipments, trip lengths, and the specific functional form used to describe the relationship. In addition, some studies measure the effect on tons shipped, and some measure the effect on ton-miles or volume. A literature survey by Christidis and Leduc (2009) reported elasticity estimates from eight studies mostly in the range of –0.5 to –1.5. A second survey (Graham and Glaister, 2004) also cites a likely range of –0.5 to –1.5 for the own-price elasticity, although it does not specify if that is for tons or ton-miles. A recent study by the Federal Highway Administration (FHWA) used own-price elasticity for ton-miles of –0.97, meaning that a 1 percent increase in the price of shipping one ton over one mile results in slightly less than a 1 percent reduction in ton-miles shipped. A study for the National Cooperative Highway Research Program cited a rail cross-price elasticity of 0.52 based on the intermodal competition model and identified a range of cross-price elasticities of 0.35 to 0.59 from an analysis of Class 1 railroads (Dennis, 1988; Jones et al., 1990). Given the wide range of estimates and the imprecision of the demand variables, it is not possible to provide a confident measure of the rebound effect.

Vehicle Class Shifting by Consumers

When manufacturers build vehicles, they make trade-offs related to various vehicle attributes in order to produce a vehicle that is most attractive to a given market segment. For example, manufacturers regularly need to balance issues of performance, cost, and fuel efficiency. In cases where regulation incentivizes a certain class of vehicles to meet a fuel efficiency standard at the expense of performance, a potential buyer may choose to purchase a larger class vehicle to offset the performance losses. This behavior would lead to less efficient vehicles on the road—exactly the opposite effect of what an efficiency standard is designed to achieve. This behavior can be called "consumer class shifting." Class shifting could also occur if the cost of different vehicle classes is affected disproportionately by the regulations. For example, requiring aerodynamic fairings on all Class 8 vehicles may cause some companies that currently use these vehicles on long-haul operations to choose smaller, less efficient vehicles rather than invest in the fairings. Others, however, will find they will have to add fairings that provide little benefit at high cost.

There is little or no literature that describes the cross-class mode shift between truck types based on cost or performance, although at least one manufacturer interviewed for this report expressed consumer class shifting as a concern (particularly for lower class vehicles). To induce consumer class shifting, regulations would need to significantly increase the cost or decrease the performance of one class of trucks relative to another.

Environmental Co-Benefits and Costs

It is often (but not always) the case that fuel efficiency improvements result in reductions of other pollutants as well. Certain technologies such as hybridization or idle reduction reduce the amount of time an internal combustion engine is operating, resulting in a direct reduction in all emissions (namely NO_x and PM) during this period. Another example occurs when engine loads are reduced (e.g., through adoption of aerodynamic or rolling resistance improvements), which in turn reduce NO_x emissions (Schubert and Kromer, 2008). Table 6-23 demonstrates trade-offs between fuel efficiency technologies and *engine-out* NO_x emissions.[15]

In interpreting Table 6-23, it is important to note that all engines have to meet stringent NO_x and PM standards at certification. In addition, new 2007-2010 exhaust emissions standards effectively require aftertreatment technologies (SCR and PM traps) that will substantially reduce engine-out emissions prior to exhaust. Therefore, although trade-offs exist between engine efficiency improvements and engine-out emissions, any engine modifications to improve fuel efficiency must ultimately comply with emission standards at certification and during in-use operation. In fact, the committee expects that new NO_x control technologies such as SCR will be so effective in reducing tailpipe emissions as to potentially allow engine manufacturers to increase engine efficiency while still complying with emission standards.

Congestion

Traffic congestion has increased dramatically throughout the world due to increased vehicle miles traveled and can result in significant loss of time, money, and fuel. Travel in congested conditions results in both longer and less predictable travel times. The Texas Transportation Institute (TTI) estimated that in 2007 the 439 urban areas in the United States experienced 4.2 billion vehicle-hours of delay, resulting in 2.8 billion gallons of wasted fuel and $87.2 billion in delay and fuel costs (Schrank and Lomax, 2005). TTI (2007) also found that the commercial vehicle cost of congestion, in both lost time and fuel, was ~$77 per vehicle-hour. The major causes of traffic congestion in the United States, as a percentage of total congestion, are (1) bottlenecks (traffic demand exceeds roadway capacity)—50 percent; (2) traffic incidents—25 percent; (3) work zones—15 percent; (4) bad weather—10 percent; and (5) poor signal timing—5 percent (AHUA, 2004). There are two questions. First, if regulations increase truck VMT, what are the potential implications for congestion? Second, if regulations cause degradation in truck performance, will slower trucks, or trucks with reduced hill-climbing ability, cause congestion to increase?

This chapter identifies how the demand for long-haul trucking could increase if regulations on reducing fuel consumption effectively reduce the net operating cost of long-haul trucking (i.e., the "rebound effect"). The effect on demand for trucking services is shown to be potentially significant, depending on the technology chosen and the own-price and cross-price elasticities of demand. There is

[15] A primary cause for the trade-off between fuel consumption and engine-out NO_x emissions is that NO_x formation is highly temperature dependent; therefore, at high combustion temperatures (which generally result in high pressures and thermal efficiency), NO_x formation increases.

TABLE 6-23 Fuel Efficiency Technology Versus NO_x Emissions Trade-off

Fuel Efficiency Technology	Effect on Efficiency	Effect on NO_x
Aerodynamic/weight reduction	+/++++	Anticipated decrease, depending on mode
Variable valve actuation	+	Depends on design
Higher cylinder pressure	++	Increases proportionally
Miller cycle	+++	May decrease due to aftercooling
Multistage turbo	++	Increases if EGR cannot be used
Mechanical turbocompound	+	Minimal
Electrical turbocompound	++	Minimal
Bottoming cycle	++++	Minimal
Thermoelectrics	+	Minimal
Enhanced exhaust insulation	Minimal	Minimal
Hybridization	+++	Reduction
Idle reduction	++++	Reduction

little information, however, about effects of potential performance degradation, although, where present, they may result in locally significant effects on congestion.

The remainder of this section focuses on basic freeway segments and is applied to the heavier, slower trucks (Classes 4 through 8). Trucks have an impact on congestion at other traffic control locations, such as signalized and unsignalized intersections, merge sections, and freeway-to-freeway off-ramps, but the data to analyze these impacts are not readily available. In the highway capacity manual (HCM), in basic freeway analysis, trucks are represented as "passenger car equivalents" (PCE; TRB, 2000). The PCE concept is meant to capture the effect that heavy vehicles have on traffic flow on a freeway because heavy vehicles occupy more space, travel more slowly up steep grades and more quickly down them, accelerate more slowly, brake more slowly, and change lanes more slowly than passenger cars. Furthermore, different passenger car operators react differently to the presence of trucks, for example, following farther behind trucks than cars.

The HCM increases the PCE for trucks based on roadway conditions such as grade, number of lanes, distance to roadside obstructions, and width of lanes. It does not, however, distinguish between truck PCEs under congested and uncongested conditions. FHWA simulated the effect of combination trucks as part of its truck size and weight study to estimate how effective truck PCEs change based on the weight-to-horsepower ratio of the truck itself, grade of the roadway, road type, geography, and congestion levels. FHWA did not estimate the impacts of other roadway characteristics such as lane width or distance to obstructions (Battelle, 2005). Generally, steeper grades, longer hills, fewer lanes, and a higher weight-to-horsepower ratio increase the number of PCEs for a single truck. Truck PCE conversions are necessary to calculate a volume-to-capacity (V/C) ratio, which can be used to estimate congestion and delay measurement.

While different formulas have been developed to estimate traffic speeds based on V/C ratios, an illustrative example can be provided through the use of the Bureau of Public Roads formula:

$$CS = \frac{FFS}{\left[1 + 0.15\left(\frac{V}{C}\right)^4\right]}$$

where CS is congestion speed, FFS is free-flow speed, V is volume of traffic, and C is traffic volume capacity. The formula implies that as a basic freeway segment approaches capacity (as the V/C ratio approaches 1.0), traffic speed will be reduced. To calculate delay, both congested and free-flow travel times must be calculated from the segment length and congested and free-flow speeds and then the difference between the free-flow travel time and the congested travel time must be measured.

An alternative perspective on congestion measurement uses estimates from the literature on the marginal cost of one combination truck on overall congestion. Parry (2006) estimates the marginal congestion cost of combination trucks to be \$0.168 per mile in urban areas and \$0.037 per mile in rural areas. The marginal congestion cost describes the cost, measured in lost travel time, that a single additional combination truck imposes on the rest of the traffic already on the roadway. Generally, as congestion increases, the marginal cost increases.

Safety Impacts

New regulations that would affect fuel efficiency of medium- and heavy-duty vehicles must consider impacts on vehicle and highway safety. The safety impacts are of several types. First, new technologies may have specific safety issues associated with them. For example, hybridization will introduce high-voltage electrical equipment into trucks; operators, service mechanics, and emergency personnel will need to be educated about appropriate handling of this equipment. Second, the rebound effect may increase overall truck traffic on the road, thereby leading to potentially higher incidences of accidents. Third, some technologies and/or approaches to improving fuel efficiency may actually lead to a safer highway system; for example, speed reductions, improved driver training, and use of side fairings might reduce hazards to other vehicles in inclement weather. Fourth, if new technologies diminish the performance of vehicles (e.g., decrease acceleration times), negative safety impacts could occur. Lastly, if new technologies or regulations have the effect of increasing payload capacity for trucks, fewer trucks may be in operation, potentially resulting in safety benefits. Of these five types of safety issues, only the second one (safety issues related to a rebound effect is discussed in detail here). A more detailed assessment would be needed on all these safety aspects based on the type of regulation that may be put forward by the U.S. Department of Transportation (DOT).

This chapter has described how the demand for long-haul trucking will increase if fuel economy regulations effectively reduce long-haul truck operating costs. Truck traffic could increase by 2.2 to 10.5 percent, depending on the technology alternatives and assumed demand elasticities. The literature shows that truck traffic has a direct correlation with injuries and fatalities. Estimates of increased truck traffic can be multiplied by the injury and fatality crash rates to estimate the range of potential deaths and injuries caused by an increase in travel. Regulations that push beyond the private cost-effective levels of fuel savings can result in a decline in VMT, and the effect would be a decline in crash rates.

Recent data for highway crashes indicate crash rates for truck tractors with trailers of 2.4 per 100 million VMT for fatal crashes and 51.1 per 100 million VMT for injury crashes (Federal Motor Carrier Safety Administration, 2007). To the extent that regulations or technologies incentivize increased VMT, accidents and fatalities could potentially increase; in contrast, if regulations or technologies decrease VMT, accidents and fatalities could potentially decrease.

To estimate the dollar costs from these additional crashes, a Federal Motor Carrier Safety Administration-commissioned study on the costs of medium and large truck crashes was used (Zoloshnja and Miller, 2006). This study provides estimates of unit costs for highway crashes involving medium/heavy trucks by severity. Crash costs are broken out by truck type and severity of the crash, including no injury crashes, crashes with nonfatal injuries, and crashes with a fatality. The injury costs represent the present value, computed at a 4 percent discount rate, of all costs over the victims' expected life span that result from a crash. They include medically related costs, emergency services costs, property damage costs, lost productivity, and the monetized value of the pain, suffering, and quality of life that the family loses because of a death or injury. As expected, fatal crashes cost more than any other crashes. The cost estimates exclude mental health care costs for crash victims, roadside furniture repair costs, cargo delays, earnings lost by family and friends caring for the injured, and the value of schoolwork lost. Table 6-24 shows the study findings for one truck type and severity of the crash. The crash costs by severity for "truck tractor and one trailer" can be used to estimate the costs resulting from the additional crashes.

TABLE 6-24 Estimated Costs for Crashes Involving Truck Tractor with One Trailer, 2006

Severity of Crash	Estimated Crash Cost ($)
Crashes with injury	200,000
Fatal crashes	3,800,000

NOTE: Weighted average calculated from the three injury categories presented in Table 2 in Zaloshnja and Miller (2006): "possible injury," "nonincapacitating injury," and "incapacitating injury."

Effects of Incremental Changes in Weight

Certain fuel-saving technologies will add to vehicle weight, affecting operators' costs in three ways. First, transporting the extra weight itself increases fuel costs, partially offsetting the fuel savings the technologies allow. This effect was discussed in Chapters 4 and 5.

Second, in medium-duty truck applications, the extra weight may increase the loaded gross weight of some present Class 2 vehicles to over 10,000 lb and of some present Class 6 vehicles to more than 26,000 lb. Surpassing these weight thresholds will subject drivers and companies operating the vehicles to federal and state motor carrier safety regulations. The federal regulations apply to vehicles over 10,000 lb in commercial use that operate across state lines. Operators are required to register with DOT. The regulations cover record keeping, driver qualifications, driver hours of service, safety equipment, and other practices (Federal Motor Carrier Safety Administration, 2008). Many states have adopted similar regulations for vehicles used intrastate. Drivers operating vehicles over 26,000 lb must have a commercial drivers license (CDL) that meets federal standards.

A truck operator who has not previously been subject to these motor carrier safety regulations or to CDL requirements and is considering whether to adopt new vehicles with fuel-saving technologies and higher weight that would trigger the regulations will have several options. The operator may acquire the heavier vehicles and comply with the regulations or specify offsetting weight-saving equipment in order to stay under the threshold, or acquire smaller trucks than previously used. Vehicle manufacturers may decide to market new vehicle designs that facilitate the latter two choices. Any of these choices will increase the operator's truck transportation costs, and the operator will select the one with the least cost. Complying with the safety regulations may have benefits to the operator and to society that at least partially offset compliance costs.

Finally, in heavy-duty operations in which trucks are sometimes loaded to the 80,000-lb legal gross weight limit that applies on most major U.S. roads, and in operations in which trucks are sometimes loaded to axle weight limits (e.g., refuse haulers, dump trucks), the added weight of some fuel-saving devices (without concomitant vehicle weight-reducing materials) will reduce cargo capacity, increasing average cost per ton-mile and necessitating more vehicle-miles of travel to carry a given quantity of freight. In an operation in which trucks are almost always loaded to the gross weight or axle weight limit, the added cost will be proportional to the loss of payload. For example, the payload of a truck loaded near the 80,000-lb limit is about 50,000 lb, so an additional 500 lb of fuel-saving devices would reduce capacity by 1 percent and increase average cost per ton-mile by 1 percent in an application in which trucks are usually loaded to the gross weight limit. Fuel savings from the devices would at least partially offset this cost increase.

Most large trucks on the road are not loaded to the gross weight or axle weight limits. For example, 70 percent of traffic of five-axle tractor semitrailer mileage is at gross weights below 70,000 lb. For operations in which trucks are never or rarely loaded to the gross or axle weight limits, the loss of cargo weight capacity would not affect costs. For a typical fleet that operates sometimes at the weight limit and sometimes below, costs would be intermediate between the two extreme cases.

Fuel-saving technologies that add weight include the following:

- Engine efficiency improvements: turbocompound systems and waste heat recovery systems
- Hybrid power systems
- Aerodynamic fairings

In hybrid systems the added weight of batteries and other components may not be offset by downsizing of the internal combustion engine. European manufacturers report payload penalties of 100 to 200 kg for their medium-duty truck hybrids (Baker et al., 2009). One manufacturer of aerodynamic fairings reports that its trailer underside fairings weigh 150 to 230 lb and a trailer rear fairing 75 lb (Freight Wing, no date; Transport Canada, 2009). A complete aero package might add 500 lb.

Manufacturability and Product Development

As a final note, the committee was tasked to determine whether the fuel consumption reduction technologies discussed throughout this report could be efficiently integrated into the manufacturing process. The committee found no current studies or analyses suggesting that manufacturability was a major barrier to the integration of gas/diesel engine, hybrid, aero, tire, or other technologies in the *vehicle manufacturing process*. (Note: There may still be issues and barriers related to the *availability* and *manufacturing* of the technologies themselves.) This finding was examined and reinforced during the committee's site visits to engine, chassis, truck, bus, hybrid, trailer, and other manufacturers/users, including Allison Transmission, ArvinMeritor, Cummins, Great Dane, Ford, Eaton, Enova, WalMart, Navistar, Azure, Peterbilt, and PACCAR. These manufacturers were familiar with the technologies that could be introduced during the 2010-2015 time period, and they did not see any significant issues for their manufacturing processes. However, several manufacturers suggested that the larger challenge was aligning new technology deployment with their product development process (PDP) or cycle. The opportunities to integrate new technology designs into a PDP are limited, and sufficient time is needed for design and validation, customer acceptance, testing, and compliance strategy development. In addition, fuel consumption reduction technologies must be integrated with emission and safety technologies in compliance with regulations.

FINDINGS AND RECOMMENDATION

Direct Costs and Benefits

Finding 6-1. Since tractor-trailer trucks have relatively high fuel consumption, very high average vehicle miles traveled, and a large share of the overall truck market, it makes sense to put a priority on fuel consumption reduction. A given percentage reduction in this vehicle category will save more fuel than a matching percent improvement in any other vehicle category. In fact, the potential fuel savings in tractor-trailer trucks represents about half of the total possible fuel savings in all categories of medium- and heavy-duty vehicles.

Finding 6-2. The fuel consumption reduction potential for the tractor-trailer application, in the 2015 to 2020 time frame, is 50.5 percent at a cost of $84,600, which results in a capital cost per percent reduction (CCPPR) of $1,674 per percent fuel consumption reduction.

Finding 6-3. The fuel consumption reduction potential for Class 6 box and bucket trucks, in the 2015 to 2020 time frame, is 47.1 percent for box trucks and 49.6 percent for bucket trucks. The resulting cost for box trucks is $43,120 with a CCPPR of $915 per percent fuel saved, and the cost for bucket trucks is $49,870 with a CCPPR of $1,005 per percent fuel consumption reduction.

Finding 6-4. The fuel consumption reduction potential for the Class 2b pickup and van application, in the 2015 to 2020 time frame, is 44.5 percent at a cost of $14,710, which results in a CCPPR of $331 per percent fuel consumption reduction.

Finding 6-5. The fuel consumption reduction potential for the refuse truck, in the 2015 to 2020 time frame, is 38.4 percent at a cost of $50,800, which results in a CCPPR of $1,323 per percent fuel consumption reduction.

Finding 6-6. The fuel consumption reduction potential for transit bus applications, in the 2015 to 2020 time frame, is 47.8 percent at a cost of $250,400 (without subsidy), which results in a CCPPR of $5,232 per percent fuel consumption reduction.

Finding 6-7. The fuel consumption reduction potential for the motor coach application, in the 2015 to 2020 time frame, is 32 percent at a cost of $36,350, which results in a CCPPR of $1,136 per percent fuel consumption reduction.

Finding 6-8. Table 6-25 summarizes, for the seven vehicle applications studied, the fuel consumption reduction potential (from Chapters 4 and 5), the capital cost, and the cost-benefit for the 2015 to 2020 time frame, as stated in Findings 6-2 to 6-7.

TABLE 6-25 Summary of Potential Fuel Consumption Reduction, Cost, and Cost-Benefit

Vehicle Class	Fuel Consumption Reduction (%)	Capital Cost ($)	Cost-Effectiveness Metric: Dollars per Percent Fuel Saved	Dollars per Gallon Saved per Year	Breakeven Fuel Price[a] ($/gal)
Tractor-trailer	51	84,600	1,674	7.68	1.09
Class 6 box truck	47	43,120	915	29.30	4.17
Class 6 bucket truck	50	49,870	1,006	37.80	5.38
Class 2b pickup	45	14,710	331	33.73	4.81
Refuse truck	38	50,800	1,323	18.90	2.69
Transit bus	48	250,400	5,232	48.05	6.84
Motor coach	32	36,350	1,135	11.59	1.65

NOTE: Values shown are for one set of input assumptions. Results will vary depending on these assumptions.

[a]Calculated assuming a 7 percent discount rate and a 10-year life, excluding incremental operating and maintenance costs associated with the technologies.

SOURCE: Adapted from TIAX (2009).

The tractor trailer offers the best cost-benefit potential, followed by the motor coach. The refuse hauler costs more than twice as much per gallon of fuel saved, and the other vehicle classes are even more expensive.

Indirect Costs and Benefits

Finding 6-9. A number of indirect effects and unintended consequences associated with regulations aimed at reducing fuel consumption in the trucking sector can be important. In particular, regulators should consider the following effects in the development of any regulatory proposals: rate of replacement of older vehicles (fleet turnover impacts), increased ton-miles shipped due to the lower cost of shipping (rebound effect), purchasing one class of vehicle rather than another in response to a regulatory change (vehicle class shifting), environmental co-benefits and costs, congestion, safety, and incremental weight impacts.

Finding 6-10. Consumer buying in anticipation of new regulations (pre-buy) and retention of older vehicles can slow the rate of fleet turnover and the rate at which regulatory standards can affect fleetwide fuel consumption. The committee believes the effects will be transient and reduced to the extent that fuel consumption savings offset incremental purchase costs. Government incentives in the form of tax credits or excise tax reductions with a sunset date could be used to help minimize anticipated pre-buy/low-buy fluctuations in the future. Regulators must be cognizant of these potential effects and should consider regulatory mechanisms that minimize these potential distortions.

Finding 6-11. Elasticity estimates vary over a wide range, and it is not possible to calculate with a great deal of confidence what the magnitude of the "rebound" effect is for heavy-duty trucks. The rebound effect measures the increase in ton-miles shipped resulting from a reduction in the cost of shipping. A rebound effect nevertheless likely exists that will partially offset fuel consumption declines due to the adoption of new cost-effective technologies. To the extent the regulation pushes beyond the private cost-effective point, the rebound effect will be reversed. Estimates of fuel savings from regulatory standards will be somewhat misestimated if the rebound effect is not considered.

Finding 6-12. Standards that differentially affect the capital and operating costs of individual vehicle classes can cause purchase of vehicles that are not optimized for particular operating conditions. The complexity of truck use and the variability of duty cycles increase the probability of these unintended consequences.

Finding 6-13. Reduced fuel consumption through fuel efficiency technologies in medium- and heavy-duty vehicles will likely reduce emissions of criteria pollutants. Efficiency improvements achieved by improved aerodynamics, tire rolling resistance, and weight reductions will translate into lower tailpipe emissions as well.

Finding 6-14. To the extent that regulations alter the number of shipments and VMT, there will be some safety and congestion impacts. A more detailed assessment of these impacts would be needed based on the type of regulation that may be put forward by the U.S. Department of Transportation.

Finding 6-15. There are potential safety issues associated with particular fuel reduction technologies. Examples are hybrids that use high-voltage batteries or aerodynamic fairings that may detach from trucks on the road.

Finding 6-16. Some fuel-efficienc-improving technologies will add weight to vehicles and push those vehicles over federal threshold weights, thereby triggering new operational conditions and affecting, in turn, vehicle purchase decisions. More research is needed to assess the significance of this potential impact.

Finding 6-17. Some fuel-efficiency-improving technologies will reduce cargo capacity for trucks that are currently "weighed out" and will therefore force additional trucks onto the road. More research is needed to assess the significance of this potential impact.

Finding 6-18. The committee found no current studies or analyses suggesting that manufacturability was a major barrier to the integration of gas/diesel engine, hybrid, aerodynamic, tire, or other technologies in the *vehicle manufacturing process*. However, there may be challenges with integrating new technologies into manufacturers' product development processes, and sufficient time is needed for design and validation, customer acceptance, testing, and compliance strategy development.

Recommendation 6-1. The National Highway Traffic Safety Administration, in its study, should do an economic/payback analysis based on fuel usage by application and different fuel price scenarios. Operating and maintenance costs should be part of any study.

BIBLIOGRAPHY

American Bus Association. 2006. Commercial Bus Emissions Characterization & Idle Reduction: Idle & Urban Cycle Test Results. Prepared by the U.S. Department of Energy, U.S. Environmental Protection Agency, and U.S. Federal Highway Administration. Available at http://www.buses.org/files/download/motorcoach_idling_study.pdf.

AHUA (American Highway Users Alliance). 2004.Unclogging America's Arteries, 1999-2004. Available at www.highways.org.

American Transportation Research Institute. 2008. An Analysis of the Operational Costs of Trucking. Total Cost Summary. P. 18, Table 4. Arlington, Va.

Baker, H., R. Cornwell, E. Koehler, and J. Patterson. 2009. Review of Low Carbon Technologies for Heavy Goods Vehicles—Annex 1. Prepared for the U.S. Department of Transportation by Ricardo PLC, West Sussex, U.K. June.

Barnitt, R. 2008. BAE/Orion Hybrid Electric Buses at New York City Transit: A Generational Comparison. Technical Report NREL/TP-540-42217. March. Available at http://www1.eere.energy.gov/vehiclesandfuels/avta/pdfs/heavy/bae-orion_he_buses.pdf.

Barth, M., and K. Boriboonsomsin. 2008. Real-world CO_2 impacts of traffic congestion. *Transportation Research Record.* Submitted March 31, 2008. Washington, D.C.

Battelle Memorial Institute. 2005. Traffic Operations and Truck Size and Weight Regulations, Working Paper 6. Prepared for the U.S. Federal Highway Administration. February. Columbus, Ohio.

Bus Engineering. 2008. Comparison of WMATA's newest bus fleet. April 16.

Calstart HTUF Dialog, October 2008. Available at http://www.calstart.org/HomePage.aspx?aspxerrorpath=/Homepage.aspx/programs/htuf/hybrid_dialog_search.php.

CARB (California Air Resources Board). 2008. Final Regulation Order to Reduce Greenhouse Gas Emissions from Heavy-Duty Vehicles. Available at http://www.arb.ca.gov/regact/2008/ghghdv08/ghgfro.pdf.

Christidis, Panavotis, and Guillaume Leduc. 2009. Longer and Heavier Vehicles for Freight Transport. JRCC52005. European Commission Joint Research Center, Institute for Prospective Technology Studies.

Cooper, C. 2009. Reducing Heavy-Duty Long Haul Combination Truck Fuel Consumption and CO_2 Emissions. Boston: NESCCAF.

Dennis, Scott M. 1988. The Intermodal Competition Model. Association of American Railroads. September.

DOE (U.S. Department of Energy), Energy Information Administration (EIA). 2009. Assumptions to the Annual Energy Outlook 2009. Available at http://www.eia.doe.gov/oiaf/aeo/assumption/transportation.html.

EPA (U.S. Environmental Protection Agency). 2006. Motor Coach Idling Field Observation Study for Washington, DC, Metro Area. Transportation and Regional Programs Division and SmartWay Transport Partnership Group, Office of Transportation and Air Quality, U.S. Environmental Protection Agency. Available at http://epa.gov/smartway/documents/420s06004.pdf.

Federal Motor Carrier Safety Administration. 2007. Commercial Vehicle Facts. November. Available at http://www.fmcsa.dot.gov/facts-research/facts-figures/analysis-statistics/cmvfacts.htm.

Federal Motor Carrier Safety Administration. 2008. A Motor Carrier's Guide to Improving Highway Safety. October.

Freight Wing, Inc. No date. Freight Wing Trailer Aerodynamics. Available at http://www.freightwing.com/.

IFEU (Institut für Energieund Umweltforschung, Heidelberg GmbH). 2003. Energy Savings by Lightweighting. Heidelberg, Germany: IFEU.

Jones, J., F. Nix, and C. Schwier. 1990. The Impact of Changes in Road User Charges on Canadian Railways. Prepared for Transport Canada by the Canadian Institute of Guided Ground Transport, Kingston, Ontario. September.

Lowe, M., et al., Center on Globalization Governance and Competiveness, Duke University. 2009. Manufacturing Climate Solutions, Carbon Reducing Technologies and U.S. Jobs. June 10. Available at http://www.cggc.duke.edu/environment/climatesolutions/greeneconomy_Ch9_HybridDrivetrainsforTrucks.pdf.

NESCCAF, ICCT, Southwest Research Institute, and TIAX. 2009. Reducing Heavy-Duty Long Haul Combination Truck Fuel Consumption and CO_2 Emissions. Washington, D.C.: International Council on Clean Transportation (ICCT). October.

Parry, I. 2006. How Should Heavy-Duty Trucks Be Taxed? Washington, D.C.: Resources for the Future. April.

Ranganathan, S. 2006. Hybrid Buses Costs and Benefits. Washington, D.C.: Environmental and Energy Study Institute. Available at http://www.eesi.org/files/eesi_hybrid_bus_032007.pdf.

Research and Markets. 2009. Lithium-Ion and Nickel-Metal Hydride Batteries, Lithium, Rare Earth Lanthanum and the Future of Hybrid Electric Vehicles, Plug-in Hybrid Electric Vehicles and Electric Vehicles 2009-2020. Brochure. Available at www.researchandmarkets.com/. Accessed November 2, 2009.

Schrank, D., and T. Lomax. 2005. The 2005 Urban Mobility Report. Texas Transportation Institute, Texas A&M University System. Available at http://mobility/tamu.edu.

Schubert, R., and M. Kromer. 2008. Heavy-Duty Truck Retrofit Technology: Assessment and Regulatory Approach. Prepared by TIAX, LLC, for the Union of Concerned Scientists, September 12.

TTI (Texas Transportation Institute). 2007. 2007 Annual Urban Mobility Report, Appendix A. College Station, Tex: TTI.

TIAX LLC. 2009. Assessment of Fuel Economy Technologies for Medium- and Heavy-Duty Vehicles. Final Report for the National Academy of Sciences. Sept. 16.

Transport Canada. 2009. Freight Wing Fleet Trial Program on Aerodynamic Fairings. Modified June 26. Available at http://www.tc.gc.ca/programs/environment/ecofreight/casestudies/freightwing-eng.htm.

TRB (Transportation Research Board). 2000. Highway Capacity Manual. Washington, D.C.: TRB.

TRB. 2009. Transportation Energy Data Book. Edition 28-2009, Table 5.14. Washington, D.C.: TRB.

Walsh, J. 2009. Mack delivers hybrid refuse truck to New York City. Press release, May 28.

Zaloshnja, E., and T. Miller. 2006. Unit Costs of Medium and Heavy Truck Crashes. Final Report. Prepared for the Federal Motor Carrier Safety Administration. December. Available from National Technical Information Service, Springfield, Va. Also available at www.ai.volpe.dot.gov/.

7

Alternative Approaches to Reducing Fuel Consumption in Medium- and Heavy-Duty Vehicles

OVERVIEW

The preceding chapters in this report focus primarily on the technical feasibility and cost-effectiveness of technologies aimed at improving fuel efficiency in medium- and heavy-duty vehicles. The technologies discussed represent options available for meeting a new federal fuel efficiency standard based on metrics previously discussed. However, technology alone is not the only approach that manufacturers, carriers, and operators have at their disposal to improve vehicle efficiency. Nor are fuel efficiency standards the only mechanisms available to policymakers aiming to incentivize more efficient movement of passengers and goods. The purpose of this chapter is to present alternative approaches for improving the fuel efficiency of medium- and heavy-duty-vehicle (MHDV) operations.

This chapter identifies the following set of alternative approaches and discusses their pros and cons:

1. Changing fuel price signals
    - Fuel taxes
    - Cap-and-trade: Implications for trucking
2. Technology-specific mandates and subsidies
    - Technology mandates
    - Equipment subsidies
    - Low-carbon fuel standards
3. Alternative/complementary regulations
    - Emissions limits
    - Size and weight limits
    - Mandatory speed limits
4. Other complementary approaches
    - Intelligent transportation systems
    - Construction of exclusive truck lanes
    - Congestion pricing
    - Driver training
    - Intermodal operations

In some cases these alternative approaches can be complementary to fuel efficiency standards. In other cases these alternatives could substitute for fuel efficiency standards. For each approach a short description is provided, as well as the advantages of the approach, the disadvantages of the approach, and any potential implementation issues associated with its application. The committee recognizes that the alternative approaches included herein are complex and require a great deal more study beyond the scope of this report, particularly with respect to the potential impact on fuel efficiency improvements that each approach may have.

CHANGING FUEL PRICE SIGNALS

Fuel Taxes

The rationale for government imposition of fuel-savings standards is that trucking firms decide to implement fuel-saving technologies based on the market price of fuel, which does not include the external costs associated with climate change and fuel security. As a result, firms do not implement technologies that are socially efficient since the private return is too low. This social inefficiency can be corrected most efficiently by imposing a tax equal to the external costs. If, for example, the problem being addressed is carbon emissions, all transportation fuels should bear a tax proportional to their carbon content. If the price of fuel were higher to reflect the external costs, more fuel-saving technologies would be adopted. In addition, the higher price of fuel would lead to more fuel-efficient operations. An example is provided by the European experience. Fuel prices in Europe are significantly higher than in the United States because of higher taxes. Higher prices have not yet led to adoption in the European truck fleet of many of the more advanced technologies such as bottoming cycle and ultralight structures. However, the committee was told by one major international engine manufacturer that it decided to develop a new engine with turbocompounding for worldwide sale based on the expected payback from the higher fuel prices in Europe. The turbocompounding in this application is expected to reduce fuel consumption by 3 to 5 percent. The European truck fleet also uses more aerodynamic fairings, relies more on driver-training for fuel-saving operations, and uses almost exclusively diesel engines, except in the lower range of Class 1. Finally, higher fuel costs would be passed on in the form of higher truck shipping rates, reducing the demand for shipping by truck and the diversion of truck shipments to other modes, particularly rail, leading to additional fuel savings.

Advantages

A tax affects the incentives associated with all of the elements in the freight transportation system. It provides incentives for technology adoption and operational efficiencies (such as reduced idling, improved driver education, etc.). These actions, in many cases, offer significant fuel savings. In addition, a tax affects the utilization of vehicles already on the road, while fuel consumption standards typically affect only new vehicles and can be implemented only slowly over time as the vehicle fleet transitions to the more fuel-efficient vehicles.

Fuel taxes would contribute toward achieving more efficient outcomes in additional ways. To the degree that demand for transportation is elastic, as discussed in Chapter 6, a fuel tax, by raising the cost of shipping, will tend to lower miles driven, thereby reducing congestion, accidents, and other driving-related negative externalities. Furthermore, increased fuel taxes would augment the highway trust fund, permitting the construction of improved transportation options, or at least offsetting the decline in revenues from reduced fuel consumption by more efficient vehicles.

Most importantly, a fuel tax economizes on the information needed by regulators. Maximizing economic efficiency requires that the marginal cost of reducing fuel consumption be the same for all vehicle manufacturers and be equal to the marginal cost of actions that vehicle operators can take to reduce fuel use. In this way, the low-cost means of reducing fuel consumption are exploited before utilizing higher-cost reduction technologies or techniques. The information needed to find the lowest-cost pattern of fuel consumption reductions places large demands on regulators when the manufacturing cost of a technology varies among manufacturers, the in-use cost varies depending on the specific use, and other measures such as driving and truck-routing procedures exist that can reduce energy consumption. A fuel tax provides incentives for private firms to take action and relies on the individual knowledge and incentives manufacturers and shippers have to reduce costs. The trucking industry is a competitive one, and the committee has found that the companies are very focused on reducing fuel costs, subject to the requirements of delivering the freight or accomplishing the particular work requirements. Furthermore, the industry is highly varied, with trucks utilized in very different tasks from long-haul freight operations to postal delivery to electric utility trucks to trash removal. Given a higher fuel price because of the tax, firms will optimize their operations to realize the greatest fuel savings while still performing the required tasks. Setting standards instead requires that regulators consider, in addition to technology options, the complexity of tasks to be accomplished, the variety of conditions under which trucks will be operated, and the changing uses over the life of the truck. A mandated fuel efficiency standard, rather than a market-based solution such as a tax, has a higher probability of counterproductive unintended consequences because of this complexity.

Finally, a fuel tax is a clear statement of the additional costs being imposed on the truck sector to accomplish societal aims, fostering transparency in the public policy process. In contrast, fuel efficiency standards can often obscure the costs to the public.

Disadvantages

Taxes involve setting a price signal and letting industry choose the most efficient means of reducing fuel consumption. In the transportation sector, setting taxes is complicated because all fuels must be appropriately priced to avoid distortions across fuel markets, for example between diesel and gasoline. The response to the tax, however, is uncertain and empirical estimates of elasticities are not precise enough to predict the resultant fuel savings. However, setting standards also involves uncertainties as to fuel savings and operational costs due to indirect effects, as discussed in Chapter 6.

In addition, a fuel tax may not provide sufficient incentive for technology development, particularly given the political difficulties associated with implementing a tax large enough to have significant incentive effects. Last, a fuel tax, while leading to immediate savings from utilization in the existing fleet, will impose costs on the fleet that were not anticipated when the investments in technology and vehicles were put in place and is likely to raise issues of equity. These issues could be accommodated by a scheduled phase-in of taxes.

Other Considerations

A variable fuel tax could be used to reduce the volatility in prices faced by trucking firms and manufacturers. For instance, a fuel tax could be implemented in a manner that would provide a price "floor" for fuel. This would reduce uncertainty and allow a clearer signal for investment in fuel-saving technologies. However, such a variable tax would create an uncertainty in the amount of dollars flowing to the highway trust fund, thus jeopardizing federal, state, and local highway construction projects. Last, a fuel tax aimed at reducing fuel consumption of heavy-duty vehicles needs to be considered in light of its impacts on the light-duty vehicle and non-road sectors.

Cap and Trade: Implications for Trucking

At the time of this writing Congress is considering enacting a "cap-and-trade" system to control the emissions of gases that contribute to climate change.[1] Such a system would cap emissions at a predetermined level and issue a number of permits equal to that cap. Any controlled entity such as electric utilities or oil refineries would have to surrender a permit for each ton of CO_2 emitted. The permits could be traded, so that an entity desiring to increase production and thereby emit additional tons of CO_2 or other global-warming gasses could buy additional permits from a permit holder willing to sell. The market price of permits will be reflected in the cost of production and passed on to the ultimate consumer. In the trucking sector the permit price would have the same effect as a tax on fuel.

Advantages

The cap-and-trade system introduces a price on CO_2 emissions, as a tax would, and provides incentives for the adoption of fuel-saving technologies as well as for the adoption of operational methods to save fuel. Applying this system over the economy as a whole it can lead to an efficient pattern of emission reduction. For example, if it is cheaper to reduce emissions from electric utilities than from another industrial plant, the electric utility will cut emissions and sell the permits it no longer needs to the industrial emitter. The industrial emitter will be willing to buy those permits as long as it is cheaper than reducing its emissions by technology or operational changes.

Once a cap is in place, regulators may have less of a need to establish fuel consumption standards for a particular covered sector. This is because any reduction in CO_2 emissions coming from trucking, for example, will result in more emissions elsewhere among covered entities, so that the total emissions remain unchanged. Similar to the case with fuel taxes, this economizes on the information regulators need about technology, operating conditions, and duty cycles for trucking operations. Under a cap-and-trade system individual firms make the decisions based on their knowledge of the operations and the price of carbon emissions.

Disadvantages

By setting a cap, the ultimate emissions are known, but the cost of achieving the cap is uncertain. The cost will emerge in the market as firms consider technological and operational changes versus the cost of purchasing permits. The price will fluctuate as the demand for permits will change in response to technological developments, to changes in expectations about future economic growth and hence the demand for permits, to weather variation, and even to interest rate changes.

Introduction of a cap-and-trade system will increase governmental administrative burdens for monitoring and policing the system, supervising markets in permits and derivatives that will emerge in financial markets. Similar to the case with fuel taxes, there may be concern that the increase in fuel prices, given political limitations on how tight a cap can be legislated, will be too small to have major impacts in generating change in technology adoption.

A cap-and-trade system is designed to cap carbon or other global-warming gases, not oil consumption. While higher carbon prices will be passed on to fuel prices and reduce oil consumption, oil security concerns may require additional measures. While fuel consumption standards would reduce oil consumption, a cap-and-trade system could accommodate an additional charge within the system so as to provide the additional incentive to save oil. For example, it could be required that 1.25 tons of CO_2 emissions coming from oil be traded for 1 ton of coal emissions.

TECHNOLOGY-SPECIFIC MANDATES AND SUBSIDIES

Technology Mandates

A technology mandate would be a regulation requiring operators of medium- and heavy-duty vehicles to purchase and use specified designs or models of vehicles or components. The required vehicles and equipment would be those embodying fuel-saving technologies. The regulator would establish a certification process to identify energy-efficient vehicles and components and would publish lists of complying models. The California Heavy-Duty Vehicle Greenhouse Gas Emission Reduction Regulation is the most relevant example of such a regulation.

In December 2008 the California Air Resources Board (CARB) adopted a regulation requiring certain operators of certain kinds of trucks to either use EPA SmartWay-certified tractors and trailers or to retrofit their vehicles with SmartWay-verified technologies. The SmartWay vehicles and equipment save fuel primarily through improved tractor and trailer aerodynamics and the use of low rolling resistance tires. The regulation applies to 53-ft or longer van trailers and to tractors that pull these trailers in California. Tractors that drive less than 50,000 miles per year are exempt, and tractors and trailers that operate within a 100-mile radius from a home base are exempt from the aerodynamics requirements.

Operators who choose to comply by retrofitting must equip their trailers with low rolling resistance tires and with aerodynamic fairings or other SmartWay-approved technologies. The technologies required will depend on a percentage greenhouse gas emissions reduction assigned to each device by CARB.

From 2010, when the rule goes into effect, through 2020,

[1]H.R. 2454 was passed by the House of Representatives on June 26, 2009. The bill sets a cap on CO_2 emissions that covers about 85 percent of total U.S. emissions and includes domestic oil refiners.

CARB expects the regulation to reduce diesel fuel consumption by 750 million gallons in the state and 5 billion gallons nationwide. For comparison, diesel motor fuel purchases in California in 2007 were 3.2 billion gallons (CARB, n.d., 2009).

Advantages

Equipment mandates are seen as a simpler alternative in circumstances where a performance standard (e.g., a gallons-per-ton-mile or gallons-per-cubic-foot-mile fuel consumption standard) would be difficult to apply or enforce. From an enforcement perspective, it is often easier for regulators to confirm a manufacturer's or user's adoption of a technology mandate (e.g., side fairings) than to determine whether a performance metric is being achieved.

Disadvantages

Under a performance standard, the regulated party is free to adopt any combination of measures that meets the standard. Each party can be expected to adopt the most cost-effective approach for the application. Equipment mandates lack this flexibility and therefore may increase regulatory compliance costs. For example, under the California regulation, the fuel-saving benefits of required trailer fairings will vary greatly from user to user. It is most likely that some users could have obtained greater fuel savings by some alternative practice at a cost equal to or lower than the cost of the fairings. In addition, a technology mandate that is not appropriately "tuned" to the characteristics or operational aspects of a particular vehicle or class of vehicles may not achieve desired benefits due to incompatibility between the technology and vehicle use. In such cases, especially when characteristics or operations are uncertain, an emissions standard may perform better.

Equipment Subsidies

Another approach to encouraging technology adoption is to offer government subsidies. The federal government and the states have offered a variety of financial incentives to firms and individuals for purchases that reduce energy consumption. The forms of incentives include tax credits, cash grants, and credit assistance. For individuals, federal incentives for the purchase of energy-saving home improvements and for of hybrid cars are well-known examples (EPA, 2009a).

Existing programs applicable to medium- and heavy-vehicle target primarily reductions in criteria pollutants rather than GHG emissions or fuel economy, but some of these programs are also intended to promote fuel savings and the program structures could be applicable to fuel economy incentive programs. Examples include the following:

- In California the Carl Moyer Memorial Air Quality Standards Attainment Program is a grant program that subsidizes replacement or retrofit of diesel engines in heavy trucks, locomotives, and other applications. Originally conceived to reduce emissions of criteria toxic pollutants, the program is described also as a means to reduce GHG emissions, mainly by subsidizing hybrid applications. The program has been disbursing about $14 million annually since 1998. Idling reduction retrofits are eligible for 100 percent funding, and fleet modernization (new vehicle purchases replacing older, more polluting equipment) for up to 80 percent funding. However, the subsidized equipment must exceed the emissions limits imposed on all vehicles by law, and as idling and emissions regulations have become more stringent, opportunities for truck operators to qualify for the grants have been reduced (CARB, 2008).
- Federal grants are available under the EPA and the Clean Fuels Grant Program for hybrid deployment in trucks and urban buses (see Chapter 6).
- A second California program is subsidizing replacement of older trucks used by drayage operators at the state's seaports, for the purpose of reducing pollutant emissions. The program is funded at $400 million (CARB, 2008).
- Several other states have offered financial incentive programs for installation of idle reduction devices (Leavitt, 2005).
- American Recovery and Reinvestment Act of 2009 (P.L. 111-5) provided $300 million for federal and state programs to pay for diesel emissions reduction. A share of this will be directly available to truck operators as financial assistance for equipment replacement (e.g., through low-interest loans). The funds are administered by EPA, in part through the SmartWay Clean Diesel Finance Program (EPA, 2009b).
- The Energy Improvement and Extension Act of 2008 (P.L. 110-343) give EPA the authority to exclude exempt idle reduction devices from the 12 percent federal excise tax on new truck purchases. EPA has certified 70 devices as eligible. The exemption is estimated to be worth $700 to $1,000 per truck to some purchasers (EPA, n.d.; Miller, 2009).

Advantages

Financial incentives may be most effective in encouraging early adoption of new technologies when the benefits are uncertain, the technology is not widely known to users, and the cost may be high because the technology is in limited production. The subsidy then transfers some of the risk from the early adopters to the public. Subsidies are also used, as the examples above illustrate, where fairness to small businesses is a concern. In the California Carl Moyers and drayage operator programs, part of the motivation was to avoid disproportionate harm to small businesses from stringent new regulations.

Disadvantages

Subsidies are best seen as a possible transition strategy rather than a major permanent feature of pollution control and energy-saving programs. If a significant U.S. greenhouse gas emissions reduction program is enacted, the cost burden of compliance will be ubiquitous and subsidizing these costs will be impractical except in very limited circumstances.

Low-Carbon Fuel Standard

The regulation of fuel quality and "chemistry" is not novel. The EPA currently regulates sulfur content for on-road diesel fuel, which was reduced from a limit of 500 ppm sulfur pre-2007 (40 CFR 80.29) to 15 ppm sulfur now (40 CFR 80.520). Building off the fuel standard approach, the State of California has moved forward with a low-carbon fuel standard (LCFS) that regulates the average carbon content of fuel used in the transportation sector. The LCFS has target carbon content values that will help the state meet GHG reduction goals over the next decade and beyond. As constructed in California, the LCFS is aimed at regulating fuel providers (e.g., oil refiners) and will ultimately require the introduction of larger percentages of alternative fuels in the transportation sector than would be otherwise expected. These alternative fuels include biofuels, natural gas, and electricity, to name a few. Other states and regions of the country (e.g., the northeastern states) are also considering implementing their own LCFSs.

Advantages

The primary advantage of an LCFS is that (if accounted for properly) it ensures a certain level of GHG reductions relative to a non-LCFS benchmark. A second advantage of a LCFS (if constructed in a similar fashion as the California approach) is that it regulates only a small body of entities (fuel providers) and so regulatory oversight is somewhat simplified. A third advantage is that an LCFS provides incentives for the research, development, and deployment of alternative fuels for transportation.

Disadvantages

The lower energy density of the fuel means that more gallons are used, larger fuel tanks will be required on trucks, road use taxes applied on a per-gallon basis will go up, and carbon emissions associated with transportation of the fuel will rise. Another disadvantage of an LCFS is the difficulty in assuring that life-cycle emissions (including those from upstream feedstock and fuel production) in fact lead to overall GHG reductions. New research in the biofuels area has asserted that direct and indirect land use changes associated with biomass feedstock production may, in fact, increase overall global GHG emissions if not done properly.

Implementation Issues

The greatest issue facing the implementation of an LCFS is in constructing an appropriate metric for measuring total fuel-cycle carbon emissions from fuel production. This includes not only biofuel production but also nonconventional fossil fuels, such as petroleum from tar sands or shale. Other implementation issues have to do with administrating, monitoring, and validating an LCFS measurement claim.

ALTERNATIVE AND COMPLEMENTARY REGULATIONS

Emission Standards

A carbon dioxide (CO_2) emissions standard may incentivize more efficient engine operations since CO_2 is directly related to the amount (and type) of fuel burned in the engine. A CO_2 emissions standard could work similarly to a fuel efficiency standard. However, one of the advantages would be to allow consideration of nonfuel-saving actions that could lower CO_2. For example, the introduction of alternative, low-carbon fuels could be an option for meeting the standard. Typical diesel fuel is ~86 percent carbon by mass, while natural gas is only ~75 percent carbon by mass. Therefore, two vehicles can achieve the same fuel efficiency, yet one operated on natural gas would have a lower CO_2 emissions rate. Biofuels could also be addressed in this manner through accounting for carbon uptake in the feedstock used to produce the fuel, and this is discussed in more detail below.

Truck Size and Weight Mandates

Motor vehicle weights and dimensions are governed by a complex mix of federal and state regulations. The main provisions of the federal regulations are as follows:

- Maximum gross weight of vehicle on interstate highways: 80,000 lb.
- Maximum axle weight on interstate highways: 20,000 lb on a single axle; 34,000 lb on a tandem axle
- Maximum weight determined by the number and spacing of axles (the "federal and state bridge formula")
- Width of vehicles: states must allow 102 in. on the National Network for Large Trucks (interstates plus 160,000 miles of other main roads)
- Trailer length and numbers: states must allow single trailers at least 48 ft in length and tractors pulling two 28-ft trailers on the national network.

Federal law forbids the states to impose more restrictive

limits on roads where the federal limits apply, but a grandfather provision allows preexisting, more liberal state limits to remain in effect. States set limits on roads not covered by federal law, issue permits exempting vehicles from the limits under specified circumstances, and are responsible for enforcement. Exemptions and exceptions from nominal limits are numerous, and enforcement often is imperfect.

The regulations have been justified as serving a variety of purposes. The original state regulations (dating from the early 20th century) and the first federal regulations (dating from 1956, when the present federal aid highway program was created) served to fix design parameters for road construction. The 1983 federal preemption of state regulations more restrictive than the federal limits on the Interstates was economically motivated, to reduce the costs of interstate commerce. The most recent federal action, a 1991 law that blocked the states from allowing expanded use of longer combination vehicles (multi-trailer vehicles longer than the federally sanctioned twin-28-ft-trailer combination) was justified as a safety measure (TRB, 2002).

The regulations have important economic consequences. They influence the cost of truck transportation to shippers and the costs of highway construction and maintenance, and probably influence highway accident losses, although in complex ways. They affect international commerce (U.S. limits differ from those of Canada and Mexico, and containers shipped in international trade often are not consistent with U.S. regulations) and railroads' profitability and market share. Proposals to change the weight regulations always are controversial. Historically, liberalization usually has been opposed by the railroads, certain safety groups, some states, unionized drivers, and some carriers. Liberalization is supported by shippers, some carriers, and some states. Several detailed studies by DOT (2000), the Transportation Research Board (TRB, 1990a,b, 2002), and others (e.g., in Canada; RTAC, 1986) have examined the costs and benefits of alternative size and weight regulations.

Advantages

Historical experience and prospective studies indicate that liberalizing size and weight regulations (i.e., allowing vehicles with greater cargo volume capacity and/or greater cargo weight capacity) could significantly reduce fuel consumption in freight transportation and also reduce total shipper costs. For example, if all loaded trucks carried the maximum legal payload weight at all times, the reduction in vehicle-miles of truck travel would be inversely proportional to the increase in payload. (Percent fuel savings, however, would be less than the percent mileage reduction because the heavier trucks would consume more fuel per mile of travel. For example, fuel consumption per mile for a class 8 truck increases by roughly 5 percent for every 10,000 lb increase in weight [Greszler, 2009]).

As a hypothetical example, consider a fleet of trucks hauling coal, fully loaded, from a mine to a rail head and returning empty. If a change in the legal weight limit allowed the operator to increase each truck's payload by 50 percent, from 50,000 to 75,000 lb, truck-miles to haul a day's output of coal would decrease by 33 percent. However, the trucks would consume about 12 percent more fuel per mile when loaded, or 6 percent more on the round trip. The fuel savings would be 100 percent × [1 – (1.06 × 50/75)] = ~29 percent.

Similarly, an increase in the maximum legal volume capacity of trailers would allow a nearly inversely proportional decrease in truck-miles of travel in a hauling operation in which all loaded vehicles carried payloads that utilized their full volume capacity. A trucking industry study estimated that a 97,000 lb six-axle tractor-semitrailer (a vehicle that industry groups have advocated legalizing in the United States) applied in a fully weight-constrained operation will consume 15 percent less fuel per ton-mile than an 80,000 lb. tractor-semitrailer and a turnpike double (twin 48-ft trailers) in a fully volume-constrained operation will consume 28 percent less fuel per ton-mile than a single-trailer combination (Tunnell, 2008).

However, an operation in which trucks frequently operate at maximum volume capacity may gain little advantage from an increase in the weight limit, and operations that normally are weight constrained may gain no advantage from an increase in legal trailer dimensions. Also, operations in which trucks frequently travel with partial loads in order to meet delivery schedules and operations in which trucks make multiple stops to partially load or unload will not be able to fully utilize an increase in the legal maximum capacity. Consequently, in practice, a hypothetical change in size and weight limits that increased both maximum volume and weight capacity by 50 percent would yield less than a 33 percent reduction in truck-miles for the entire fleet.

A high percentage of trucks on the road at any time are empty or are loaded to less than either their weight or volumetric capacity limit. In the 1990s less than 50 percent of VMT among all five-axle tractor-semitrailers was driven by trucks with 55,000 lb or less operating weight (see Figure 7-1).

The effect of a change in size and weight limits on fuel consumption in a particular trucking operation will depend on the characteristics of the operation and on the details of the regulatory change. Consequently the impact of a proposed change on VMT or fuel consumption is difficult to project, and past estimates have varied widely. Table 7-1 shows some illustrative projections of fuel savings.

Each of the studies noted in Table 7-1 considered induced freight traffic and diversion from rail in their estimates. The relatively small impacts estimated in the TRB studies reflect those studies' less liberal hypothesized limit changes and their conclusion that short heavy double trailer vehicles would be attractive only in a limited range of applications.

A more recent study (Woodrooffe et al., 2009) surveyed 100 companies that operate private fleets about the potential

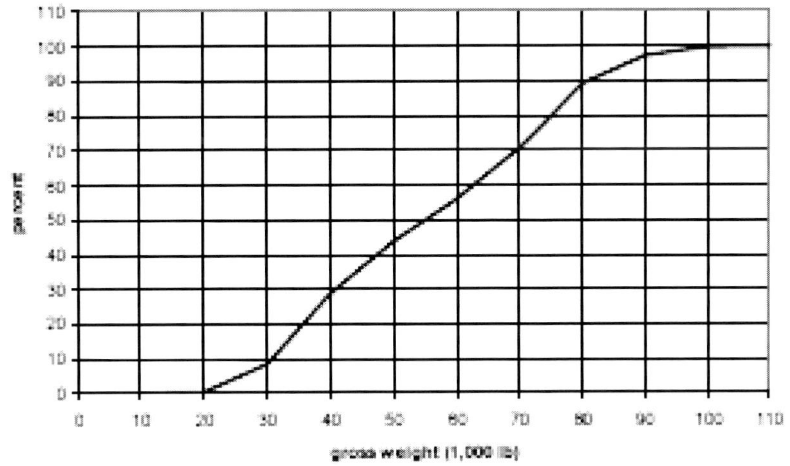

FIGURE 7-1 Five-axle tractor-semi vehicle-miles traveled by operating weight (cumulative percentage). SOURCE: FHWA (1997).

benefits of liberalized size and weight limits. In-depth interviews were conducted with seven companies as case studies. Two of the companies reported that because they carry low-density cargo, their trucks always reach volume capacity at a gross weight lower than the present limit. The other five companies at least occasionally load their trucks to the weight limit. These five were asked to report the percentage fuel consumption reduction they would expect if the gross weight limit were increased sufficiently to add 14,000 lb to cargo capacity. The median estimate of fuel consumption reduction was 23 percent. This prediction appears optimistic, since it could be attained only if all trucks began all their runs fully loaded to the weight limit and the heavier trucks had the same rate of fuel consumption per mile as trucks in present operations.

The impacts of past U.S. limit revisions have never been systematically monitored. In the United Kingdom the consequences of a recent change in the weight limit have been measured (McKinnon, 2005). In 2001, the U.K. weight limit was raised from 41 to 44 metric tons, allowing a 12 percent increase in maximum payload weight. Dimensional limits were not changed. Extrapolating the first two years of data on trends in truck travel and weight distributions, the study estimated that the eventual net effect would be a reduction of 170 million vehicle-kilometers/year by 2007, representing a 0.6 percent reduction in travel of freight vehicles. The percentage change in fuel consumption would be somewhat less. The estimate includes an allowance for road traffic generated by the reduced cost of truck transport. The impact has been small because a large fraction of freight vehicles do not operate at the weight limit. The characteristics of truck travel in the United Kingdom are different from those in the United States, but this example illustrates that the increase in fuel efficiency from a weight limit increase can be much less than the increase in maximum cargo capacity.

A 2002 TRB committee (TRB, 2002) reviewed the estimates of costs and benefits of revisions to size and weight limits in the past DOT, TRB, and Canadian studies. It concluded that liberalizing the regulations would reduce shippers' freight transportation costs. Highway agencies costs for constructing and maintaining roads might increase or decrease, depending on the details of the regulatory changes and on how agencies changed their practices to accommodate the traffic changes; however, properly designed revisions to limits would yield freight cost savings exceeding any added extra infrastructure costs.

TABLE 7-1 Some Illustrative Projections of Fuel Consumption Savings

Study	Limit Change	Change in Truck Vehicle Miles Traveled (%)	Change in Fuel Consumption (%)
DOT (2000)	97,000-lb six-axle semis and 131,000-lb short doubles nationwide	−11	−6
DOT (2000)	148,000-lb turnpike doubles on interstates nationwide; 124,000-lb short doubles on most roads	−23	−13
TRB (1990a)	89,000-lb 6-axle semis and 96,000-lb short doubles nationwide	−3	−2
TRB (1990b)	110,000-lb short doubles nationwide where compatible with bridges	−3	−2

NOTE: Short doubles in the studies were twin 28- or 33-ft trailers; turnpike doubles were twin 53-ft trailers. The TRB studies did not estimate fuel savings; the values above are consistent with the studies' traffic change estimates.

Disadvantages

The main arguments against increasing the limits have been that highway safety would be degraded, that diversion of freight from rail to truck would increase the social cost of freight transportation, and that highway agencies could not afford the cost of upgrading infrastructure to accommodate larger trucks. The 1990 TRB studies concluded that the safety impact of sensibly liberalized limits would be positive, because the dominant influence on safety would be a reduction in truck VMT. The studies found that the inherent safety differences between the old and new vehicles would be slight. The 2002 TRB study acknowledged that understanding of the factors that determine the safety performance of large trucks is incomplete and therefore called for regulatory changes to be tested through rigorously monitored large-scale pilots.

Regarding other public costs, raising limits would increase air pollutant emissions as a result of induced freight demand and diversion of freight from rail to truck. The studies' projections imply that the change in emissions will be small in comparison with total truck emissions and its significance will diminish as truck emissions regulations become more stringent. Traffic impacts are projected to be positive on net because of the reduction in truck VMT.

Regarding highway agency costs, all the TRB studies recommended that truck fees be adjusted to cover the cost of providing infrastructure for them. The 2002 TRB committee concluded that DOT studies had overstated the probable cost of bridge repairs and replacements that would be required to accommodate larger trucks and recommended that limit revisions be accompanied by improvements in the states' asset management programs.

All the studies predict that liberalizing limits would divert some freight from rail to truck. This diversion would not increase the social cost of freight transportation provided trucks paid fees that covered their infrastructure costs and provided that pollution, safety, and congestion effects are small or positive, as the TRB studies predicted. Regulatory changes that allowed widespread use of longer combination vehicles were projected to have a greater effect on rail traffic than changes in weight limits alone.

Implementation Issues

The regulatory changes most commonly proposed include the following:

- Raising the federal weight limit to 97,000 lb. This would allow about a 25 percent increase in payload for weight-limited shipments on Interstates and other main roads. A bill introduced in Congress in 2009 (H.R. 1799: Safe and Efficient Transportation Act of 2009) would accomplish this.
- Giving states the option of allowing operation of longer combination vehicles (primarily turnpike doubles, a configuration with a tractor pulling two full-sized trailers) on roads with high design standards. Several western states have from time to time advocated this change.
- Allowing operation of a new kind of configuration, double trailers that would be longer and heavier than the twin-28-ft double now in use nationwide but shorter and more maneuverable than turnpike doubles. The DOT and TRB studies described above considered this kind of vehicle, which today is little used.

More ambitious proposals call for construction of new exclusive right-of-way for larger trucks. Four states, with the partial support of a DOT planning grant, are studying construction of truck-only lanes on segments of Interstate-70 through Ohio, Indiana, Illinois, and Missouri, together with staging areas for assembly and disassembly of longer combinations (FHWA, n.d.). Enactment of more restrictive limits also has been proposed, and a bill for this purpose (S.779, Safe Highways and Infrastructure Preservation Act) was introduced in Congress in 2009.

The 2002 TRB committee emphasized that changes in size and weight regulations made in coordination with complementary changes in highway management would offer the greatest potential for improving system performance. Specifically, the committee recommended adjusting truck fees to cover highway agency costs, improved bridge management, systematic monitoring of truck traffic, reform of enforcement methods, and vehicle safety regulations governing the performance of larger trucks.

Mandatory Speed Limits (Road-Speed Governors)

Road-speed governors have been standard features on trucks with electronic engine controls for many years. In the U.S. market at this time, it is up to the vehicle owner to decide whether or not to use the road-speed governor at all and what speed to select if the governor is used. Most large truck fleets do use road-speed governing today, with typical governed speeds in the 65 to 70 mph range. A few fleets set their governors as low as 60 mph, while many smaller fleets and owner-operators do not use the governor feature at all.

In Europe all trucks have their road-speed governors set by the factory to a specified value which is determined by law. The approach evaluated in this section would be in implementing a European-style mandatory road-speed governor regulation in the U.S. market.

Advantages

The NESCCAF/ICCT report (2009) projects a fuel savings of 0.7 percent per mph speed reduction for an aerodynamically optimized tractor/trailer combination truck on a simulated long-haul duty cycle. Other sources put the fuel savings at up to 1 percent per mph for tractor-trailers with today's standard aerodynamics, when cruising at 65 mph. Most studies determine the benefit of lower road speeds using 65 mph cruise as a baseline, and they do not take into account the fact that well under 100 percent of the truck's duty cycle is spent at cruise speed. This leads to a tendency to overestimate the potential benefit.

A fleet that operates long-haul tractors in areas with little congestion can gain significant benefits. For example, a fleet that governs today at 65 mph could see a 3.5 to 5 percent benefit by lowering governed speed to 60 mph, while a fleet that runs 70 mph could see a 7 to 10 percent fuel savings by cutting speed to 60.

A universal truck road-speed-governor requirement would almost completely eliminate issues with speeding by trucks, possibly providing significant safety benefits. Only in the case of tampering would speeding be possible, and tampering would be easy to detect. Any truck running significantly over the required governed speed setting could be assumed to be tampered with.

The cost of implementing mandatory road-speed governors is very low. For new vehicles the cost would be in engineering development only, with no manufacturing cost unless features need to be added to make tampering more difficult. For existing vehicles with electronic engine controls, the retrofit cost would be limited to development cost and the cost of a service stop.

Disadvantages

A number of disadvantages must be taken into account before making decisions regarding mandatory implementation of road-speed governors:

- If the regulation is applied to existing trucks, many of them will need changes to the rear axle ratio to match cruise engine speed to the new, lower road speed.
- Governors will only save fuel in situations where a truck would otherwise run faster than the governed limit. Vehicles that operate in urban or congested areas will normally see little or no benefit from governing. This means that the overall fleet fuel savings will be significantly less than projections derived from open-road driving scenarios.
- In situations where the fuel savings is significant, so is the increase in trip time. Higher trip time decreases the distance a driver can cover during a workday, meaning that more trucks would be required to move a given amount of freight. This has three undesirable effects: increased shipping costs, increased traffic congestion, and increased opportunity for accidents because of the increase in the number of trucks on the road. The lower the governed speed is set, the bigger these issues become.
- Larger fleets today are relatively sophisticated in balancing fuel cost and trip time through their use of road-speed-governor settings. These fleets would lose the benefit of being able to determine their own trade-offs. On the other hand, most smaller fleets and owner-operators are not very sophisticated in their cost-benefit analysis, and these operators might benefit from a mandatory requirement.
- If governed speeds are set significantly below the typical travel speeds of light vehicle traffic, the result will be a significant increase in traffic congestion and an increased risk of accidents because of increased speed differentials between trucks and light vehicles. Light-duty vehicle drivers in the United States are not accustomed to the sort of lane discipline required to achieve good traffic flow and safety in situations where large speed differentials exist. (Preventing excessive speeds by trucks could be a safety benefit.)
- Having all trucks governed to the same speed will result in a situation where all trucks operate at nearly, but not exactly, the same speed. Inevitable tolerance differences will result in slight speed differences. Thus, when one truck passes another, it will take a long time and create a potential for rolling roadblocks that impede light-duty vehicle traffic. This disadvantage could be reduced by allowing drivers to override the governed speed for brief periods to enable faster passing. Current road-speed governors (and current European regulations) do not allow for this override feature.
- Tampering might become a significant issue. Vehicle and engine manufacturers have gotten pretty sophisticated in their techniques for making tampering difficult, but some operators will have a significant financial and personal incentive to tamper.
- Many long-haul truck drivers are paid by the mile. A road-speed governing regulation would amount to a direct pay cut for these drivers. For many owner-operators, implementation of a road-speed-governor requirement could make the difference between making the monthly truck payment and becoming unemployed. The incentive to run longer (illegal) driving hours would become stronger.
- Engine and vehicle makers are developing increasingly sophisticated control features aimed at changing driver behavior in ways that save fuel. One feature used today allows drivers a slightly higher road-speed-governor setting if they follow other operating requirements aimed at saving fuel. This gives the fleet what it wants (fuel savings) and the driver what he wants (higher pay and a shorter trip time). Allowing features like this could greatly complicate a regulation.

Implementation Issues

All electronic engines today already have a road-speed-governor feature built in. The feature would need to be modified to prevent owner or user changes to the speed-limit setting, and to prevent the feature from being turned off. These changes would be easy for vehicle and engine manufacturers to implement. Making these features sufficiently tamper-proof might prove to be a much greater challenge.

Road-speed governors on new trucks would be easy to implement in a relatively short time frame. Manufacturers will need to modify and validate their existing road-speed-governor features to meet the requirements of the new regulation. It would also be relatively easy to develop calibrations that could be retrofitted to existing vehicles with electronically controlled engines. Getting owners to bring in their vehicles for a retrofit calibration that includes a new road-speed governor might be very difficult, however. Most owners would try to put this off as long as possible, preferably for the life of the truck.

Older vehicles that have mechanical fuel systems could in theory be retrofit with road-speed-governors, but several issues would need to be overcome. First, systems would need to be developed for this market, and they would probably not be low cost. There were road-speed-governor systems for these vehicles many years ago, but they were not low cost or widely used. Second, some way to force implementation by owners would be required. These older vehicles tend to travel few miles per year, so the potential fuel savings is limited. The owners of older trucks often lack the money to pay for an upgrade. On the other hand, if older vehicles were exempt from the speed-governor regulation, this would increase the value of older vehicles and encourage these trucks to be maintained rather than scrapped. Since older trucks have much higher emissions, any incentive to prolong their life would not be desirable. Like many other good-sounding fuel-saving ideas, the unintended consequences of road-speed governing can outweigh the benefits if great care is not taken in implementation.

OTHER COMPLEMENTARY APPROACHES

Intelligent Transportation Systems

Intelligent transportation systems (ITS) encompass a broad range of wireless and wire-line communications-based information, control, and electronics technologies. When integrated into the transportation system infrastructure, and in vehicles themselves, these technologies help monitor and manage traffic flow, reduce congestion, provide alternate routes to travelers, and enhance productivity—all to improve mobility and safety. DOT has developed the National ITS Program Plan for ITS, which provides a new vision for surface transportation in the United States in the following areas:

- Travel and transportation management
- Travel demand management
- Public transportation operations
- Electronic payment
- Commercial vehicle operations
- Emergency management
- Advanced vehicle control and safety system

The national ITS architecture (see Figure 7-2), provides a common structure for the design of ITS. It is not a system

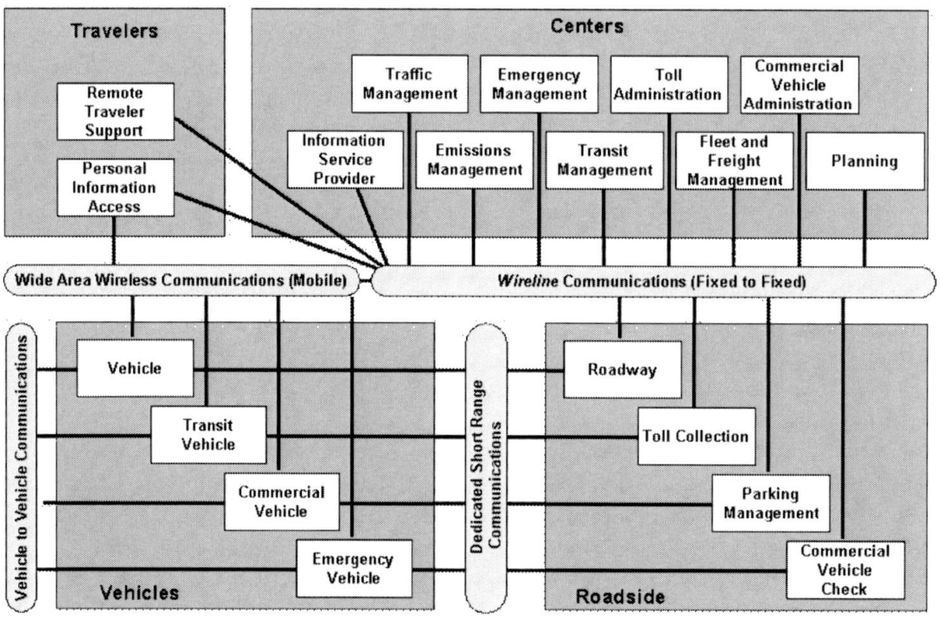

FIGURE 7-2 U.S. national ITS architecture. SOURCE: FHWA (2008).

design nor design concept but rather a framework around which multiple design approaches can be developed, each one specifically tailored to meet the individual needs of the user.

ITS is very broad in scope. This section limits the discussion to a sampling of the technologies that can play a significant role in reducing the fuel consumption of medium- and heavy-duty trucks. The focus is on technologies and applications in the infrastructure that help reduce the bottlenecks that truckers often experience—namely, congestion, toll booths, weigh stations, and inspection stations. In addition, Chapter 5 (under "Intelligent Vehicle Technologies") discusses applications for reducing fuel consumption using technologies of ITS that reside primarily on the vehicle. However, several of these applications cannot operate exclusive of the infrastructure. For example, those that use real-time traffic information require technologies on the infrastructure side to sense and communicate this information to the vehicle.

As noted in Chapter 6, traffic congestion in the United States produces significant cost in terms of wasted fuel and vehicle-hours of delay. There is a general consensus among transportation planners that we cannot "build" our way out of congestion but instead need to utilize existing capacity more efficiently to improve mobility. Some ITS technologies that can contribute to utilizing capacity more efficiently are as follows:

- Historical or real-time traffic information provided to travelers via Internet Websites during pretrip planning or via dynamic message signs or highway advisory radio while en route.
- Adaptive traffic signal control and coordinated signal timing.
- Ramp control such as ramp meters that use sensor data to optimize freeway travel speeds and ramp wait times.

Real-time traffic data can come from imbedded inductive loop detectors in the highway, such as the Performance Measurement System (PeMS) in California, or from traffic probe vehicles that carry special cell phones that are able to communicate the vehicle's position and velocity in real-time to a traffic management center (TMC). The TMC integrates a variety of ITS applications to facilitate coordination of information and services within the transportation system.

Electronic toll collection (ETC) is one of the most successful ITS applications with numerous benefits, including delay reductions, improved throughput, and reduced fuel consumption. ETC systems support the collection and processing of toll plaza transactions without requiring the driver to stop and pay manually, thereby increasing operational efficiency and convenience for travelers. ETC systems operate as either integrated multistate systems such as the E-Z Pass system, or single-state or single toll authority systems such as the Oklahoma Turnpike system. In most existing charging schemes, vehicles are identified via an in-vehicle transponder or by a video image of the license plate if the vehicle does not have a transponder. For traditional ETC systems, vehicles must pass through a gate at speeds less than 5 mph to allow time for the vehicle to be recognized and the gate lifted or a light to change from red to green. With newer technologies, such as open-road tolling, toll transactions can be processed at freeway speeds, thereby reducing the need for fuel-wasting speed fluctuations and for toll booth barriers.

ITS/Commercial Vehicle Operations (CVO) applications are designed to enhance communication between motor carriers and regulatory agencies, particularly during interstate freight movement. Commercial vehicle clearance, automated roadside safety inspection, on-board safety monitoring, hazardous materials incident response, automated administrative processing, and commercial fleet management are some of the key functions that ITS can provide for commercial vehicles. The Commercial Vehicle Information System and Networks (CVISN) program, created by the Federal Motor Carrier Safety Administration, is a nation-wide framework of communication links that State agencies, motor carriers, and stakeholders can use to conduct business transactions electronically to support CVO.

An example of the use of CVISN for supporting CVO is electronic screening that includes safety screening, border clearance, weight screening, and credential checking. Communications equipment at the roadside can query trucks equipped with in-vehicle transponders as they approach a station and issue a red or green light on the transponder so drivers know whether to continue on the mainline (bypass) or report to the station for possible inspection.

In the United States there are currently two major national electronic screening programs, the North American Preclearance and Safety System (NORPASS) and PrePass. As of March 2008, NORPASS was available in 11 states and Canadian provinces and had an enrollment of more than 93,000 trucks, and PrePass was available in 28 states and had an enrollment of more than 423,000 trucks (Maccubbin et al., 2008).

Development of the Comprehensive Modal Emissions Model (CMEM) which can predict second-by-second vehicle fuel consumption based on different traffic operations is important for developing and evaluating transportation policy for reducing fuel consumption (Barth and Boriboonsomsin, 2008). CMEM is comprehensive in that it covers 30 vehicle/technology categories from the smallest light-duty vehicle to Class 8 heavy-duty diesel trucks. In their congestion research, Barth and Boriboonsomsin (2008) worked with the California Department of Transportation's (Caltrans) Freeway Performance Measurement System. The PeMS collects real-time speed, flow, and density data from loop detectors embedded in freeways and makes the data available for transportation management, research, and commercial research.

Advantages

Evaluations of traveler information services, including real-time traffic information, show that these systems are well received by those who use them. Benefits are found in the form of improved on-time reliability, better trip planning, and reduced early and late arrivals. Studies show that drivers who use route-specific travel time information instead of area-wide traffic advisories can improve on-time performance by 5 to 13 percent (Maccubbin et al., 2008).

ITS applications for traffic control using both adaptive signal control and coordinated signal timing to smooth traffic can lead to corresponding safety improvements through reduced rear-end crashes. Studies of signal coordination in five U.S. cities and one Canadian city have shown reductions in stops from 6 to 77 percent, while 2 statewide studies have shown average improvements from 12 to 14 percent (Maccubbin et al., 2008). Reducing the number of stops reduces fuel consumption because the trip time is shorter and there are fewer energy-consuming speed fluctuations.

Ramp metering reduces the number of acceleration-deceleration cycles and smooths traffic flow. Traffic signals on freeway ramp meters alternate between red and green to control the flow of vehicles entering the freeway. A study in Minneapolis-St. Paul, MN (Maccubbin et al., 2008) showed a 21 percent crash reduction and 10 percent higher freeway volumes compared to when the ramp metering was shut down. A simulation study of two sections of freeway of that same system, each about 12 miles long, showed a 2 to 55 percent fuel savings compared to when the ramp metering was shut down. Data were collected over a three-day period, and the performance of ramp metering depended on the daily fluctuations of the demand patterns (Hourdakis and Michalopoulos, 2001).

Ninety-five percent of toll plazas in the 108 largest metropolitan areas in the United States are equipped with ETC. In Florida, ETC decreased delay by 50 percent for manual cash customers and by 55 percent for automatic coin machine customers (Maccubbin et al., 2008). On the Tappan Zee Bridge toll plaza near New York City, the ETC lane more than doubles vehicles per hour compared to the manual lanes.

Electronic screening will reduce the number of stops and starts that commercial vehicles must make for weight and safety inspections, thus reducing fuel consumption and time spent idling in lines. The Oregon Department of Transportation's Green Light Program, a weigh-in-motion system, indicates a 36 to 67 percent reduction in pollutants—particulate matter, carbon dioxide, nitrogen oxides, carbon monoxide, and hydrocarbons—when trucks stayed at highway speed past a weigh station. Trucks that avoided deceleration to enter a station and then acceleration to exit also experienced over a 50 percent reduction in fuel consumption during this deceleration/acceleration event (see http://oregon.gov/ODOT/COMM/greenlight).

Disadvantages

Disadvantages, in terms of counterbenefits, are few if the ITS technologies described above are deployed. Maccubbin et al. (2008) rated the impact of ITS deployment in six key goal areas: safety, mobility, efficiency, productivity, energy and environment, and customer satisfaction. For all ITS deployments, he gave one of the following impact ratings was given for each goal area:

- Substantial positive impact
- Positive impact
- Negligible impact
- Mixed result
- Negative impact
- Not enough data

For all of the ITS technologies described above, none received a "negative impact" in any of the goal areas. Only one, ETC, received a "mixed result" in the safety goal area. In Florida the addition of open-road tolling to an existing ETC mainline toll plaza decreased crashes by an estimated 22 to 26 percent. However, an earlier experience in Florida found that driver uncertainty about toll plaza configuration and traffic speeds contributed to a 48 percent increase in crashes at plazas with traditional ETC lanes.

Although freeway ramp metering may result in higher freeway volumes, it does require an additional stop before entering the freeway if the light is red. The additional time spent accelerating from a stop to freeway speed increases fuel consumption.

Implementation Issues

A number of implementation issues arise with ITS that make it unique to other approaches:

- Deployment of ITS is almost always regional, often covering several states, and rarely locally confined to a single city. As a result, the planning, funding, operation, and maintenance of ITS is multijurisdictional and requires cooperation at many levels of government, Federal, State and local governments.
- Interoperability is important when planning an ITS deployment that borders similar ITS deployments in adjacent jurisdictions.
- Advanced traveler information system deployments in rural and/or remote areas present special challenges. Often a remote location makes equipment more susceptible to vandalism. Also, available power to the equipment may not be nearby and may require installation of power lines.
- One of the largest and most common hurdles when deploying ITS is to make the systems compatible with existing systems already deployed. This can have a

significant impact on ITS costs and deployment schedules.
- Privacy issues can present particular challenges in ITS projects, as new ITS technologies can often raise concerns about intrusive, "Big Brother"-type surveillance.

Construction of Exclusive Truck Lanes

The idea of exclusive truck lanes covers several types of designs and how each type can be used to better improve efficient use of the highways, reduce traffic congestion, improve safety for all highway vehicles, and reduce the cost of moving goods. Truck-only lanes allow for the possibility for future technologies such as ITS to be used to improve all of the aforementioned items. Construction of these lanes also offers the opportunity to upgrade the current highway designs for increased weight and traffic of the future. During the Missouri Department of Transportation study of Interstate-70 between Kansas City and St Louis, the supplemental environmental impact statement team chose the truck-only lanes strategy as the preferred alternative, instead of the widen existing I-70 strategy. With that selection, the next step was to apply the strategy across the corridor as alternatives. The study team assessed several alternatives before recommending a preferred one that, at a minimum, provides two truck-only lanes on the inside and two general-purpose lanes on the outside for both eastbound and westbound travelers.

From the perspective of traffic and engineering, the truck-only lanes strategy compared more favorably than the widen existing I-70 strategy in the key areas of freight efficiency, safety, constructability, and maintenance of traffic.

The design that is the most prominent uses two lanes in each direction for truck-only traffic. These lanes are placed on the inside of the current lanes of the federal highways such as interstate highways. The design fits best in the rural and country areas, so that the width of the road right-of-way does not become a problem.

In areas where that the road right-of-way does not allow for construction of the lanes on the same level plane, another design is considered as a possibility. This design places the truck lanes over the current auto traffic lanes. There is a third design that employs underground tunnels for getting past the problem of clearance or the lack of property for the extra lanes. Several studies have been done in the United States by state transportation departments, but to date no lanes have been built for the purpose of moving only truck traffic for any long distances. Figures 7-3 to 7-5 show the various designs that have been considered.

The provision of access points to/from the truck-only lanes depends on the nature of the corridor. For corridors serving long-haul/through trips, access points can be limited to key interchanges and staging areas (if long combination vehicles [LCVs] are permitted to operate). On the other hand, in urban corridors, where most trips are a relatively short distance, more access points would be required. In this case, the cost and financial analyses should consider the tradeoffs among capital costs, usage/toll revenues, and safety. The use of tolls to offset some of the costs to build and maintain these truck-only lanes must be cost effective for the vehicle owners, or they will be bypassed by drivers.

The California State Route 60 and Interstate 710 corridor studies demonstrated the importance of providing frequent access points to increase truck traffic demands in urban truck-only toll corridors that serve primarily short-haul trips. In the State Route 60 study, the tradeoff between limiting access points and generating high demand was a major issue, especially because high demand is desired to maximize possible toll revenues. Yet adding access points increases the capital costs for the corridor.

For LCVs to be effective, staging areas are needed to make up and break up the trailer combinations. The cost of these staging areas might be borne by the owner/operator of the toll facility or by the private sector. In Oregon, staging facilities are privately owned.

Advantages

The major advantage to truck-only lanes is that freight can move faster and more efficiently along these corridors. Longer and heavier loads on highways built for the extra loads and length would make the movement of goods more efficient. In addition, it is expected that congestion should be reduced by separating truck traffic from small-vehicle traffic. With proper planning for cross-lane traffic and intersections, along with access and egress, car and truck accidents will be

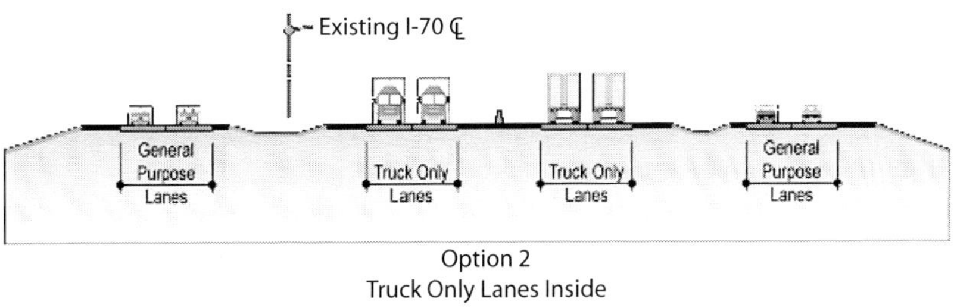

FIGURE 7-3 Example of truck-only lanes. SOURCE: FHWA (2005).

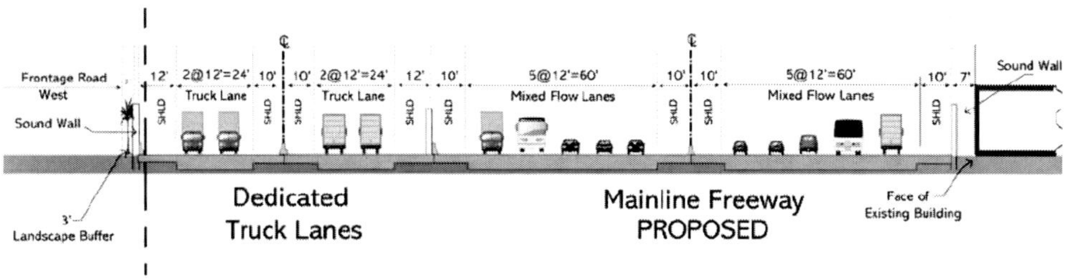

FIGURE 7-4 Concept for reducing the need for additional road right-of-way. SOURCE: FHWA (2005).

reduced and possibly there will be lower insurance costs as the accident rates drop.

Last, during the construction of the truck-only lanes, there would be the ability to update and repair the present roadways at a cheaper cost than going out to maintain or repair the current lanes. This is due to being able to use the material, equipment, and workers to do both jobs.

Disadvantages

Financing for truck-only lanes will be difficult to obtain. Tolls from trucks will not pay for construction and maintenance, and public funds and additional taxes would be needed to meet the construction cost of these lanes. The social return on investment has not yet been established.

Furthermore, adequate right-of-way (ROW) is not currently available for the construction of these lanes, so additional land will need to be purchased, and the widening of the right-of-way and the clearing of land may have a negative impact on the environment.

Last, the time that it takes to construct a usable network of truck-only lanes will be several years before any benefits of a better transportation system will be realized. Due to the long time that will be needed, new technologies may not be put into place until some of the lanes are completed.

Congestion Pricing

Congestion pricing refers to variable road tolls (higher prices under congested conditions and lower prices at less congested times and locations) intended to reduce peak-period traffic volumes to optimal levels. Congestion pricing could take different forms, such as area-wide network pricing on freeways and possibly arterials, "cordon" or area pricing in central business districts, or truck-specific congestion pricing such as the varying time-of-day gate fees implemented at the ports of Los Angeles and Long Beach.

Area-wide congestion pricing is applicable to freeways and major arterials where there is significant congestion. Cordon pricing strategies are only applicable in major urban areas with significant congestion. The limited geographic applicability of these two scenarios limits the fuel reduction potential. Area-wide congestion pricing has greater potential since it is estimated that nearly 30 percent of urban vehicle miles travelled (VMT) occurs at the level of service E (unstable flow) or F (forced or breakdown flow; TRB, 2000). Cordon pricing of metropolitan area central business districts, however, is estimated to affect only 3 percent of total VMT nationwide. Furthermore, evidence suggests that there will be little, if any, overall impact on total truck traffic (as the added costs are likely to be marginal, or the option of moving to the off-peak period is unacceptable), but

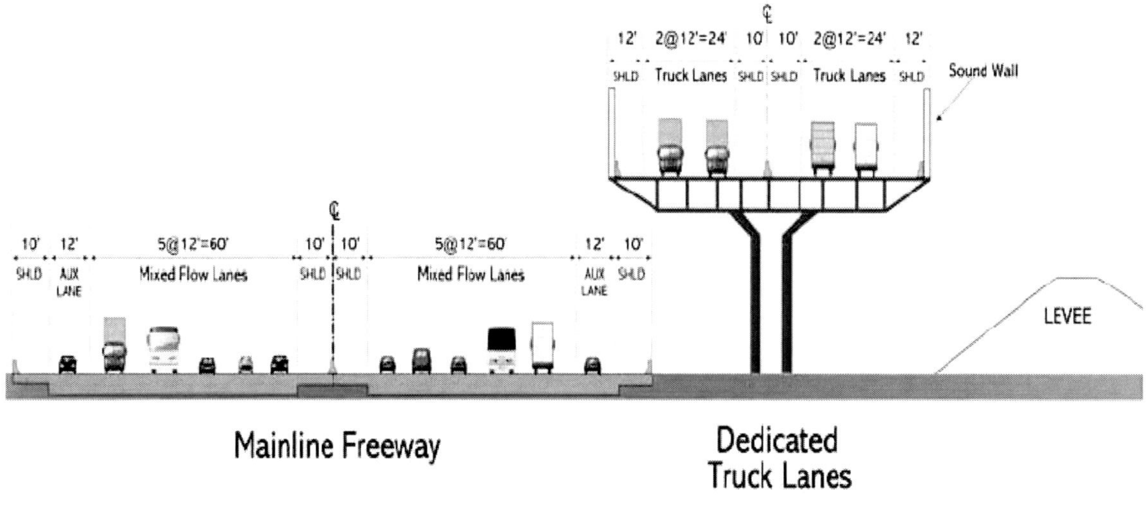

FIGURE 7-5 Elevated truck lanes. SOURCE: FHWA (2005).

rather that the benefits will occur from trucks operating under improved flow conditions and therefore using less fuel due to idling or stop-and-go operations. This will have a larger impact on smaller urban trucks since larger long-distance trucks operate mostly on uncongested highways.

Advantages

Congestion pricing could affect truck fuel consumption by:

- Shifting trips to less congested off-peak hours;
- Reducing congestion for trucks continuing to operate during peak periods, thereby improving their fuel economy and productivity (and offsetting the congestion pricing);
- Reducing the overall movement of goods and related truck traffic due to higher costs; and
- Increasing the shift in logistics patterns—for example, leading firms are establishing consolidation centers on the edges of urban areas to reduce truck activity within the congested area.

Most studies of the impact of congestion pricing have focused on all traffic, rather than distinguishing impacts on personal versus commercial vehicle traffic. A study for the U.S. Department of Energy used travel demand models in Minneapolis-St. Paul and Seattle, in conjunction with speed-fuel efficiency relationships, to evaluate the combined benefits of travel reductions and operating efficiencies from area-wide systems of managed lanes.[2] The results from different scenarios ranged from a 0.1 to 2.5 percent impact on fuel consumption and greenhouse gas (GHG) emissions depending upon the scenario. Extrapolating these results to a national level based on projected 2030 congestion levels in different urbanized areas led to an overall estimated reduction in national fuel consumption ranging from 0.5 to 1.1 percent (EEA, 2008). Another national study of GHG emission reduction strategies estimated that cordon pricing could potentially reduce VMT on the order of 3 percent if applied to all metropolitan areas in the United States (Cambridge Systematics, 2009). These are rough estimates for all vehicles, however, and may not be transferable to truck traffic.

Evaluations of a cordon pricing scheme implemented in London examined effects specifically on truck traffic (Transport for London, 2006). The experience suggests that the reduction in overall vehicle-kilometers of travel has come almost exclusively from passenger vehicles rather than trucks. However, the trucks benefited from reduced queuing and, subsequently, reduced truck idling and fuel consumption. Once the scheme was introduced, excess delays were reduced by 26 percent, from 2.3 to 1.7 minutes per kilometer. For 70,000 truck-kilometers traveled and a reduction in excess idling delay of 0.6 minutes per kilometer, the scheme reduced truck idling by a total of 700 hours. With each truck-hour of idling consuming 0.8 gallons, the truck fuel consumption reduction from congestion pricing would have been 560 gallons annually (EPA, 2004). For trucks whose average fuel consumption is 40 liters per 100 km (10.4 gal/100 km), this represents a reduction in fuel consumed of about 7.7 percent.

Disadvantages

As congestion is reduced, average speed increases because speed variability declines and less time is spent at idle. Therefore, fuel consumption declines with increasing average speed—up to a point. When speeds average greater than 40 to 55 mph and approach free-flow highway speeds, fuel consumption rates increase. Congestion is likely to affect urban service and delivery movements more than long-haul freight, and therefore it is the fuel consumption characteristics of smaller trucks that are most important.

If congestion pricing is implemented only on a limited basis (e.g., only freeways), diversion of traffic to other non-tolled facilities is likely to be a significant concern because of the impacts on neighborhoods and local traffic. Increases in VMT on alternate routes could offset the fuel savings achieved from reductions in VMT and congestion on the facility itself. Therefore, congestion pricing will be most effective at reducing fuel consumption if it is implemented universally (on all major roads in an area). In addition, HOT lane implementation could potentially reduce available conventional lanes, particularly if separation lanes are needed between HOT lanes and conventional lanes. The impacts of the loss of conventional lanes on traffic flow and costs need to be evaluated in decision making regarding HOT lanes.

Implementation Issues

Congestion pricing has been experimented with in a number of areas, primarily on existing tolled facilities, but has not yet gained widespread popularity. From a technical standpoint, congestion pricing is relatively easy to implement on facilities that already are tolled. The broader-scale application of this strategy beyond existing or proposed toll highway facilities, however, is likely to require universal deployment of electronic toll collection technologies. This will require coordination by a state or regional transportation agency. The U.S. Department of Transportation (DOT) is encouraging greater experimentation in this area. In 2007, DOT awarded $853 million in funding to five metro areas for urban partnership agreements to reduce congestion, including a significant focus on tolling/strategies.

[2]These systems included high-occupancy/toll (HOT) lanes on freeways, in which drivers of single-occupancy vehicles can use the lanes if they pay a fee, which depends on the congestion on the untolled travel lanes. Depending on the scenario, either existing/planned high-occupancy vehicle (HOV) lanes were converted to HOT lanes, or a new HOT lane was constructed alongside an existing/planned HOV lane to form two HOT lanes.

Driver Training and Behavior

Driver training requires relatively low initial investment and appears to be a highly cost-effective strategy for improving fuel efficiency and lowering operating costs and harmful emissions. One option includes training designed to educate drivers about operating practices that influence fuel consumption and improve their driving skills.

Professional drivers of heavy-duty vehicles must provide proof to insurance companies of a minimum number of training hours. There are three different types of truck driver training programs: private schools, public institutions, and training programs run by the motor carriers themselves. Most private truck driving schools and publicly funded truck driving programs provide a certificate or diploma upon graduation, which is generally recognized and accepted by some carriers as proof of acceptable training. Most larger fleets require that all new drivers go through a company training course that includes a driving course and test before they are allowed to drive for the company.

A review of tuition requirements for various commercial driver's license (CDL) schools in the United States found that tuition can range from approximately $2,000 to $4,000 per driver. Some of these schools have already incorporated fuel-efficient driving instruction into their curricula. Drivers who have already obtained their CDL, however, would only receive targeted instruction for fuel-efficient driving at a lower tuition rate (Latty, 2009).

There are several fundamental principles and techniques each driver should know in order to minimize fuel consumption, as described below.

- *Minimize speed fluctuation.* Smooth acceleration reduces inertial effects as well as wear on the engine and equipment, especially in hilly or mountainous terrain. Rapid acceleration causes undue wear on the engine, drivetrain, and tires as well as requiring more fuel to achieve the same end result. In addition, braking results in a loss of energy as vehicle momentum is converted to heat. Braking also activates the air compressor, which draws power from the engine, further increasing fuel consumption. Smooth braking saves fuel, reduces brake wear, and reduces engine load.
- *Engine braking.* Use of the engine brake allows for smooth deceleration, reduces brake wear, and saves fuel.
- *Shift optimization and gear selection.* One gear down may increase fuel consumption by approximately 15 percent from optimal conditions. For example, for 10 to 15-liter engines found in Class 8 trucks, a constant operation below 1,300 rpm significantly improves fuel economy, with the target range being between 1,200 and 1,500 rpm. A 20 percent difference in the time spent in top gear could improve fuel efficiency by as much as 4 percent (M. England, personal communication, 2009).
- *Idling.* An average heavy diesel engine uses about one gallon of fuel per hour while idling (CARB, 2005). To reduce fuel consumption, certain types of idling can be minimized or eliminated altogether. For example, electronically controlled engines do not require significant warm-up or cool-down periods; a driver can reach 70 percent throttle as soon as oil pressure is up.[3] Moreover, the cost of turning the engine off and starting it again is frequently less than the cost of idling, since excessive idling leads to increased maintenance and engine wear.
- *Tires.* Tire condition and inflation are just as important on trailer tires as on tractor tires.
- *Speed.* Road speed has a direct impact on aerodynamic drag. Higher speeds also cause extra wear on the engine and transmission systems.
- *Cruise control.* Cruise control optimizes the electronic control system's fuel delivery and improves fuel efficiency.
- *Clutch control.* Double clutching increases clutch wear and reduces fuel efficiency. Double-clutching is not necessary on synchromesh gear boxes. Many drivers have learned to shift gears without using the clutch except to stop and start.
- *Trip planning.* Total fuel usage should be considered in trip planning. Also, each full stop requires approximately one-third of a gallon to return to highway speed. Accordingly, drivers should consolidate stops for food, fuel, and so forth to increase fuel efficiency (personal communication, L. Harvey, Natural Resources Canada, 2009).
- *Block shifting/skipping gears.* Fewer gear changes results in greater fuel efficiency. The quicker a driver moves up the gearbox to top gear, the more fuel that is saved. Each gear shift up improves instantaneous fuel consumption by 10 to 30 percent (personal communication, L. Harvey, Natural Resources Canada, 2009).
- *Aerodynamics.* Vehicles with adjustable roof-mounted air deflectors can improve fuel efficiency. Covering trailers, whether loaded or empty, ensuring curtains are tear-free, and correctly positioning a load all help reduce aerodynamic drag and improve fuel efficiency.
- *Overfilling the fuel tank.* Overfilling the fuel tank causes fuel to be lost through the breather vent when it is heated and expands, resulting in lower fuel efficiency.
- *Maintenance.* Changing air and fuel filters when vacuum specifications are exceeded can improve fuel

[3]Personal communication, Juan Ortega, Longhorn International Trucks LTD.

efficiency. Ensuring proper wheel alignment also has a substantial impact on fuel efficiency, up to 3 to 4 percent in some cases, as well as reducing tire wear (TIAX, 2009).

Advantages

According to a staff member at Natural Resources Canada, initial case studies indicate approximately 2 to 8 percent reduction in fuel consumption and associated GHG emissions due to driver training (personal communication, L. Harvey, Natural Resources Canada, 2009; DOE, 2008). These results are consistent with values reported in the Freight Best Practice case studies. In numerous case studies, companies and drivers commonly reported an average fuel efficiency improvement of 5 percent, with actual results across all case studies reviewed ranging from 1.9 to 17 percent improvement (Freight Best Practice, 2009).

All things being equal, driver training is expected to be more effective for high-load operations, since greater energy requirements are at stake for each acceleration, gear shift, and braking event. In addition, urban drive cycles are expected to be more sensitive to improvements in driver behavior than line-haul cycles. For example, the high number of starts, stops, and transient operation events associated with urban drive cycles could provide frequent opportunities for smoother braking and acceleration. On the other hand, free-flow highway driving is usually extremely uniform in nature, providing little opportunity for driver modifications. This is especially true for vehicles with a speed governor in place, which effectively limits the one meaningful operational parameter (in terms of efficiency improvements) over which the line-haul driver has control.

Another advantage is that driver training programs work with existing and new equipment. Therefore, the typical lag times associated with market penetration of new technology standards are avoided, and the trucking sector could see immediate benefits based on the existing truck fleet. Drivers who use the proper driving methods are also accepted by the public as responsible providers for the transportation of the nation's goods. Last, driver training requires relatively low initial investment and appears to be a highly cost-effective strategy for improving fuel efficiency, lowering operating costs and harmful emissions.

Implementation Issues

Three major challenges to implementing and promoting driver training have been identified by FleetSmart personnel (personal communication, L. Harvey, Natural Resources Canada, 2009): (1) licensing requirements that do not require additional training or testing; (2) drivers' schedules that may prevent them from attending training programs; and, (3) accessibility of training programs to all drivers, especially those in remote areas. Companies will need to be convinced of the cost savings and the need for this new way of managing fleets and drivers. These challenges could be overcome through appropriate policy incentives and/or mandates.

The financial and environmental impacts of driver training provide strong incentives for companies to adopt aggressive driver training programs. The promotion of such programs is fully consistent with the goals of the U.S. Environmental Protection Agency's (EPA) SmartWay program. Development and/or validation of training programs in the United States could be facilitated through SmartWay, with third-party training services joining the program as partners.

Intermodal Transport

Intermodal transport involves the movement of goods by more than one mode on a single journey (Corbett and Winebrake, 2007; Winebrake et al., 2008). Commonly, intermodal transport combines a truck mode with either ship or rail to improve shipping efficiency, reduce costs, or achieve some other desirable performance attribute. Because rail and ship are significantly less energy-intensive than truck, incentivizing the movement of goods from truck to rail or ship is one way to improve the overall efficiency of the freight transportation system.

Estimates of total potential freight mode shifting have been hypothetical in nature, rather than based on empirical data, due in large part to the complex nature of competition between trucks and rail. The potential for mode shifting is limited to certain types of commodities—those that are heavy, of low value, and do not have an acute need for reliable and timely delivery—for example, building stone and waste, as well as certain movements—in particular, long-haul movements where the efficiency benefits of rail outweigh the additional handling/logistics costs and time at either end, generally shipments longer than 1,000 miles. Furthermore, market demand both affects and is dependent on the quality of service. Rail service improves significantly as demand between market pairs increases—increased traffic (trains per day) increases the level of service that railroads provide to customers and means that improved access is possible (shippers need access to rail facilities to ship via rail). In short, shippers choose a mode that minimizes their total logistics cost.

Increasing intermodal freight shipments requires significant investment in rights-of-way, in rolling stock, and in overcoming infrastructure-induced capacity constraints. The investment should be justified based on the overall economics of the investment in the delivery system, and not just the fuel consumption savings that would result from diversion of freight from truck to rail. Nevertheless, there are fuel savings to be realized in some transport corridors where economically sound investments can be undertaken. For example, the Mid-Atlantic Rail Operations study (Cambridge Systematics, 2010a) estimated the potential for a 4-11 percent diversion based on $12 billion in infrastructure investment in the midtlantic rail network.

Advantages

Reductions in fuel consumption on the order of 60 percent per ton-mile are typical for shifts from trucking (trailers or containers) to long-haul intermodal rail, with reductions decreasing with shorter distances (Cambridge Systematics, 2010b). Savings can vary significantly, however, depending on the distance of the movement and type of cargo.

Disadvantages

Intermodal movement of goods requires large investments in infrastructure that can only be accomplished over a long time period. In addition, freight delivery service and performance may be sacrificed in the shift from truck to rail and/or ship.

Implementation Issues

Many different policy mechanisms exist to promote intermodal transport. The approach that could have the greatest impact would be investments into intermodal facilities such that delays and costs associated with cargo transfer and logistics are minimized. Actions that can affect a truck-rail mode shift include investment in rail and intermodal terminal infrastructure, direct operating subsidies for railroads, land use regulations (e.g., to preserve rail sidings for rail-oriented businesses), and taxes to increase the cost of truck travel, as previously discussed.

Improvements to intermodal transport, such as rail capacity improvements and bottleneck relief, intermodal (truck-rail) terminals, and financial/pricing incentives, could potentially encourage shippers to make greater use of rail in place of trucks, increasing the efficiency of freight movement on a ton-mile basis. The government could promote rail diversion through the promotion of freight "villages" that include intermodal terminals, transload facilities, and bulk storage facilities; expanded market reach for regional railroads; and continued improvement in rail infrastructure, including signal, track, bridge, terminal, and clearance upgrades.

While freight rail infrastructure investment has traditionally been left to the private sector, the federal government and a number of states have increasingly become involved in this issue for purposes of economic development and road traffic reduction. There are several state and federal programs that will fund rail improvements to help bridge the gap between investment needs and the availability of private capital. The federal-aid highway funding program also allows some flexibility in using funds for nonhighway freight transportation projects.

To date, most of the easier rail capacity improvement projects have been built, leaving primarily the more difficult and expensive projects. In addition to being expensive, many of the remaining critical needs are set in urban environments where there are substantial constraints on rights-of-way as well as added costs for mitigation of impacts. These barriers will pose challenges to large-scale improvements in freight infrastructure sufficient to leverage significant truck-rail mode shift.

FINDINGS AND RECOMMENDATIONS

Finding 7-1. The committee examined a number of approaches for reducing fuel consumption in the trucking sector and found suggestive evidence that several approaches—particularly driver training and longer combination vehicles (LCVs)—offer potential fuel savings for the trucking sector that rival the savings available from technology adoption for certain vehicle classes and/or types. Any government action taken to reduce fuel consumption in the trucking sector should consider these alternatives.

Finding 7-2. Fuel taxes offer a transparent and efficient method for internalizing the potential societal costs of climate change and oil imports (e.g., energy security) and reducing fuel consumption in road transport. Fuel taxes operate to make fuel-saving technologies more attractive and provide incentives for saving fuel in operations, while involving fewer unintended consequences than standards.

Finding 7-3. Fuel taxes can be designed to lessen the uncertainties facing the truck sector and provide a market signal for investments in fuel-saving technology.

Recommendation 7-1. Although the committee recognizes the political difficulty associated with increasing fuel taxes, it strongly recommends that Congress consider fuel taxes as an alternative to mandating fuel efficiency standards for medium- and heavy-duty trucks.

Finding 7-4. A cap-and-trade system for carbon emissions would provide market signals for truckers to adopt fuel-saving technology and operations. The signal, however, is more uncertain and volatile than would be provided by fuel taxes.

Finding 7-5. A cap-and-trade system, such as is being considered by Congress that would limit total CO_2 emissions by primary energy producers, would have implications for the trucking sector. Regulators would then not need to develop standards for CO_2 emissions that apply to specific trucks and trucking operations, avoiding the complexity of different classes and duty cycles of trucks. On the other hand, the cap-and-trade system would likely involve new administrative burdens for monitoring emissions from the primary producers and policing the system.

Finding 7-6. Methods to encourage the adoption of specific technologies—mandates or subsidies—are best utilized when options are limited and the compatibility with truck usage and duty cycle are clear.

Finding 7-7. When there are several fuel-saving options and complex truck operating conditions, performance standards are likely to be superior to specific technology requirements.

Finding 7-8. Increasing vehicle size and weight limits offers potentially significant fuel savings for the entire tractor-trailer combination truck fleet. This approach would need to be weighed against increased costs of road repair. Example case studies explored in this report demonstrate fuel savings of up to 15 percent or more. These savings are similar in size but independent and accumulative of other actions that may be taken to improve fuel consumption of vehicles; therefore the net potential benefit is substantial. To achieve these savings would require the federal government to:

- Change regulatory limits that currently restrict vehicle weight to 80,000 lb and that freeze LCV operations on the Federal Interstate System.
- Establish a regulatory structure that assures safety and compatibility with the infrastructure. One possible regulatory structure has been proposed by the Transpotation Research Board in *Regulation of Weights, Lengths, and Widths of Commercial Motor Vehicles*, Special Report 267 (TRB, 2002).
- Consider the necessary changes that would be required to permit reasonable access of LCVs to vehicle breakdown yards and major shipping facilities in close proximity to the interstate.

Recommendation 7-2. Congress should give serious consideration to liberalizing weight and size restrictions and should consider how the potential fuel savings and other benefits of such liberalization can be realized in a way that maintains safety and minimizes the cost of potential infrastructure changes.

Finding 7-9. Mandatory road-speed-governor settings have long been used in Europe. Most large U.S. fleets already use speed governors, and they could be implemented more generally in the U.S. market. The committee found that the benefit of these governors is significant only for vehicles that spend a large amount of time at high-speed cruise, where one might expect ~1 percent fuel savings for each mile per hour reduced (e.g., reducing speeds from 65 mph to 60 mph may lead to 3 to 5 percent fuel savings. Road-speed governors have a number of disadvantages and potential unintended consequences.

Finding 7-10. Intelligent transportation systems enable more efficient use of the existing roadway system by improving traffic flow and reducing or avoiding congestion. This results in a reduction of large variations in speed, idle time, and periods of high acceleration, which have a considerable impact on fuel consumption. Many ITS applications are now being tested or deployed throughout the country. Although the cost of deployment is not insignificant, it may allow deferment or as an alternative to expanding the existing roadway system.

Finding 7-11. Congestion pricing offers several potential benefits: reduced congestion increases overall efficiency in the freight delivery system, and increases fuel savings on the order of 0.1 to 7.7 percent based on the examples described herein.

Finding 7-12. There are significant opportunities for savings in fuel, equipment, maintenance, and labor when drivers are trained properly. Indications are that this could be one of the most cost-effective and best ways to reduce fuel consumption and improve the productivity of the trucking sector. For example, cases evaluated herein demonstrate potential fuel savings of ~2 percent to 17 percent with appropriately trained drivers.

Recommendation 7-3. The federal government should encourage and incentivize the dissemination of information related to the relationship between driving behavior and fuel savings. For example, one step in this direction could be to establish a curriculum and process for certifying fuel-saving driving techniques as part of commercial driver license certification and to regularly evaluate the effects of such a curriculum.

Finding 7-13. Intermodal transport offers significant environmental and energy advantages compared to trucking alone on an individual cargo movement basis.

Finding 7-14. The system-wide opportunities for intermodal transport are currently limited based on existing infrastructure, customer demands, cargo compatibility, and economic feasibility.

REFERENCES

Barth, Matthew, and Kanok Boriboonsomsin. 2008. Real-world CO_2 impacts of traffic congestion. Transportation Research Record, submitted March 31, 2008.

Cambridge Systematics. 2009. Moving Cooler: An Analysis of Transportation Strategies for Reducing Greenhouse Gas Emissions. Washington, D.C.: Urban Land Institute. July.

Cambridge Systematics. 2010a. Mid-Atlantic Rail Operations Study. Prepared for the I-95 Corridor Coalition. Forthcoming.

Cambridge Systematics. 2010b. Transportation's Role in Reducing Greenhouse Gas Emissions. Prepared for the Federal Highway Administration. Forthcoming.

CARB (California Air Resources Board). No date. Heavy-Duty Vehicle Greenhouse Gas Measure. Available at http://www.arb.ca.gov/cc/hdghg/hdghg.htm.

CARB. 2005. Proposal to Reduce Idling Emissions from On-Road Heavy-Duty Diesel Vehicles. Available at www.arb.ca.gov/msprog/truck-idling/workshop-03-23-2005.ppt.

CARB. 2008. The Carl Moyers Program Guidelines: Approved Revision 2008. April 22.

CARB. 2009. Facts About Heavy-Duty Vehicle Greenhouse Gas Emission Reduction Regulation: Reducing Climate Change Emissions from Tractor-Trailers. June 30.

Corbett, J.J., and J.J. Winebrake. 2007. Sustainable movement of goods: Energy and environmental implications of trucks, trains, ships, and planes, Environmental Management (November): 8-12.

DOE (U.S. Department of Energy). 2008. Energy Market and Economic Impacts of S. 2191, the Lieberman-Warner Climate Security Act of 2007. Energy Information Administration. Available at http://www.eia.doe.gov/ oiaf/service_rpts.htm.

DOT (U.S. Department of Transportation). 2000. The U.S. Department of Transportation's Transportation Comprehensive Truck Size and Weight Study. August. Available at http://www.fhwa.dot.gov/reports/tswstudy/TSWfinal.htm.

EEA (Energy and Environmental Analysis). 2008.Market-Based Approaches to Fuel Economy: Summary of Policy Options. Prepared for National Energy Technology Laboratory, U.S. Department of Energy.

EPA (U.S. Environmental Protection Agency). No date. Federal Excise Tax Exemption. Available at http://www.epa.gov/smartway/transport/what-smartway/idling-reduction-fet.htm.

EPA. 2004. A Glance at Clean Freight Strategies: Idle Reduction. February. Available at http://nepis.epa.gov/Exe/ZyPURL.cgi?Dockey=P1000S9K.txt.

EPA. 2009a. Federal Tax Credits for Energy Efficiency. Updated September 14. Available at http://www.energystar.gov/index.cfm?c=tax_credits.tx_index#s3.

EPA. 2009b. Innovative Financing: SmartWay Clean Diesel Finance Program. August. Available at http://www.epa.gov/otaq/smartway/transport/what-smartway/financing-clean-diesel-info.htm.

FHWA (Federal Highway Administration). No date. Corridor: Interstate 70 (I-70): Dedicated Truck Lanes—Missouri to Ohio. Available at http://www.corridors.dot.gov/i70.htm.

FHWA. 1997. Federal Highway Cost Allocation Study.

FHWA. 2005. *I-710 Major Corridor Study, Final Report.* March.

Freight Best Practice. 2009. Examples of Cost Savings. Available at www.freightbestpractice.org/examples-of-cost-savings.

Greszler, A. 2009. Heavy duty vehicle fleet technologies for reducing carbon dioxide: An industry perspective. Pp. 101-116 in Reducing Climate Impacts in the Transportation Sector, D. Sperling and J.S. Cannon, eds. Springer.

Hourdakis, J., and P.G. Michalopoulos. 2001. Evaluation of Ramp Control Effectiveness in Two Twin Cities Freeways, Transportation Research Board 2002 Annual Meeting, Washington, D.C.

Latty, J. 2009. JCL Driver Training, Southport, Merseyside, England. Available at www.jcldrivertraining.co.uk.

Leavitt, W. 2005. Idle cash. FleetOwner. April.

Maccubbin, R., B. Staples, F. Kabir, C. Lowrance, M. Mercer, B. Phillips, and S. Gordon. 2008. Intelligent Transportation Systems Benefits, Costs, Deployment, and Lessons Learned: 2008 Update. Report No. FHWA-JPO-08-032. Federal Highway Administration, U.S. Department of Transportation.

McKinnon, A. 2005. The economic and environmental benefits of increasing Maximum Truck Weight: The British Experience. Transportation Research Part D: Transport and Environment 10(1): 77-95.

Miller, E. 2009. EPA exempts 70 idle-reduction devices from federal excise tax for first time. Transport Topics, February 2, p. 4.

NESCCAF/ICCT, Southwest Research Institute, and TIAX. 2009. Reducing Heavy-Duty Long Haul Combination Truck Fuel Consumption and CO_2 Emissions. Washington, D.C.: International Council on Clean Transportation (ICCT). October.

RTAC (Roads and Transportation Association of Canada). 1986. Vehicle Weights and Dimensions Study—Technical Steering Committee Report. December.

TIAX LLC. 2009. Assessment of Fuel Economy Technologies for Medium- and Heavy-Duty Vehicles. Final Report for the National Academy of Sciences. Sept. 16.

Transport for London. 2006. Central London Congestion Charging: Impacts Monitoring, Fourth Annual Report. June.

TRB (Transportation Research Board). 1990a. Special Report 225: Truck Weight Limits: Issues and Options. Washington, D.C.: TRB.

TRB. 1990b. Special Report 227: New Trucks for Greater Productivity and Less Road Wear: An Evaluation of the Turner Proposal. Washington, D.C.: TRB.

TRB. 2002. Special Report 267: Regulation of Weights, Lengths, and Widths of Commercial Motor Vehicles. Washington, D.C.: TRB.

TRB. 2000. Highway Capacity Manual. Washington, D.C.: TRB.

Tunnell, M. 2008. Energy and Emissions Impacts of Operating Higher Productivity Vehicles: Update: 2008. American Trucking Research Institute. March.

Winebrake, J.J., J.J. Corbett, A. Falzarano, J.S. Hawker, K. Korfmacher, S. Ketha, and S. Zilora. 2008. Assessing energy, environmental, and economic tradeoffs in intermodal freight transportation. Journal of the Air and Waste Management Association 58(8): 1004-1013.

Woodrooffe, J., B. Belzowski, J. Reese, and P. Sweatman. 2009. Analysis of the Potential Benefits of Larger Trucks for U.S. Businesses Operating Private Fleets. Prepared for the National Private Truck Council. University of Michigan Transportation Research Institute. May.

8

Approaches to Fuel Economy and Regulations

This chapter examines the broad variations in medium- and heavy-duty vehicles and explains how the complex nature of trucks influences regulatory options. It explores metrics that capture the work task of the vehicle, thereby providing a means for comparing the relative fuel consumption performance of vehicles on the basis of task. The chapter concludes with a discussion of regulatory approaches and includes examples of fuel consumption regulatory instruments that may be suitable for implementation.

Thirty years ago, the National Highway Traffic Safety Administration (NHTSA) of the U.S. Department of Transportation was faced with the question of how to design a new fuel economy regulation for passenger vehicles. For many important elements, NHTSA was able to build off an existing exhaust emissions program. The regulated parties (vehicle manufacturers), test method (chassis dynamometer), and test cycle (federal test procedure) were adopted by NHTSA for its new fuel economy program. Of course, there were many important elements of the program that were unique to the new fuel economy regulation, such as the method of allowing compliance by means of a corporate average of annual sales and the development of a second test cycle to reflect highway operation.

Today, while there is an existing heavy-duty vehicle exhaust emissions program with its own regulated entities (engine manufacturers), test method (engine dynamometer), and test cycles (Federal Test Procedure [FTP], Supplemental Emissions Test [SET], Ramped Mode Cycle [RMC], and in-use tests), there are factors associated with the U.S. vehicle market that make fuel consumption regulations more difficult and complicated than the design of fuel economy standards for passenger vehicles. Consider the following three examples:

- The heavy-duty vehicle market is extremely diverse, with a wide range of vehicle types, sizes, and duty cycles.
- Heavy-duty vehicle manufacturing is driven by customer specifications, which often leads to a far greater variety of pairings between major components (e.g., engine, transmission, chassis, axles, wheels, body shape).
- Unlike passenger vehicles, vehicle manufacturing is often split between two different manufacturers: the producer of the chassis and a second manufacturer that purchases the chassis, adds a body and special equipment, and ultimately sells the vehicle to the consumer (see Figure 8-1). The exception is pickup trucks and truck tractors, which are completely assembled by the final manufacturer.

PURPOSE AND OBJECTIVES OF A REGULATORY PROGRAM

The purpose and structure of a regulatory program should be as follows: (1) generate cost-effective reductions in fuel consumption from medium- and heavy-duty vehicles, maximizing the savings of fuel at a justifiable cost imposed on the industry and society; (2) accelerate the research, development, and market penetration of new and existing energy saving technologies; (3) reduce the amount of energy consumed per movement of freight or passengers; (4) build on existing market incentives and company practices to lower fuel consumption; and (5) minimize additional administrative burden upon the regulated industry.

There are a handful of major technical and policy questions that must be addressed when developing a new regulatory program. Each is discussed in turn throughout this chapter:

- *Regulated vehicle types.* What types of vehicles should be regulated?
- *Regulated parties.* Who should the regulated parties be?
- *Metrics for fuel consumption.* What metric should be used to measure performance?

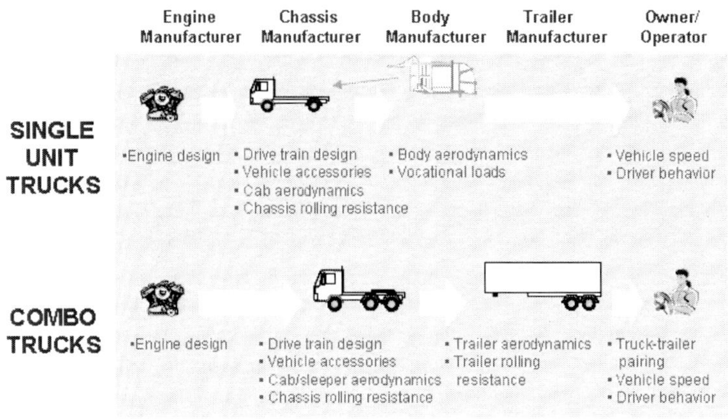

FIGURE 8-1 Shared responsibility for major elements that affect heavy-duty-vehicle fuel efficiency. SOURCE: Bradley and Associates (2009).

- *Methods for certification and compliance.* What methods will be used to determine compliance and overall program effectiveness?
- *Regulatory model.*

REGULATED VEHICLE TYPES

The committee has considered a broad range of vehicles. These include pickup trucks, transit buses, motor coaches, school buses, delivery vans, straight trucks, and combination vehicles such as tractor trailers. The largest fuel use from the heavy-truck fleet is associated with the vehicles that move the vast majority of the freight: Class 8 tractor trailers with gross combined weight (GCW) ranging from 80,000 lb on the interstate and in excess of 130,000 lb on some state highways (GCW varies considerably, as it is governed by federal and state size and weight regulations). This is not surprising, considering the huge jump in weight hauling capacity between Class 8 (in excess of 130,000 lb) and the rest of the heavy-duty fleet (Class 2b through 7 weight capacity ranges from 8,500 to 33,000 lb). Class 8s are about 20 percent of the fleet in total number of vehicles, but 61 percent of the fuel use of all heavy-duty vehicles. The second largest fuel use segment is the Class 2B, which makes up the majority of heavy-duty vehicles (53 percent) and which is responsible for just under 20 percent of fuel consumption. The third largest class is Class 6. These are considered medium heavy-duty and generally have only a single rear axle, while Class 8 vehicles typically have tandem drive axles. Class 6 vehicles make up about 16 percent of the heavy truck population and consume 11 percent of the fuel. Table 8-1 gives more detail.

Most Class 8 vehicles are combination trucks for which several trailer options are available to complete the vehicle system (see Figure 8-2), adding another dimension to an already complex regulatory challenge. For example, the type of trailer used will influence the vehicle's overall aerodynamic drag coefficient and the projected frontal area, both of which influence aerodynamic losses and directly affect fuel consumption. The tires, on both the tractor and the trailer, will influence the rolling resistance. In addition, weight and dimension regulations define the "legal" GVW, which also influences fuel consumption. An added complication is that the size and weight regulations for a given vehicle vary depending on the jurisdiction—federal or state.

The problems are compounded further in that vehicles

TABLE 8-1. Mileage and Fuel Consumption by Vehicle Weight Class

Vehicle	Population (millions)	Annual Miles (million)	Annual Fuel Use (million gallons)	Percent of Population	Percent of Annual Miles	Percent of Fuel Use
Class 2B	5.800	76,700	5,500	52.8	35.1	19.3
Class 3	0.691	9,744	928	6.3	4.5	3.3
Class 4	0.291	4,493	529	2.6	2.1	1.9
Class 5	0.166	1,939	245	1.5	0.9	0.9
Class 6	1.710	21,662	3,095	15.6	9.9	10.9
Class 7	0.180	5,521	863	1.6	2.5	3.0
Class 8	2.154	98,522	17,284	19.6	45.1	60.8
Total	10.992	218,580	28,444	100.0	100.0	100.0

SOURCE: DOT (2002).

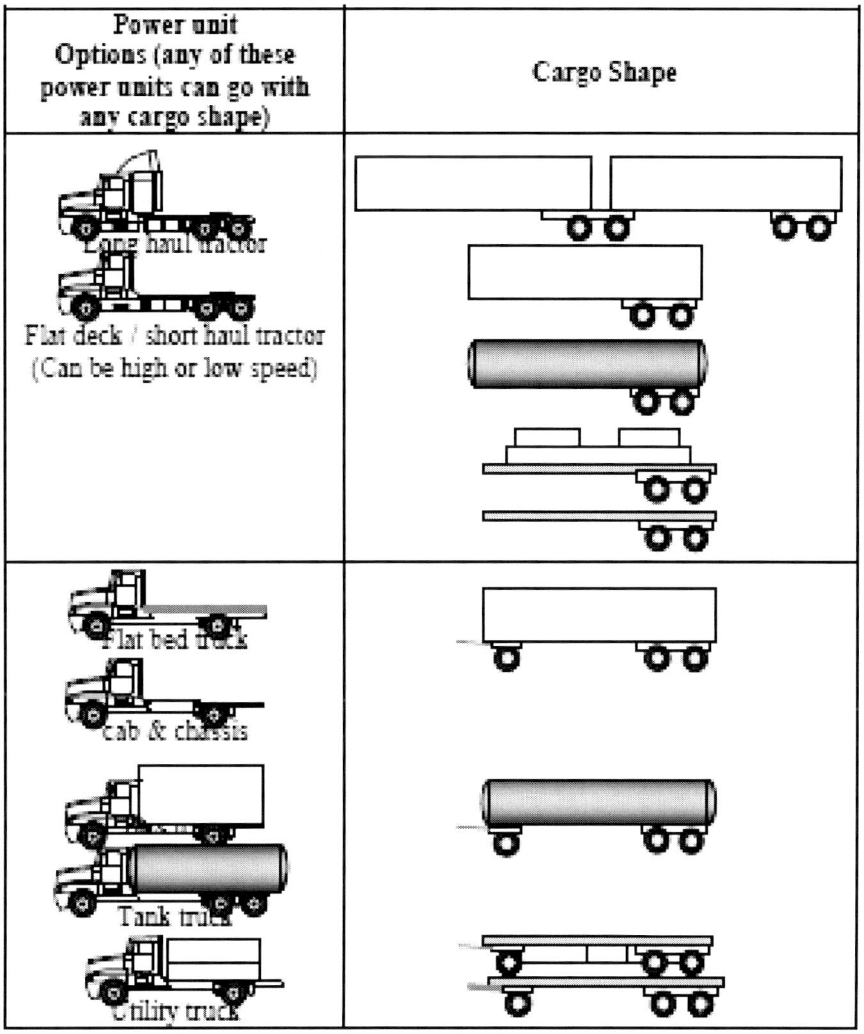

FIGURE 8-2 Illustration of diversity of trailer and power unit (tractor) options.

haul freight of various shapes, sizes, and densities and are often loaded below capacity. While van trailers present a convenient and predictable space envelope for aerodynamic evaluation, other trailers such as flat-decks have no such fixed outer shell, and therefore the shape of flat deck trailers will change with each load transported, which in turn influences aerodynamic drag and thus fuel consumption.

Given the complexity and challenge of establishing a new regulatory policy for medium- and heavy-duty vehicles, there could be a tendency to narrow the scope of a regulation, or at least to focus early implementation, on the largest fuel users in the fleet. This would suggest focusing on Class 8, Class 6, and Class 2b vehicles to cover roughly 90 percent of the fuel use in the medium- and heavy-duty vehicle fleet.

While starting with a subset of heavy-duty vehicles is tempting, there are several drawbacks that should be considered. Uneven policy application will cause disruptions in the marketplace and create the potential for reclassifying various classes of vehicles, as has been done in light-duty vehicles (LDVs), in which cars were reclassified as light trucks to achieve less stringent regulatory standards. Related to "reclassification" is a potential for change in market behavior to avoid higher prices due to regulation (i.e., if Class 2b is regulated but not Class 3 then buyers might buy more of the larger Class 3 trucks because they would become less expensive relative to 2b trucks).

Second, while regulation of medium- and heavy-duty vehicles is complicated, there could be further complications created by seeking to draw artificial lines between various segments. For example, Class 8 straight/vocational trucks typically have two drive axles and Class 7 trucks have one. Class 5 and Class 6 trucks are similar to each other, and Classes 2b and 3 tend to be more similar to each other.

Third, when considering a regulatory framework, there are a number of important parameters beyond fuel use, such as cost effectiveness, equity among manufacturers, potential for gaming, minimizing unintended outcomes, and technology potential. Finally, Congress instructed NHTSA to establish fuel economy standards for medium- and heavy-duty vehicles—not a portion or subset of medium- and heavy-duty vehicles.

REGULATED PARTIES

There are at least two principal considerations to be evaluated when seeking to determine the most effective point of regulation. The first practical consideration is that the number of regulated entities must be a manageable number (in the tens rather than the hundreds) of parties to limit compliance and administrative burdens. Second, to be effective, the regulation must affect the corporate parties with the greatest control and authority over vehicle design and over those components that offer the potential for substantial reductions in fuel consumption.

Market Concentration

As discussed in Chapter 2, the majority of production of commercial trucks is concentrated in about 12 major corporations that control different portions of the market. Manufacture of class 8 trucks (tractors and straight) is dominated by four companies (Daimler AG, Volvo, PACCAR, and Navistar) that account for more than 90 percent of U.S. truck registrations. The smallest heavy-duty vehicles—Class 2b to Class 4—are dominated by the Big 3 U.S. auto manufacturers with 89 percent of registrations.

Large original equipment manufacturers (OEMs) with significant engineering capability design and manufacture almost all Class 2b, 3, and 8b (semitractors) vehicles. These OEMs have design control over features that determine the completed vehicle aerodynamics. In some cases, these OEMs make the engine and driveline components, while in others these are outsourced to specialist suppliers, who could be given responsibility for regulatory compliance of the power train. For Classes 4 through 8a, there is a mix of vehicles made by large and small OEMs and smaller final-stage manufacturers. The vehicle mix includes box trucks, bucket trucks, school buses, transit buses, motor coaches, refuse haulers, and dump trucks (see Chapter 6). These small entities purchase the chassis, engine, driveline, and in some cases complete cab and chassis units from suppliers who could be made responsible for regulatory compliance. In some cases the supplier content will determine aerodynamic characteristics, but in many cases, the final-stage manufacturer will significantly influence aerodynamic characteristics. Given the very limited engineering resources available to these smaller OEMs, it is unreasonable to assume that such small companies can conduct the necessary aerodynamic tests for each specialty vehicle produced.

Control Over Design and Important Components

A vehicle may be considered as consisting of three major energy-consumption-related components that are separable from one another. These are (1) the wheels and tires yielding tire rolling resistance, (2) the body-yielding aerodynamic losses and (3) the power train. Arguably there is a fourth notable category of energy consumption-auxiliaries, that be considered, but is not as clearly separable. Vehicle chassis represents a structure that simply connects the body and the power train, and may be regarded as part of the power train or part of the body for testing purposes. These are the key features that affect fuel consumption in some way. The vehicle weight and aerodynamics are affected by the hood, cab, skirts, bumpers and overall shape and design of the vehicle body. The power train consists of the engine, hybrid components, transmission, differentials, and drive axles. In considering the entity to regulate, those responsible for the total vehicle or for each/any of these major components are among the options.

Table 8-2 lists the advantages and disadvantages of each choice of regulated party. Regulating at the point of the engine manufacturer is likely to impose the lowest additional administrative burden but limits the program to a small subset of potential reductions in fuel use. A second option is to impose regulation on the power train integrator, which could be an engine company partnered with a drive train component supplier (e.g., hybrid component manufacturers), or a power train integrator could be an integrated truck OEM with selected suppliers. This option would benefit from accounting for engine and power train improvements, but would, in some cases, require greater integration of the existing industry. Regulation of the vehicle manufacturers offers the greatest amount of potential improvements while limiting the number of regulated parties, but the administrative burden will be substantially higher, particularly for smaller vehicle manufacturers.

In the final analysis, the concept of addressing the power train, aerodynamics, and tires seems to have strong potential for success as it would maintain focus on the dominant fuel-consumption-related components of vehicles. The suppliers of these components are arguably in the best position to control future improvements of the components that they manufacture. The final stage manufacturers need some means of assurance that they receive accurate and meaningful data from the suppliers in order to evaluate the final vehicle fuel consumption. Therefore, the point of regulation would need to be at the final-stage vehicle manufacturer, supplemented by the provision of consistent component performance data by the component manufacturers. Annex 8-1 presents a more detailed analysis of a methodology that might underlie a component-based regulatory program.

A perplexing problem for any option, regarding Class 8 vehicles, is what to do about the trailer. The trailer market represents a clear barrier with split incentives, where the owner of the trailer often does not incur fuel costs, and thus has no incentive to improve aerodynamics of the trailer itself or to improve the integration of the trailer with the tractor or truck. Furthermore, legal authority is tenuous, given that trailers are not self-propelled vehicles. One option could be

TABLE 8-2 Advantages and Disadvantages of Each Choice of Regulated Party

Regulated Entity	Advantages	Disadvantages
Engine manufacturer	• Utilizes existing regulatory framework for criteria pollutants: test cycle (though current cycles may need updating), engine tests, compliance testing • Manageable number of regulated parties • Low administrative burden	• Misses the bulk of potential improvements in drivetrain, hybrids, tires, aerodynamics, vehicle accessories, component integration, improved design • Does not include trailer
Power train integrator	• Captures hybrid systems and transmission packages when the dynamic power train system is broader than engine • Builds on existing regulatory framework of engine tests and cycles • Allows vehicle and trailer attributes to be covered by simulation with test cycles • Reduces need for full vehicle testing	• May require two or more industry entities to define the power train hardware as team; new business model in some cases • Will require upgrades to certification engine cell controls to accommodate range of vehicle load inputs and hybrid drive train components
Final stage vehicle manufacturer	• Includes nearly all vehicle parameters that affect fuel use in single heavy-duty vehicles • Manageable number of regulated parties	• Class 8 trailers and bodies of vocational trucks not included • Higher administrative costs to develop test cycles, conduct vehicle testing, perform certification and compliance testing
Fleet owner Vehicle owner	• Allows for greater range of operational improvements (driver training, intermodalism)	• Unmanageable number of regulated entities (hundreds of fleets: half of heavy-duty vehicles in fleets of less than 10 trucks) • Would still require mandatory fuel efficiency testing of HDVs to provide fleet owners with information required to make smart compliance decisions

to allow manufacturers to certify for additional credits if an improved trailer design and/or integration is satisfactorily incorporated as a complete vehicle. However, in many cases tractors change semitrailers frequently, making integration difficult without standardization of design.

METRICS FOR FUEL CONSUMPTION

Considering the complexity of heavy-duty vehicles and the highly specialized nature of vehicle design and operation with respect to vehicle task, the following advisory principles were developed by the committee:

- The metrics should incentivize subcomponent and total vehicle development.
- The metrics should relate to the transport task or vehicle vocation.
- The metric should encourage energy conservation for a given task.
- The metric should be based on energy or fuel consumption—e.g. equivalent diesel gallons/cargo ton-mile. (See discussion in Chapter 2.) Normalizing to equivalent diesel fuel permits fair comparison across fuel type as energy density varies with fuel type and specification.

The committee recognized that an equipment specification regulation was an option, considering the ongoing SmartWay program as an example. However, a performance-based metric is strongly recommended as it will more adequately address the advisory principals above.

The practical effect of using a gallon per mile metric is that it will result in improvements only to the vehicle itself and in all likelihood encourage smaller vehicles with smaller payloads, resulting in serious erosion of transportation efficiency. On the other hand, the load specific fuel consumption (LSFC) metric such as gallons per cargo ton-mile will promote technical improvements and configuration development that increase the amount of cargo that can be carried for a given amount of fuel consumed. Improvements can be achieved in two ways under an LSFC metric: (1) by improving the efficiency of the vehicle (power train, tires, aerodynamics, etc.), the vehicle can move a given amount of freight with lower fuel consumption; (2) by increasing the cargo capacity of the vehicle, the regulated party will also be able to improve its fuel efficiency rating—independent of any change in truck subcomponent fuel efficiency. In combination these two distinct approaches will provide the greatest potential for energy conservation and savings.

Smaller-class single-unit trucks and buses have design and operating characteristics that are different from larger vehicles. For example, utility trucks used by electric power companies may be equipped with a bucket crane and not carry any substantial cargo. Clearly it would not be practical to evaluate the performance of such a vehicle in terms of the mass of transported cargo. Single-unit trucks may also be placed into service towing trailers with a drawbar hitch as shown in Figure 8-2, which is common in the West and which further complicates assigning an operating weight to the truck. Considering the multiplicity of factors influencing fuel consumption and the complexity of larger vehicle systems and operations, the committee concludes that the notion of a single metric being applied identically to all classes of vehicles appears to be problematic. However, the committee is confident that a standard measurement protocol coupled with different standards and metrics will provide a means of assessing fuel consumption on the basis of work task for medium- and heavy-duty vehicles.

Class 2b and 3 vehicles tend to be higher volume general-purpose vehicles with less custom built content. The high production volume of this vehicle class is conducive to a more general metric such as gallons per mile, gallons per mile per person weight, or gallons per ton mile. Buses also have substantial variability. Of the bus categories, the long-distance motor coach not only transports passengers but they also transports freight, and therefore the task-based metric would need to consider both freight (baggage and package cargo) and passenger mass. Passenger mass can be estimated using "typical passenger" mass multiplied by the number of available seats. Freight mass can be estimated by using a "typical" freight density term multiplied by the cubic capacity of available cargo space.

Implications of Cargo Density for Fuel Consumption Evaluation

For most truck transportation, the nature of the freight task can be classified as volume limited or mass limited. Mass-limited freight is of sufficiently high density that the GVW will be reached before the volumetric capacity of the vehicle is fully utilized. Volume limited freight is of sufficiently low density that it occupies the available cargo space before the GVW is achieved. It is estimated that the split between volume-limited and mass-limited freight on the U.S. highway network is approximately 50/50. Vehicles are often designed on the basis of mass or volumetric capacity, and the characteristics of these vehicles are somewhat sensitive to the methods used to calculate fuel consumption. The following example illustrates practical considerations that will be necessary when developing a fuel consumption regulatory instrument.

Consider the real-world example of two tractor trailers having identical power units but with trailers of different cargo mass capacity and identical volumetric capacity (Figure 8-3). Vehicle A has a GVW of 80,000 lb and a cargo capacity of 48,000 lb. Vehicle B has a GVW of 97,000 lb and a cargo capacity of 61,000 lb (allowing 4,000 lb for the extra axle, suspension and additional trailer structure). Both trailers have identical cargo volume capacity of 3,650 ft^3.

It is clear that Vehicle A is better suited to cargo weighing 48,000 lb or less, and vehicle B is better suited to cargo weighing more than 48,000 lb. There is no difference in the volumetric capacity of these vehicles; therefore, the cargo mass dictates the vehicle choice. On the surface this case appears to be ideally suited to the vehicle mass fuel consumption metric (gal/ cargo ton-mile). However, when examined more closely, it is apparent that the identical tractor would have different fuel consumption values for each of the two cases given the difference in the GVW and cargo mass capacity. It is likely that if the mass metric were applied, Vehicle B would always outperform Vehicle A (assuming that a proportionate cargo mass is used). This would be counterproductive for low-density, volume-limited freight applications because the mass metric would encourage heavier capacity vehicles with higher tare weight. In such cases, this problem can be offset by considering an alternate metric such as gallons per cargo ft^3-mile. Further discussion of alternative metrics can be found in Annex 8-2 to this chapter.

METHODS FOR CERTIFICATION AND COMPLIANCE

The choices of possible methods for certification and compliance of fuel consumption standards for medium and heavy duty vehicles involve some of the most challenging regulatory design issues.

One broad choice pertains to whether it would be possible to establish average standards by corporate entity, as is done under the light duty vehicle CAFE program, or whether the breadth and diversity of the medium and heavy-duty vehicle market precludes such an option. In general there are important benefits associated with a corporate average standard in that it allows corporations flexibility to focus improvements on vehicle types within the retooling cycle. The challenge with a corporate average standard is that the medium- and heavy-duty vehicle market is extremely diverse and would require establishing categories of vehicles by type and application. In addition, the light-duty vehicle CAFE program placed full line, largely domestic, manufacturers at

FIGURE 8-3 Identical tractors used to pull trailers of different mass capacity but identical volume capacity.

a competitive disadvantage due to the form of the regulatory standard. The committee therefore strongly urges NHTSA to proceed with caution if it considers a corporate average standard. Another type of regulatory flexibility is to allow manufacturers to average emissions across engine families. This approach could be used to allow manufacturers to average fuel consumption across vehicle lines. NHTSA would need to perform a separate analysis on the market structure of various truck manufacturers to understand the pros and cons of setting corporate average standards within vehicle types and categories, as well as examining other regulatory flexibilities.

It is clear that the regulatory system should incentivize the subcomponent manufacturers to make real gains in efficiency, but this could be achieved even if the point of regulation is at the OEM. The engagement of a purchaser in seeking highly efficient vehicles is very different from what is typical in the passenger car market. One of the main thrusts of a fuel consumption regulation for medium- and heavy-duty vehicles should be to ensure that the customer has access to reliable data that are based on performance metrics related to the intended function of the vehicle.

Approaches to characterizing or certifying heavy-duty vehicle fuel consumption toward a standard are summarized in Table 8-3. The options range from testing assembled vehicles to modeling and simulating assembled vehicles with most testing at only the power train, tires, and aerodynamics component levels. Any procedure must characterize the consumption or efficiency using a duty cycle that is reasonably representative of real use. The greater the degree of representation, the greater will be the number of test cycles required to cover the applications, but the greater will be the accuracy of the process in reflecting real-world data. A high level of fidelity between regulatory cycles and the real world is required to enable regulators to make the correct decisions and drive the market in the desired direction. Vehicle pur-

TABLE 8-3 Options for Certification of Heavy-Duty Vehicles to a Standard

Method	Equipment	Advantages	Disadvantages
In-use test (complete vehicle)	400- to 600-mile test course(s) on public roads	• Easy to conduct • Relatively inexpensive • Well-developed procedure • Familiar with HDV fleets • SAE procedures	• High test-to-test variation including driver differences, ambient conditions, and traffic variations • Best for comparing one truck to another • Requires the use of a "reference truck" to limit test-to-test variability
Test track (complete vehicle)	Closed, 1- to 5-mile oval or circular test track	• Easy to conduct • Good repeatability	• Facilities are limited and expensive • Complexity of test cycles limited • Best for high-speed steady-state test cycles • Cannot incorporate changes in grade to test cycle • Affected by ambient conditions • Requires "reference truck" to reduce test-to-test variability
Chassis dynamometer	Heavy-duty chassis dynamometer with data from a coast-down test track	• Well-developed procedure • Computerized drivers' aids ensure very good compliance with transient test cycles • Very good repeatability	• Facilities are limited and expensive • Accuracy depends on accurate input data from coast-down test • Coast-down data not reliable • Inability to handle variable grade
Engine test plus vehicle simulation modeling	Engine dynamometer Vehicle simulation model	• Well-developed test • Minimal additional burden • Lowest total cost to vehicle manufacturers • Ability to run large number of vehicle test cycles off a single engine test	• Accuracy depends on complexity of simulation model and "accuracy" of model inputs • Development of vehicle-specific modeling parameters likely to require additional vehicle/component testing (i.e., dynamic wind tunnel tests for aerodynamic drag, tire tests)
Power train test plus vehicle simulation modeling	Engine dynamometer that will accommodate hybrid power train hardware and model/cycle control (CIL) Vehicle simulation model	• Builds on current practice of engine dynamometer tests • Ability to accommodate many cycles and vehicles via models • Facilitates harmonization with pollutant emission certification	• New business model may be needed to integrate engine and other power train components • Process development required for integration of simulation into regulatory framework (see above).
Simulation of entire vehicle	Vehicle simulation model	• Ability to accommodate many cycles and vehicles via models	• Still requires substantial testing for model development and validation • Models not adequate to cover regulated pollutants, so emissions test still required

SOURCE: Modified from Bradley and Associates (2009).

chasers require an even higher level of fidelity to make the best decisions when specifying a new vehicle. Sophisticated larger fleets will often change engine, transmission, tire, or even OEM selection to gain a 1 or 2 percent fuel consumption reduction. Achieving this level of fidelity is a major challenge.

As discussed previously, using the results from existing engine dynamometer testing for heavy-duty vehicles would allow for accurate, repeatable comparisons, but there are substantial drawbacks to limiting the scope of the rule to only engine technologies. For example, there is a potential lack of fidelity between the dynamometer test cycle and real world performance. Table 8-3 describes the four major widely used test methods—in-use testing, test track testing, chassis dynamometer testing, and simulation modeling—and identifies the advantages and disadvantages of each approach. A final method, generally used in power train development to test the combined engine and drive train, would require the engine dynamometer test cycle to utilize the load characteristics of real trucks over real duty cycles. The load on the engine would be determined by a vehicle simulation, an approach that bears some similarity to the approach used in Japanese regulations. The truck and trailer simulation (aerodynamics, tires, mass) could include applicable fuel-saving features to represent a range of truck models, unlike the fixed truck characteristics defined in Japan's model. To carry this concept to hybrid vehicles, the "engine" would need to be augmented with the hybrid components and thus become a "power train" in the test cell (also called component in the loop; CIL). As the interaction between engines and conventional (nonhybrid) transmissions becomes increasingly emphasized, the concept of evaluating the performance of a complete power train retains merit.

Three of the methods listed in Table 8-3 for determining vehicle fuel consumption require the use of complete vehicles: in-use testing, test track testing, and chassis dynamometer testing. Vehicles that operate with trailers would need to have standard trailers for the testing. In the case of Class 8b vehicles, at least two trailer types would be required: a standard box van trailer to be tested on tractors intended for this type of application, and a low frontal area trailer for trucks to be used with other trailer types.

Two of the methods for determining vehicle fuel consumption that are listed in Table 8-3 require only component-level testing. These are the engine test plus vehicle simulation approach, and the power train test plus vehicle simulation. The final method listed in Table 8-3 is pure simulation, although even this approach will require some testing to validate the data used in the simulation model.

Sample Applications of Methods to Vehicle Classes

For Class 2b vehicles and Class 3 pickup trucks, a chassis dynamometer test for fuel consumption similar to the test used in light-duty vehicles is a viable option. These vehicles are often used in ways similar to light duty vehicles, and existing test facilities could be used. Many of these vehicles are also made in relatively high volume, making a full-vehicle test less difficult to manage for the manufacturer. An ability to rely on existing industry and regulator experience and capability makes this approach attractive.

In medium- and heavy-duty vehicles from Class 3 up through Class 8a, manufacturing volumes are often low and many different configurations are built on a given platform. This makes chassis dynamometer fuel consumption testing much more difficult and expensive than in smaller vehicles. For these larger vehicles it makes sense to combine engine or power train test data with vehicle simulation models. Particularly in the case of hybrid vehicles, it will be important to have high-fidelity data for the fuel consumption and performance of the power train. This data may come from testing, simulation, or a blend of simulation such as hardware-in-the-loop (HIL) or CIL. Even if a pure simulation approach is used, some level of test data is required to validate the models. A wide range of vehicle duty cycles may need to be simulated or tested in order to achieve adequate fidelity with real-world fuel consumption data.

Tractor trailers (Class 8b) are also available in dozens of configurations, many of which are produced in low volume. Once again, chassis dynamometer fuel consumption testing would be difficult and expensive, and so some level of vehicle simulation modeling is likely to be required. Engine or power train test data can be provided as needed. Again, many vehicle duty cycles may need to be simulated or tested in order to achieve adequate fidelity. Regulators will want to reinforce rather than impede the fuel consumption sensitivity in Class 8b, where purchasing decisions are often based on differences as small as 1 or 2 percent in fuel consumption. For example, a buyer will want to know which of 10 or more available aerodynamic treatments will perform best in the buyer's particular application. Defining tests or simulations that can provide an accurate answer to a question like this will not be easy. Defining a regulatory process and standards that do not drive incorrect decisions by vehicle manufacturers will require considerable care.

For the first iteration of the new regulatory program for medium- and heavy-duty vehicles, the committee recommends that regulators consider test methods that minimize the administrative burden on those vocational vehicles that are not the large fuel users (that is, all vehicles not included in Class 2b, Class 6, or Class 8b line-haul tractor trailers). For these numerous types of vehicles that account for less than 10 percent of commercial truck fuel consumption, the committee recommends two options: (1) pure vehicle simulation, and (2) engine-in-the-loop, also called CIL (engine connected to a dynamometer that emulates the rest of the vehicle). The pure simulation modeling approach allows the regulated entity to piggyback off of existing engine tests and data for other components (e.g., transmission, tires, electric machine). The CIL approach takes advantage of the existing engine test procedure while also incorporating existing component data. At this point, it appears that pure simulation would be the less expensive option.

When simulation models are used, inputs are required to represent components. Some of these inputs may come from standardized tests, such as a test for tire rolling resistance or a wind tunnel test for aerodynamic drag. These inputs may also come from simulation models, if the models are validated and sufficiently accurate. For example, a CFD model may be used to determine the Cd of a vehicle in place of wind tunnel testing. Data used in a model for regulatory approaches will need to come from a standard test or analysis process that is recognized by the entire industry. Many new test and analysis procedure standards will be needed. It may be necessary to approve both codes and experimental facilities to insure quality control and uniformity in the determination of wind drag.

Figure 8-4 shows an outline of how a power train test can be combined with a vehicle simulation model to determine the fuel consumption of a vehicle. This is the CIL approach. This figure shows a hybrid electric power train with a diesel engine including exhaust aftertreatment. The power train is tested in a test cell, where the dynamometer load is determined by a vehicle model. The vehicle model, in turn, uses input data for parameters such as rolling resistance, mass, and aerodynamic drag. The vehicle model is exercised over a specified route, and the resulting power demands are applied on the power train by the dynamometer. The resulting fuel consumption is experimentally measured.

The choice of test cycle is a critical part of any vehicle fuel consumption test or simulation. Test cycles selected for regulatory use will need to reflect real-world duty cycles to the extent possible. Parameters of importance include maximum speed, average speed, speed fluctuation, number of stops, and amount of idling. It will not be possible to faithfully reproduce the duty cycle to be experienced by every vehicle, so similar applications will be represented by one or a few duty cycles for regulatory purposes.

Overall Regulatory Structure

Introduction

Applying a regulatory system that pushes technology in the drive train, tires, and vehicle shape (aerodynamics) ensures that incentives are applied at the foundation of the major vehicle systems that influence fuel consumption.

Given the high fuel consumption sensitivity of some medium- and heavy-duty vehicle purchasers, it appears that one priority should be to ensure that accurate information on the fuel consumption characteristics of a completed vehicle is available to the purchaser. Having such information would help drive the selection of vehicles with the lowest fuel consumption for the task performed. The notion of regulating the final-stage manufacturer and including a requirement on the component manufacturers to provide relevant performance data to the purchaser will be an important part of the regulation.

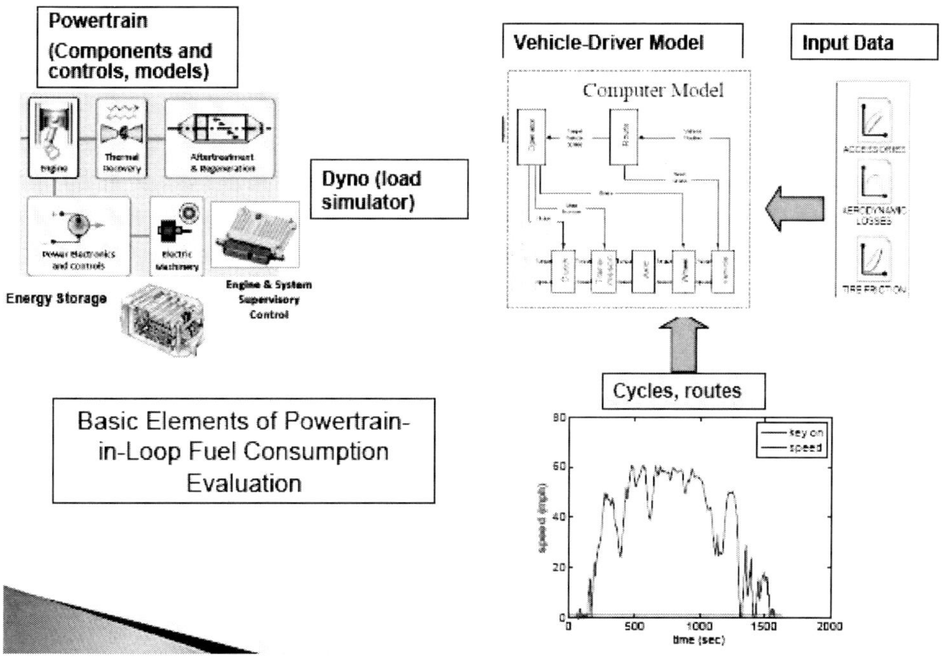

FIGURE 8-4 CIL test of a hybrid vehicle power train to determine vehicle fuel consumption on a specific test route.

Focusing on the power train, aerodynamics, and tires provides a means of incentivizing these three important areas. Measuring and documenting the performance of these key components in a constant and transparent way will be important for the final-stage manufacturer. NHTSA may wish to require suppliers to provide information in standardized form at the power train, aerodynamics and tire levels. The final-stage manufacturer would have the responsibly of combining the performance of these components to achieve the lowest cost for the intended vehicle task while complying with the regulatory fuel consumption requirement. The task-based fuel consumption metrics would be used by the final stage manufacturer to inform the customer as part of a labeling requirement.

In summary, the concept of this example regulatory model is as follows:

1. Major components such as power train, tires, and aerodynamics (including factors for accessories and auxiliaries) would each be tested or simulated, and efficiency data made available to OEMs in a common format using industry standard metrics and procedures.
2. The completed vehicle would be regulated, and in addition customer specific data would be provided that would inform the vehicle purchaser about the fuel consumption performance of the particular vehicle in relation to the intended task. This form of regulation would simplify the regulatory task compared to other alternatives and provide the flexibility needed to address the complex nature of the industry.

Compliance, Audit, Enforcement

Chapter 3 describes the compliance audit process for heavy-duty engines and for passenger car emissions. The committee believes that similar methods would be adequate to audit compliance for heavy-duty vehicle fuel consumption. The use of models and simulation in the certification process could cause a complication in auditing in that the audit test on a real vehicle might need to be assessed against a model output generated for certification. This concern would need to be addressed in the design of the regulation.

Pilot Program

The committee recommends that NHTSA conduct a pilot program to "test drive" the certification process and validate the regulatory instrument proof of concept. There are two broad purposes for such a pilot program. In the first element, the agency would gain experience with certification testing, data gathering, compiling, and reporting. There needs to be a concerted effort to determine the accuracy and repeatability of all test methods and simulation strategies that will be used with any proposed regulatory standards and a willingness to fix issues that are found. There are numerous technical challenges related to implementation of this program (e.g., reliable and accurate methods to determine tire rolling resistance and vehicle aerodynamic drag coefficients, incorporation of simulation modeling with hardware, integrating a hybrid drive train within the standard test cell, characterizing subcomponents for use in simulation modeling). This trial period will serve as a means for developing and refining the regulatory processes before the official start date of the program.

A second element would include gathering data on fuel economy from several representative fleets of commercial trucks (e.g., long-haul, delivery vans, specialty vehicles, and large pickups). These data would continue to be collected once the program was established in order to provide a real world check on the effectiveness of the regulatory design on the fuel economy of trucking fleets in various parts of the marketplace and in various regions of the country. As this program will place an additional administrative burden on NHTSA and private operators, the committee recommends that Congress consider an annual funding allocation for this program.

Concluding Comments

This is an important juncture. The choices that will be made over the course of the next few years will establish the regulatory design for medium- and heavy-duty-vehicle fuel consumption standards for the next several decades at least. Although the stringency of the standards themselves may be revisited from time to time, the regulatory design elements (regulated parties, certification tests and procedures, compliance methods)—once established—are far more difficult to modify.

In many cases the commercial vehicle market is sophisticated, driven by knowledgeable purchasers who focus on the efficiency of their operations, including the fuel costs associated with accomplishing their tasks. Thus, one of the most important challenges facing NHTSA is how to enhance and improve upon the commercial truck industry's existing incentive to maximize fuel economy of its trucks and fleets.

At the same time, there are commonly acknowledged characteristics in the commercial marketplace for trucks and buses that may be improved by a regulatory approach, such as split incentives between owners and operators (e.g., trailers), and the short payback period of 18 months to 2 years, that create barriers to the adoption of efficiency technologies for many purchasers. The existence of technology packages for some vehicle classes that offer significant fuel consumption reduction potential at reasonable costs suggests that well-designed policies to overcome problems such as split incentives or too short a payback period may yield important benefits (see Table 6-19).

Due to the complexity of the vehicle market the committee was not able to give adequate consideration to the non-commercial markets such as personal pickup trucks, school buses, personal motor homes. NHTSA should consider these applications in their regulatory proposal.

A fundamental concern raised by the committee and those who testified during our public sessions was the tension between the need to set a uniform test cycle for regulatory purposes, and existing industry practices of seeking to minimize the fuel consumption of medium and heavy-duty vehicles designed for specific routes that may include grades, loads, work tasks or speeds inconsistent with the regulatory test cycle. This highlights the critical importance of achieving fidelity between certification values and real-world results to avoid decisions that hurt rather than help real-world fuel consumption.

Because regulations can lead to unintended consequences, either because the variability of tasks within a vehicle class is not adequately dealt with or because regulations may lead to distortions between classes in the costs of accomplishing similar tasks, the committee urges NHTSA to carefully consider all factors when developing its regulatory proposal.

FINDINGS AND RECOMMENDATIONS

Regulated Vehicle Types

Finding 8-1. While it may seem expedient to focus initially on those classes of vehicles with the largest fuel consumption (i.e., Class 8, Class 6, and Class 2b, which together account for approximately 90 percent of fuel consumption of medium- and heavy-duty vehicles), the committee believes that selectively regulating only certain vehicle classes would lead to very serious unintended consequences and would compromise the intent of the regulation. Within vehicle classes, there may be certain subclasses of vehicles (e.g., fire trucks) that could be exempt from the regulation without creating market distortions.

Regulated Parties

Finding 8-2. Large OEMs, which have significant engineering capability, design and manufacture almost all Class 2b, 3, and 8b vehicles. Small companies with limited engineering resources make a significant percentage of vehicles in Classes 4 through 8a, although in many cases they buy the complete chassis from larger OEMs. Regulators will need to take the limitations of these smaller companies into account.

Finding 8-3. Commercial trailers are produced by a separate group of about 12 major manufacturers that are not associated with truck manufacturers. Trailers, which present an important opportunity for fuel consumption reduction, can benefit from improvements in aerodynamics and tires (see Chapter 5 for details).

Recommendation 8-1. When NHTSA regulates, it should regulate the final-stage vehicle manufacturers since they have the greatest control over the design of the vehicle and its major subsystems that affect fuel consumption. Component manufacturers will have to provide consistent component performance data. As the components are generally tested at this time, there is a need for a standardized test protocol and safeguards for the confidentiality of the data and information. It may be necessary for the vehicle manufacturers to provide the same level of data to the tier suppliers of the engines, transmissions, and after-treatment and hybrid systems.

Recommendation 8-2. Separate regulation of trailer manufacturers will be necessary to promote more fuel-efficient trailers, including integration of the trailer design with the tractor for improved aerodynamic performance, lower tare weight, and a requirement for low-rolling-resistance tires.

Fuel Consumption Performance Metrics

Finding 8-4. Since the primary social benefit of the medium- and heavy-vehicle sector is the efficient and reliable movement of freight, movement of purpose-built integrated equipment, or performance of a task, it is necessary to establish a metric that includes a factor for the work performed (e.g., gallons per cargo ton-mile rather than simply gallons per mile) to ensure that the regulatory instrument meets societal goals.

Finding 8-5. Choosing a metric associated with the movement of freight will promote improvements that increase the amount of cargo that can be carried per unit of fuel consumed, and thus provide a means of quantifying the benefits of more productive vehicles that move the same amount of freight with fewer trips and fewer vehicle-miles traveled, such as longer combination vehicles (LCVs).

Finding 8-6. Setting a metric based exclusively on gallons per cargo ton-mile (gal/ton-mile) may not adequately address light-density freight that is limited by volume.

Recommendation 8-3. NHTSA should establish fuel consumption metrics tied to the task associated with a particular type of medium- or heavy-duty vehicle and set targets based on potential improvements in vehicle efficiency and vehicle or trailer changes to increase cargo-carrying capacity. NHTSA should determine whether a system of standards for full but lightly loaded (cubed-out) vehicles can be developed using only the LSFC metric or whether these vehicles need a different metric to properly measure fuel efficiency without compromising the design of the vehicles.

Methods for Certification and Compliance

Finding 8-7. The regulatory certification and compliance options discussed in this report are the product of much discussion and thought by committee members, supported by input from industry, government, and other organizations. Some certification and compliance methods seem more practical than others, and the committee acknowledges that there may be other options or variations that have yet to be identified. Nevertheless, the committee has determined that regulating the total vehicle fuel consumption of medium- and heavy-duty vehicles will be a formidable task due to the complexity of the fleet, the various work tasks performed, and the variations in fuel-consumption-related technologies within given classes, including vehicles of the same model and manufacturer.

Finding 8-8. A certification test method must be highly accurate, repeatable, and identical to the in-use compliance tests as is the case with current regulation of light-duty vehicles tested on a chassis dynamometer, and for heavy-duty engine emission standards tested on engine dynamometers.

Finding 8-9. Using the process and results from existing engine dynamometer testing for criteria emissions to certify fuel economy standards for medium- and heavy-duty vehicles would build on proven, accurate, and repeatable methods and put less additional administrative burden on the industry. However, to account for the fuel consumption benefits of hybrid power trains and transmission technology, the present engine-only tests for emissions certification will need to be augmented with other power train components added to the engine test cell, either as real hardware or as simulated components. Similarly, the vehicle attributes (aerodynamics, tires, mass) will need to be accounted for, one approach being to use vehicle-specific prescribed loads (via models) in the test cycle. This will require close cooperation among component manufacturers and vehicle manufacturers.

Finding 8-10. At present there is no established federal test method for heavy-duty vehicle fuel consumption. Empirical testing (from components in an emulated vehicle environment to the whole vehicle), simulation modeling, or both may be used for the characterization and certification of regulated equipment. Each approach involves uncertainties that can affect certification and compliance. This finding underscores the need for a pilot regulation program.

Finding 8-11. Significant segments of the medium- and heavy-duty-vehicle purchasing process are highly consumer driven, with many engine, transmission, and drive axle choice combinations resulting in a wide array of completed vehicles for a given vehicle model. From a regulatory standpoint, the use of expensive and time-consuming chassis testing on each distinct vehicle variation is impractical. However, by knowing the effects of the performance of major subcomponents on fuel consumption, it may be practical to demonstrate compliance certification with vehicle standards by aggregating the subcomponents into a specified virtual vehicle for computers to evaluate fuel consumption of the completed vehicle.

Finding 8-12. Further research will be required to underpin the protocol used to measure key input parameters, such as tire rolling resistance and aerodynamic drag forces and to ensure the robustness of simulations for evaluating vehicle fuel consumption. These major components may be assembled through simulation to represent a whole-vehicle system, and models benchmarked to reliable data may be used to extend the prediction to a variety of vehicle types, by changing bodies (aerodynamic measures), tires, and operating weights associated with the power trains.

Recommendation 8-4. Simulation modeling should be used with component test data and additional tested inputs from power train tests, which could lower the cost and administrative burden yet achieve the needed accuracy of results. This is similar to the approach taken in Japan, but with the important clarification that the program would represent all of the parameters of the vehicle (power train, aerodynamics and tires) and relate fuel consumption to the vehicle task. Further, the combined vehicle simulation/component testing approach should be supplemented with tests of complete vehicles for audit purposes.

Finding 8-13. There is an immediate need to take the findings and recommendations in this report and begin the development of a regulatory approach. Significant engineering work is needed to produce an approach that results in fuel efficiency standards that are cost-effective and that accurately represent the effects of fuel-consumption-reducing technologies. The regulations should fit into the engineering and development cycle of the industry and provide meaningful data to vehicle purchasers.

Recommendation 8-5. Congress should appropriate money for and NHTSA should implement as soon as possible a major engineering contract that would analyze several actual vehicles covering several applications and develop an approach to component testing and related data collection in conjunction with vehicle simulation modeling to arrive at LSFC data for these vehicles. The actual vehicles should also be tested by appropriate full-scale test procedures to confirm the actual LSFC values and the reductions measured with fuel consumption reduction technologies in order to validate the evaluation method.

Recommendation 8-6. NHTSA should conduct a pilot program to "test drive" the certification process and validate the regulatory instrument proof of concept. It should have the following elements.

1. Gain experience with certification testing, data gathering, compiling, and reporting. There needs to be a concerted effort to determine the accuracy and repeatability of all the test methods and simulation strategies that will be used with any proposed regulatory standards and a willingness to fix issues that are found.
2. Gather data on fuel consumption from several representative fleets of vehicles. This should continue to provide a real-world check on the effectiveness of the regulatory design on the fuel consumption of trucking fleets in various parts of the marketplace and various regions of the country.

BIBLIOGRAPHY

Bradley, M.J. and Associates LLC. 2009. Setting the Stage for Regulation of Heavy-Duty Vehicle Fuel Economy and GHG Emissions: Issues and Opportunities. Washington D.C.: International Council on Clean Transportation. February.

Davis, S., S. Diegel, and R.G. Boundy. 2009. Transportation Energy Data Book, Edition 28. Rept. No. ORNL-6984. Knoxville, Tenn.: U.S. Department of Energy, Energy Efficiency and Renewable Energy.

DOC (U.S Department of Commerce), Census Bureau. 2005. 2002 Economic Census, Industry Product Analysis. March. Washington, D.C.

DOE, EIA (U.S. Department of Energy, Energy Information Administration). 2009a. Annual Energy Review 2008. DOE/EIA-0384(2008). Washington, D.C.: EIA.

DOE, EIA. 2009b. Annual Energy Outlook 2009. DOE/EIA-DOE/EIA-0383(2009). March. Washington, D.C.: EIA.

Federal Highway Administration, Highway Statistics Summary to 1995, Table VM-201A and Highway Statistics (annual releases), Table VM-1. Washington D.C.

NESCCAF, ICCT, SwRI. 2009. Regulation of Heavy-Duty Vehicle Fuel Economy and GHG Emissions: Issues and Opportunities. October. Available at www.nescaum.org/.../heavy-duty-truck-ghg_report_final-200910.pdf.

TIAX, LLC. 2009. Assessment of Fuel Economy Technologies for Medium- and Heavy- Duty Vehicles. Final Report. Report to the National Academy of Sciences. Cambridge, Mass. September.

ANNEX 8-1: COMPONENT-BASED FUEL CONSUMPTION ASSESSMENT METHOD

Introduction

The objective of this annex is to present, in broad terms, an example of a credible methodology for characterizing the fuel consumption of a medium- or heavy-duty vehicle by tests and/or simulation of the major components. This concept considers a certification and labeling approach for the main fuel consumption components that would be aggregated to provide fuel consumption performance of the completed vehicle. This approach recognizes that many fuel consumption improvements occur at the sub component level and provides a means of quantifying the performance of these components so that the final stage manufacturer will have reliable performance data to determine vehicle fuel consumption.

For advanced heavy-duty vehicle designs, the physical power train is not readily classed into separately operating sub-components such as engine and transmission, because there can be a high level of communication between these sub-components and because they are mutually controlled to achieve their function. For example, an engine and a hybrid drive train are inseparable in operation because the control system commands both major subcomponents to achieve propulsion. Even with the lowest technology option of a manual transmission attached via a clutch to an engine, the average human driver determines the engine operating envelope based on the duty cycle, the engine performance map, the transmission ratios and the drive axle ratio. The realistic engine operating envelope is not defined uniquely without these sub-components attached and in use. With some qualification, present day engine testing may not necessarily reflect how the engine will be used in a specific application insofar as the torques and speeds of the engine in use, and the nature of transients in use may not reflect the torques, speeds and transients employed in the test cell.

If simulation and physical testing are equally verifiable as facsimiles of the real-world operation of a vehicle, its components or its sub-components, then simulation and physical testing should be equally valid techniques for use in certification. For example, aerodynamic drag on a body may be found either by wind-tunnel testing or by computer-aided aerodynamic modeling, provided both methods can be shown to be accurate and repeatable. As a further example, the performance characterization of an engine and intelligent transmission combination may be found using a dynamometer in a test cell, or by fully characterizing the engine and transmission separately and combining them with a model for their controller through a simulation exercise. On the one hand, when simulation is used, each component in the simulation must be fully characterized, usually necessitating a larger number of measurements of subcomponents. On the other hand, a simulation proves economical in modeling the effect changes in controls or architecture, or in examining different trucks employing some similar components, since these changes would otherwise necessitate a large number of tests of a whole component or major component subsystem.

The accepted methods used to measure rolling resistance of wheels (with tires), the efficiency of a power train, and the drag characteristics of a body all differ substantially. The existing test methods may need further development to achieve fidelity with real world vehicle operating results.

- Tire rolling resistance is determined independently and physically in accordance with accepted test protocol. Power train performance is determined physically either by attaching a power train to a dynamometer or by operating a chassis or mule containing a power train with wheels on the drive axle on a chassis dynamometer: in the latter case the wheels and dynamometer rollers are merely connection components between the power train and a rotating dynamometer. Aerodynamic body performance is determined in a wind tunnel. These three test approaches are substantially different and independent from one another, and employ different apparatus.
- An admixture of these characterizations may be obtained through the coast down of a whole vehicle on the road, but the metrics required are intertwined and the accuracy of the process is challenged by surface and atmospheric effects.
- Tire rolling resistance may be simulated through computer-aided finite element design models relying on fundamental materials stress and deformation equations. Power train performance may be simulated by using models currently available for whole vehicle simulation, by assembling accurate sub-models of sub-components with links that rely on basic physics (torques and speeds) or precisely defined control algorithms. Aerodynamic body performance may determined using finite element fluid flow models which may vary in their level of empirical tuning. These three modeling processes are substantially different from one another, and employ different types of code for their execution.

The Component-Based Procedure

The terms "measured" and "measurement" below are intended to reflect either output from testing activities or output from modeling or simulation activities, with no preference for either, but with the assertion that either approach requires verified fidelity with regard to accuracy and precision.

The three components, namely body, power train and wheels (with tires), of a vehicle proposed for sale should be measured separately. Certain accessories may also require independent measurement if they are not easily included in the other three.

Wheels (Including Tires)

The wheels of a vehicle are usually characterized with respect to rolling resistance. The rolling resistance, as a coefficient, represents the horizontal force which must be applied to overcome the internal energy losses of the tire when it supports a given vertical weight on a horizontal surface. The coefficient of rolling resistance (C_{rr}) is reasonably constant with respect to speed and vertical load. The rolling resistance is influenced largely by the tire material construction, wear, and to a lesser degree the road surface.

Aerodynamics

The aerodynamics of the vehicle should be determined by installing a body on a suitable facsimile of wheels and chassis in a wind tunnel, or else by installing a whole vehicle in a wind tunnel. A scaled model may be used where it can be demonstrated by similarity analysis that the resulting measurements may be applied to a full scale measurement. The measurement of interest is the drag force on the body as a function of wind speed over the body. Drag force varies in close proportion to the square of the air speed, and is influenced by air density and by yaw (which results in real vehicle operation from the presence of wind which is not in the direction of travel and which is not very low in speed with respect to the vehicle speed.) Computer simulation of air flow over the body may be used to infer the drag force, provided appropriate controls are used.

Certain trucks and tractors simply would not benefit from some aerodynamic accouterments, either because they are not intended to be driven at high speed, or because they do not carry a box body or tow a box trailer with a large frontal area. For vehicles which are not intended to be driven at high speed, determination of C_dA would be purposeless.

Power Train

The power train of the vehicle consists typically of an engine, a transmission, which may include hybrid hydraulic or hybrid electric components, one or more drive axles, possibly an energy storage system, and a control system to manage the components in response to driver commands under constraint of road load. The exhaust aftertreatment system, often considered part of the engine, may evolve to the status of a separate component in future versions of testing.

A power train may be used with a variety of tires and bodies in real vehicle applications, and may be configured or optimized differently for each application, or configured generically for use in several applications.

Vehicle

Different vehicle types, categorized by weight class and use, may each be associated with one or more drive cycles, reasonably corresponding to real use of those vehicles. For example, an over-the-road tractor may be associated with a high-speed cycle, indicative of freeway behavior, and a low-speed cycle, indicative of transient behavior in an urban environment. Consider that the efficiency of a power train is to be measured or modeled, and that its associated application, body and wheels are all defined. These drive cycles exemplifying the application can be translated to a set of hub speed (versus time) and hub torque (versus time) target values, provided that the power train is considered to be propelling this well-defined vehicle. Data which are required to formulate a power train test (physical or simulated) are largely the same as those required to execute a light-duty vehicle chassis test and are as follows:

- Vehicle mass
- Tire rolling diameter
- Effective C_{rr} value for the wheels
- Effective C_dA value or the aerodynamic drag
- Value for air density
- Test cycle, as a set of speed versus time values

The power train should then be exercised through the speed-torque target values either physically in a test cell or through simulation. A human driver will be needed, although physical testing and simulation may otherwise employ a driving algorithm, provided that the algorithm reasonably represents a human driver. The choice of driver or driving algorithm must be addressed carefully, because it may impact engine transient behavior and manual transmission behavior substantially. For rapid decelerations, it will be necessary to use friction brakes or a retarder to provide deceleration torque, and for hybrid vehicles the decelerations may be used for energy capture. A physical power train test may also be accomplished with a mule or complete vehicle on a chassis dynamometer, but the dynamometer coefficients measured or projected at the drive hubs must be set to reflect the values of C_{rr} and C_dA required for the designated power train test, and not necessarily for the vehicle on which the test is being performed.

Assembled Components

The tires, aerodynamics and power train might all be separately regulated. However, they might also be combined to mimic a completed vehicle. Experimentally, a power train test will provide whole vehicle fuel consumption data if the power train test uses values for aerodynamic drag and for tire rolling resistance which represent that vehicle. However, it is inappropriate to expect that each variant of a vehicle should require a separate power train test. If modeling is used in a verifiable fashion to mimic the power train and provide efficiency data for a specific vehicle, then the same model may be modified with reasonable confidence to accommodate varying values for aerodynamic drag, rolling resistance, and vehicle weight. In this way, a wide variety of vehicles which use the same power train may be simulated economically.

For any particular power train configuration, the power train performance may be confirmed on two or three cycles using weight, aerodynamic drag and tire rolling resistance suited to a reasonable vehicle type. Data of this kind may be extended through modeling to reveal the power train performance for any test cycle. If several cycles are executed, the variation of fuel consumed with respect to average speed can be computed, and these data would be available for consumer information and as regulatory metrics. As feedback controls become common, power trains will use many sensor inputs to ride up against NTE (Not to Exceed) limits for NO_x. This means that actual fuel consumption may vary significantly as a function of ambient temperature, humidity, intake manifold temperature, coolant and oil temperatures, barometric pressure, aftertreatment temperature, aftertreatment aging, and other factors. Some reasonable consensus on standard test conditions will be important for reporting against a fuel consumption standard.

ANNEX 8-2: ALTERNATIVE METRICS

The measures listed in Table 8-3 address the vehicle task and are based on fuel or energy consumption. This list is not complete as there are many vehicle tasks not covered. This annex elaborates on some of these tasks.

One alternative approach to regulating tractor trailer fuel consumption would be to simply regulate the tractor based on standard loads tied to engine power rating. Tractors could be grouped into power ranges such as low-, medium-, and high-horsepower categories, and corresponding GVW values applied. Each of these power categories would have separate fuel consumption targets. See Figure 8-2-1.

Consider the single vehicle units shown in Figure 8-2-2 with the same GVW rating and the identical power train and chassis. For the purpose of this discussion it is assumed that the flat bed unit is used for heavy loads such as steel or lumber and the box truck is used for lighter density cargo such as courier packages. If the mass metric (gal/cargo ton-mile) were used then there would be strong incentive to minimize vehicle tare weight so that the cargo mass term in the metric could be increased thereby improving performance. This would have a clear benefit for the flat bed truck as the cargo is of sufficient density to benefit from the lighter vehicle. However for the box truck with low density cargo, the incentive for reduced tare weight may not provide any direct benefit to the shipper. In addition, when the mass metric is applied to the box truck, it would give preference to a lighter smaller box which would undermine the volumetric value of the vehicle. A metric based on volume (gal/cargo ft^3-mile) would resolve this particular application.

Operating characteristic	Candidate Power Units	Standard Trailers	Performance metric
High speed: Aero Med speed Aero + hybrid Low speed hybrid	 Long haul tractor  Flat deck / short haul tractor (Can be high or low speed)	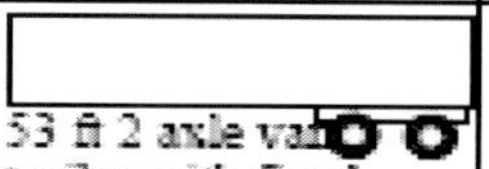 53 ft 2 axle van trailer with fixed tare weight and variable cargo mass summing to GVW 80,000 lbs 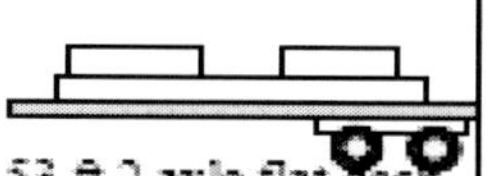 53 ft 2 axle flat deck trailer with fixed tare weight, with variable cargo shape and mass summing to GVW 80,000 lbs	Gal/ton-mile (kWh/cargo ton-km) Gal/cargo ft^3 -mile (kWh/cargo m^3-km)
Med speed Aero + hybrid Low speed hybrid			Gal/cargo ft^3 -mile (kWh/cargo m^3-km)
Med speed Aero + hybrid Low speed hybrid	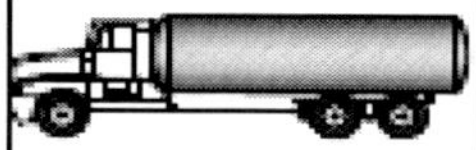		Gal/cargo ft^3 -mile (kWh/cargo m^3-km) or Gal/cargo ton -mile (kWh/cargo ton-km)
Service duty cycle	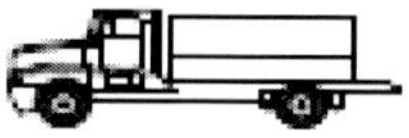		Gal/hr (kWhr / hr) or Gal/cargo ton -mile (kWh/cargo ton-km)
High speed: Aero Med speed Aero + hybrid	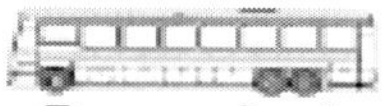 Buses: school, shuttle, transit, motorcoach		Gal/passenger, cargo ton mile (kWh/passenger cargo ton-km)
Low speed hybrid			
Corporate Average fuel Economy (CAFE)	Small trucks, pickups, and vans		MPG

FIGURE 8-2-1 Options for performance metrics.

FIGURE 8-2-2 Identical GVW rated straight trucks for high- and low-density commodities.

Appendixes

A

Statement of Task

The committee will conduct an assessment of fuel economy technologies for medium- and heavy-duty vehicles. According to the Energy Independence and Security Act of 2007, Section 108, the study is to develop a report evaluating medium-duty and heavy-duty truck fuel economy standards. Based on the language in Section 108, the NRC committee formed to carry out this study will address the following tasks, all of which will be discussed with the DOT/NHTSA representatives, as well as any relevant Congressional staff, at the committee's first meeting. The purpose of these discussions will bring the benefit of the expertise of the committee to bear on what information and data could be made available to the committee, which will determine the extent to which the tasks can be addressed. Thus, the committee will:

(1) consider appropriate approaches to measuring fuel economy for medium- and heavy-duty vehicles that would be required for setting standards. Given the diversity of vehicles and applications, consideration of classification of vehicles will likely be required in this review. In addition, the committee will likely have to work with DOT/NHTSA, EPA, and others, as appropriate, to identify a reasonable approach, which would then lay the basis for what technologies to consider for potential improved fuel economy.

(2) assess current and potential technologies and estimate improvements in fuel economy for medium-duty and heavy-duty trucks that might be achieved. The committee will need to decide on what time frame is appropriate to consider for the technology assessment. In addition, the committee should try to estimate the costs of technologies for fuel economy improvements. Costs may be difficult to estimate given the proprietary nature of the business and an approach may need to be worked out with NHTSA/DOT and the industry to collect information that would provide the committee with enough confidence in estimating a range of costs.

(3) an analysis of how the technologies identified in Task 2 above may be used practically to improve medium-duty and heavy-duty truck fuel economy. This will likely entail a discussion by the committee of barriers, time frames, competitive pressures, and other factors that may inhibit or accelerate the adoption of technologies for improved fuel economy.

(4) an analysis of how such technologies may be practically integrated into the medium-duty and heavy-duty truck manufacturing process. Again, the committee will likely identify barriers, timing, competitive pressures, and other factors that may inhibit or accelerate the practical implementation into the manufacturing of the various vehicles under consideration.

(5) an assessment of how such technologies may be used to meet fuel economy standards to be prescribed under section 32902(k) of title 49, United States Code, as amended by this subtitle.

(6) identify the potential costs and other impacts on the operation of medium-duty and heavy-duty trucks. For those technologies that can be integrated into the design of engines, vehicles, and trailers, it is likely that the committee would focus on any incremental costs and whether there are any special requirements for these technologies that might affect operation of such vehicles. The language in Section 108 refers to "congestion" as well, and this may be an issue for those systems engineering technologies, for example, integrated intelligence systems that may provide the opportunity to control traffic flow. It is anticipated that the committee would review any studies conducted on this subject to address this congestion issue.

(7) write a report documenting its conclusions and recommendations.

B

Presentations and Committee Meetings

COMMITTEE MEETING, WASHINGTON, DC DECEMBER 4-5, 2008

Opening Remarks
David Strickland, Senate Commerce Committee

Overview of the NHTSA Program, Including Sponsor's Expectations for the Study
Stephen Kratzke, National Highway Traffic Safety Administration

SmartWay Partnership and Heavy Vehicles
Mitchell Greenberg, U.S. Environmental Protection Agency

U.S. Department of Energy Truck Technology
Kenneth Howden, U.S. Department of Energy

Engine Manufacturers Association
Timothy Blubaugh, Engine Manufacturers Association

Reducing Emissions in Heavy Vehicles
Anthony Greszler, Volvo Powertrain

Freightliner
David Kayes, Daimler Trucks North America

Heavy-Duty Vehicle Fuel Economy and Emissions Improvement Project of the Northeast States Coalition for a Clean Air Future and the International Council for Clean Transportation
Coralie Cooper, The Clean Air Association of the Northeast States (NESCAUM)

Southwest Research Institute Studies to Support the NESCCAF Heavy-Duty Vehicle Project
Thomas Reinhart, Southwest Research Institute

EPA SmartWay Truck Emissions Test Protocol
Mitchell Greenberg, U.S. Environmental Protection Agency

Commercializing Hybrid and High-Efficiency Trucks: Hybrid Truck Users Forum: Program Status, Update, and Next Steps
Bill Van Amburg, CALSTART

Fuel-Efficiency Technologies for Heavy Vehicles
Gurpreet Singh, U.S. Department of Energy

Heavy-Duty Hybrid Technology
Willam A. Batten, Eaton Corporation

Heavy-Duty Trucks Fuel Economy Technology
K.G. Duleep, Energy and Environmental Analysis

COMMITTEE MEETING, WASHINGTON, DC FEBRUARY 4-5, 2009

Hybrid Status
Mike Roeth, Navistar

Heavy-Duty Trucks
Terry Penney, National Renewable Energy Laboratory

Truck Tires and Rolling Resistance
Calvin Bradley, Michelin Tire North America

Medium-Duty and Heavy-Duty Commercial Vehicle Business
Mike Roeth, Navistar

Fuel Efficiency Study
Jeff Seger, Cummins, Inc.

Heavy-Duty Fuel Economy Computer Models and Simulation
Daniel Kieffer, Kenworth Truck Company

Medium- and Heavy-Duty Vehicle Models: State-of-the-Art and Challenges
Aymeric Rousseau, Argonne National Laboratory

A Modular Approach to Fuel Efficiency
Stefan Larsson, European Auto Manufacturers

Japanese Fuel Efficiency Regulation
Akihiko Hoshi, Ministry of Land, Infrastructure, Transport, and Tourism

Simulation Method
Akihiko Hoshi, Ministry of Land, Infrastructure, Transport, and Tourism

Final Results Summary
Tom Reinhart, Southwest Research Institute

Cost Effectiveness of Heavy-Duty Vehicle Technologies
Bob Wilson, TIAX LLC

SmartWay-Certified Tractor and Trailers
Mitchell Greenberg, U.S. Environmental Protection Agency

COMMITTEE MEETING, DEARBORN, MI APRIL 6-7, 2009

Fuel Consumption as a Metric
John Johnson, Michigan Technological University

International Council on Clean Transportation Research
Drew Kodjak, International Council on Clean Transportation Research

Modeling and Simulation Issues
Ron Graves, Oak Ridge National Laboratory

Vehicle Classification and Mapping
John Woodrooffe, University of Michigan Transportation Institute

Technology Matrix
Dave Merrion, Detroit Diesel (retired)

Indirect Costs
James Winebrake, Rochester Institute of Technology

Industrial Perspectives of the 21st Century Truck Partnership
Vinod Duggal, Cummins, Inc.

Army Ground Vehicles
Paul Skalny, U.S. Army

EPA Hybrid Technology
John Kargul, U.S. Environmental Protection Agency

Report on Results
Michael Jackson, TIAX LLC

COMMITTEE MEETING, ANN ARBOR, MI JUNE 18-19, 2009

LTL Carrier's View of Past and Current Technologies
Duke Drinkard, Southeastern Freight Lines (retired)

Effective Practices and Programs for Improving FE
Dave Miller, Con-Way Freight, Inc.

Passenger Car and Light Truck CAFE Analysis and Tech Inputs
Ryan Harrington, NHTSA

EPA Vehicle Standards and GHG Reductions
Byron Bunker, EPA

Measuring and Modeling Traffic Congestion Impacts on Heavy-Duty Trucks
Matthew Barth, UC Riverside

Safety and Productivity Issues Long Combination Vehicles
John Woodrooffe, UMTRI

The Potential of Intermodalism for Meeting Energy and Environmental Goals
James Winebrake, Rochester Institute of Technology

Assessment of FE Technologies for Medium- and Heavy-Duty Vehicles
Joe Morris, NAS

ArvinMeritor's Advanced Safety Technologies for Commercial Vehicles
Brad Hicks and Alan Korn, ArvinMeritor

Vehicle Integration Team, NRC
Ron Graves, Andrew Brown, Jr., Aymeric Rousseau, Garrick Hu, NRC

The Aluminum Advantage: Exploring Commercial Vehicle Applications
Randall Scheps, Alcoa

Truck Aerodynamics
Charles Salter, Consultant

COMMITTEE MEETING, SAN ANTONIO, TX AUGUST 6-7, 2009

Intelligent Vehicle Applications
Chelsea White, Georgia Tech

Framework for the Regulation of Greenhouse Gases from Commercial Vehicles
John Wall, Cummins, Inc.

Wind Tunnel Studies of the Aerodynamics of Heavy Vehicles
Fritz Marinko and Mitch Camosy, Auto Research Center, LLC

SWRI Overview
Walt Downing, SwRI

Medium-Duty HEDGE
Chris Chadwell, SwRI

Fuel-Saving Opportunities with Intelligent Highway Systems
Ryan Lamm, SwRI

NO_x/BSFC Trade-offs in Modern Diesel Engines
Tom Ryan, SwRI

SmartWay Testing
Cheryl Bynum, EPA

COMMITTEE MEETING, WASHINGTON, DC SEPTEMBER 23-24, 2009

Modeling Fuel Consumption and CO_2 Emissions from Heavy-Duty Vehicles
Stefan Larsson, ACEA

Reducing Heavy-Duty Long-Haul Combination Truck Fuel Consumption and CO_2 Emissions
Tom Reinhart, SwRI

C

Committee Biographical Sketches

Andrew Brown, Jr. (NAE) is executive director and chief technologist at the Delphi Corporation, where he reports to CEO/president and division presidents on matters of innovation and technology. He represents Delphi globally in outside forums on matters of technology and innovation, including government and regulatory agencies, customers, alliance partners, vendors, contracting agencies, and academia. Previously, he was responsible for leading Delphi's engineering community of 17,000 engineers and scientists with a budget of nearly $2 billion annually. Dr. Brown came to Delphi from the General Motors Research and Development Center in Warren, Michigan, where he served as director of strategic futures and research and was responsible for managing GM's envisioning process and directing various research projects. He served as manager of Saturn car facilities from 1985 to 1987. At Saturn he was on the Site Selection Team and was responsible for the conceptual design and engineering of this innovative manufacturing facility. Dr. Brown began his GM career as a project engineer in manufacturing development in 1973. He progressed in the engineering field as a senior project engineer, staff development engineer, and manager of R&D for the manufacturing staff. During this period he worked on manufacturing processes and systems with an emphasis on energy systems, productivity improvement, and environmental efficiency. Before joining GM he supervised process development at Allied-Signal Corporation, now Honeywell, Inc., in Morristown, New Jersey. He earned a bachelor of science degree in chemical engineering from Wayne State University (WSU) in 1971. He received a master of business administration in finance and marketing from WSU in 1975 and a master of science degree in mechanical engineering with a focus on energy and environmental engineering from the University of Detroit, Mercy, in 1978. He completed the Penn State Executive Management Course in 1979. A registered professional engineer, Dr. Brown earned a doctorate of engineering in 1992. He is currently or has served on the boards of the following organizations: Society of Automotive Engineers, Engineering Society of Detroit, Convergence Education Foundation, National Inventors Hall of Fame, Convergence Transportation Electronics Foundation, National Council of Engineering Examiners, State of Michigan Board of Professional Engineers, and the WSR College of Engineering Board of Advisors. Dr. Brown has been an adjunct professor at WSU, the University of Michigan, and Tsinghua University (Beijing, China). As an ordained deacon, he has conducted religious/medical missions in Jamaica and South Africa.

Dennis N. Assanis (NAE) is the Jon R. and Beverly S. Holt Professor of Engineering in the Mechanical Engineering Department; director of the Michigan Memorial Phoenix Energy Institute, co-director of the General Motors Engine Systems Research Collaborative Research Laboratory; and director of the W.E. Lay Automotive Laboratory at the University of Michigan. Dr. Assanis is recognized internationally for his innovative development of modeling methodologies and experimental techniques that have shed light on complex thermal, fluid, and chemical phenomena that occur in internal combustion engines and exhaust aftertreatment systems, with applications to design integration and control of complex and hybrid power train systems. He was elected to the National Academy of Engineering in 2008 for scientific contributions to improving fuel economy and reducing emissions of internal combustion engines and for promoting automotive engineering education. His Ph.D. (power and propulsion) was awarded by the Massachusetts Institute of Technology in 1985. He holds three MIT M.S. degrees: one in management, from the Sloan School of Management, 1986; one in mechanical engineering, 1982; and one in naval architecture and marine engineering. He also earned a B.Sc. in marine engineering from Newcastle University in 1980.

Roger Bezdek is president of Management Information Services, Inc., a Washington, D.C.-based consulting firm. Dr. Bezdek has 30 years experience in consulting and management in the energy, utility, environmental, and regulatory areas. He has served as corporate director, corporate president and CEO, university professor, research director at the Energy Research and Development Administration/Department of Energy, senior advisor on energy in the Office of the Secretary of the Treasury, and as a participant in the U.S. State Department's AMPART program for Asia. He has been a consultant to the White House, federal and state government agencies, and various corporations and research organizations, including the National Academy of Sciences. His consulting background includes energy technology and market forecasting, estimating costs and benefits of environmental legislation and regulation, assessment of energy and environmental R&D programs, and energy price and market forecasting. A recent study he conducted evaluated the costs and benefits of light-duty fuel economy standards. He also served as a member of the National Research Council's Committee on Energy Futures and Air Pollution in Urban China and the United States. Dr. Bezdek received his Ph.D. in economics from the University of Illinois, Urbana.

Nigel N. Clark is professor and director of the Center for Alternative Fuels, Engines, and Emissions, Mechanical and Aerospace Engineering, and holds the George Berry Chair of Engineering, College of Engineering and Mineral Resources, University of West Virginia (UWV). He has also held assistant and associate professor positions there. Previous positions include factory survey engineer with the Water Research Commission, Durban, South Africa, and contract researcher for the Council for Mineral Technology, Durban, South Africa. Dr. Clark's areas of interest include vehicle design, advanced vehicle concepts, alternative fuels, and measurement and reduction of vehicle emissions. He has published extensively in the areas of particle science and multiphase systems. He has conducted research and published in the areas of fuel economy and emissions from heavy-duty vehicles, including buses and heavy hybrid vehicles. He received the College of Engineering and Mineral Resources Outstanding Researcher Award and the Researcher of the Year Award from the UWV, College of Engineering and Mineral Resources. He has a Ph.D. in engineering and a B.S. in chemical engineering from the University of Natal, South Africa.

Thomas M. Corsi is professor and co-director of the Supply Chain Management Center, Robert H. Smith School of Business, University of Maryland. He joined the Robert H. Smith School of Business in 1976 as a professor of logistics and transportation. He served as chairperson of the Logistics and Transportation Group from 1986 through 1994. During that time the group received recognition from the *Transportation Journal* as the most prolific faculty group in the nation based on published research in the field. He is an associate editor of the *Logistics and Transportation Review* and serves on the editorial review board of the *Transportation Journal*. He has authored more than 100 articles on logistics and transportation. He has consulted for such organizations as the Interstate Commerce Commission, the Maryland State Department of Transportation, the National Science Foundation, the U.S. Department of Transportation, the National Truck Stop Operators, United Parcel Service, the U.S. Department of Energy, and the U.S. Army Logistics Agency. He has authored three books: *The Economic Effects of Surface Freight Deregulation* (Brookings Institution, 1990), *Logistics and the Extended Enterprise: Benchmarks and Best Practices for the Manufacturing Professional* (John Wiley & Sons, 1999), and *In Real Time: Managing the New Supply Chain* (Praeger Books, 2004). He holds a Ph.D. in geography from the University of Wisconsin, Milwaukee.

Duke Drinkard is retired and works part-time as president of the 21st Century Driver and Truck Alliance and as an energy consultant. After high school he worked for Southeastern Freight Lines for 48 years in various jobs, including; dock worker, all jobs in the maintenance shop, pickup and delivery driver, road driver, dock foreman, driver trainer, building maintenance, real estate, field engineer, director of maintenance, and vice president of maintenance. He was a member of the North Carolina Maintenance Council, founding member of the South Carolina Maintenance Council, and past member of the RCCC and ATA's TAG committee. He served in all positions, including chair of the TMC and the South Carolina Maintenance Council, chair of the Future Truck Committee and VMRS Committee, and vice chair of the S-6 and S-12 study groups of TMC, and he currently cochairs the Far Horizons Committee and is a member of the S-11 Energy Conservation Study Group. He has made presentations before many groups, including the Society of Automotive Engineers and the New Zealand Safety Council. He holds an associate degree in mechanical engineering from Midland Technical College.

David E. Foster is professor of mechanical engineering, University of Wisconsin, Madison, and director of the Engine Research Center. A member of the faculty at the University of Wisconsin since completing his Ph.D., Dr. Foster teaches and conducts research in thermodynamics, fluid mechanics, internal combustion engines, and emission formation processes. His work has focused specifically on perfecting the application of optical diagnostics in engine systems and the incorporation of simplified or phenomenological models of emission formation processes into engineering simulations. He is a recipient of the Ralph R. Teetor Award, the Forest R. McFarland Award, and the Lloyd L. Withrow Distinguished Speaker Award of the Society of Automotive Engineers and is an SAE Fellow. He has served on a number of National Research Council committees including the Committee on Review of the Research Program of the Partnership for a New Generation of Vehicles and the FreedomCAR and Fuel Partnership. He is a registered professional engineer in the State of Wisconsin and has won departmental, engineering society, and university awards for his classroom teaching. He received a B.S. and an M.S. in mechanical engineering from the University of Wisconsin and a Ph.D. in mechanical engineering from the Massachusetts Institute of Technology.

Roger D. Fruechte retired from General Motors in 2003 as director of the Electrical and Controls Integration Lab at GM's R&D center in Warren, Michigan, and as co-director of the Collaborative Research Laboratory at Carnegie Mellon University. He was responsible for research in the areas of active safety, including crash avoidance, vehicle electrical architecture, chassis and power train control, hybrid vehicles, and telematics. He began his career with GM as a development engineer with the Delco Electronics Division in Milwaukee, Wisconsin. He then spent 31 years at the GM R&D center working on various automotive control system projects and intelligent transportation systems. He currently serves as a member of the Vincent Bendix Automotive Electronics Engineering Award Board for SAE, as a design judge for the Intelligent Ground Vehicle Competition, and as a member of Kettering University's ECE Industrial Advisory Board. He received the B.E.E. from Kettering University, an M.S.E.E. from the University of Toledo, and a Ph.D. in electrical engineering with a specialty in automatic control from the University of Wisconsin, Madison.

Ron Graves is director of the Fuels, Engines, and Emissions Research Center (FEERC) with programmatic, technical, and strategic responsibility for this U.S. Department of Energy (DOE) User Facility and the numerous projects conducted therein. He joined Oak Ridge National Laboratory (ORNL) in 1976. He was national project manager for the DOE Alternative Fuels Utilization Program from 1984 to 1990, during which he started the fuels-engine laboratory at ORNL that grew to be FEERC. He was technical manager of DOE's earliest projects in diesel emission controls. He was chosen by DOE to be technical coordinator for the Diesel Crosscut Team in 1997 and continues in that role. He was a member of the DOE/Industry Advanced-Petroleum Based Fuels Steering Committee, the DOE program that contributed heavily to the U.S. Environmental Protection Agency rule for lowering sulfur in diesel fuel in December 2000. Dr. Graves is ORNL's representative to the 21st Century Truck Partnership "Lab Council" and is responsible for facilitating the engine-fuels efforts in that government-industry initiative. He was a major contributor to DOE's heavy vehicle R&D plans from 1983 to 1997 and then authored the emission-control sections of the 21st Century Truck Technical Roadmap in 2000. He is an invited member of the FreedomCAR Advanced Combustion and Emission Control Tech Team and also a member of the Coordinating Research Council Working Group on Advanced Vehicle Fuels and Lubricants. He has a record of over 55 publications and reports that encompass subjects in fossil energy, internal combustion engines, fuels, and materials. He is a fellow of the Society of Automotive Engineers and has organized or chaired over 20 technical sessions at technical conferences. He has three patents, with an additional one in progress. He is a licensed professional engineer in the State of Tennessee and has a Ph.D. in mechanical engineering from the University of Tennessee.

Garrick Hu retired in 2008 after 36 years in the commercial vehicle industry and now works as a consultant in the area of strategic technology related to heavy trucks. He last served as vice-president of global engineering for ArvinMeritor Commercial Vehicle Systems. While at ArvinMeritor he directed the concept and development of a plug-in battery electric vehicle program with Unicell and Purolator, as well as the concept and development of a dual-mode hybrid electric Class 8 vehicle in partnership with International Truck and Engine, Cummins Engine Company, and WalMart. Prior to joining ArvinMeritor, he was group vice-president of advanced engineering for Volvo Global Trucks. He also served as a group vice-president for the Renault/Mack Group and as senior vice-president for Mack Truck Company. He has also worked as director of advanced vehicle systems concepts and development for International Truck and Engine Company. He was director of engineering at Kenworth Truck Company and general manager of the Paccar Technical Center. He has served as vice-chair of the Truck Manufacturers Association and as chairman of the Society of Automotive Engineers Vehicle Dynamics subcommittee. He is on the external advisory board of the University of Michigan Transportation Research Institute and serves on the visiting committee of the University of Michigan Dearborn College of Engineering. He holds undergraduate and graduate degrees in mechanical engineering from the University of Michigan and an M.B.A. from Chapman College.

John H. Johnson is a presidential professor with the Department of Mechanical Engineering-Engineering Mechanics, Michigan Technological University (MTU), and a fellow of the Society of Automotive Engineers and the American Society of Mechanical Engineers. His experience spans a wide range of analysis and experimental work related to advanced engine concepts, diesel and other internal combustion engine emissions studies, fuel systems, and engine simulation. He was previously a project engineer with the U.S. Army Tank Automotive Center and chief engineer at Applied Engine Research, the International Harvester Co., before joining the MTU mechanical engineering faculty. He served as chair of the MTU mechanical engineering and engineering mechanics department from 1986 to 1993. He has served on many committees related to engine technology, engine emissions, and health effects—for example, committees of the Society of Automotive Engineers, the National Research Council (NRC), the Combustion Institute, the Health Effects Institute, and the U.S. Environmental Protection Agency, and he serves as a consultant to a number of government and private-sector institutions. In particular, he served on the NRC Committee on Fuel Economy of Automobiles and Light Trucks, the Committee on Advanced Automotive Technologies Plan, and the Committee on Impact and Effectiveness of Corporate Average Fuel Economy (CAFE) Standards and chaired the Committee on Review of DOE's Office of Heavy Vehicle Technologies. He recently served as the chair of the NRC Committee on Review of the 21st Century Truck Partnership and is a member of the NRC Committee on Technologies for Improving the Fuel Economy of Light-Duty Vehicles. He received his Ph.D. in mechanical engineering from the University of Wisconsin.

Drew Kodjak is executive director of the International Council on Clean Transportation (ICCT), a group of government environmental regulators and international experts from around the world who participate as individuals with a common purpose of improving the environmental performance and efficiency of vehicles and fuels. Prior to joining the ICCT in 2005, Mr. Kodjak served as program director for the D.C.-based National Commission on Energy Policy, a bipartisan 16-member commission of energy experts that released a highly influential report, *Ending the Energy Stalemate*, in December 2004. Earlier, Mr. Kodjak spent several years as an attorney-advisor to the U.S. Environmental Protection Agency's Office of Transportation and Air Quality in Ann Arbor, Michigan. During his tenure with the EPA, Mr. Kodjak was awarded the Gold Medal for his work on the heavy-duty diesel rule. Mr. Kodjak is a member of bar associations in Minnesota, New Jersey, and the District of Columbia Court of Appeals.

David F. Merrion is chair of David F. Merrion, LLC, chair of Green Vision Technology, and a member of the board of directors of Clean Diesel Technologies, Inc. and Hy-Drive Technologies, Ltd. He retired as executive vice president of engineering for Detroit Diesel Corporation (DDC). His positions at DDC included staff engineer, emissions and combustion; staff engineer, research and development; chief engineer, applications; director, diesel engineering; general director, engineering (engines and transmissions); and senior vice president, engineering. He has extensive expertise in the research, development, and manufacturing of advanced diesel engines, including alternative fueled engines. He is a Society of Automotive Engineers fellow and member of the American Society of Mechanical Engineers. He served as former president of the Engine Manufacturers Association and as a member of Environmental Protection Agency's (EPA) Mobile Sources Technical Advisory Committee, the Coordinating Research Council, and the U.S. Alternate Fuels Council. He served on the National Research Council's Standing Committee to Review the Partnership for a New Generation of Vehicles program and more recently as a member of the Committee on Review of the 21st Century Truck Partnership. He is a consultant to the DDC, which included compliance auditor for the consent decree signed with EPA/California Air Resources Board/Department of Justice in 1998. He has a B.S. in mechanical engineering from General Motors Institute/Kettering University and an M.S. in mechanical engineering from the Massachusetts Institute of Technology.

Thomas E. Reinhart is program manager, Engine Design & Development, Engine, Emissions, and Vehicle Research Division, Southwest Research Institute. His previous positions were with Cummins, Inc., Columbus, Indiana, 1980-2000 (Noise, Vibration and Harshness [NVH] engineer, 1980-1984; senior engineer, Midrange Engine NVH, 1984-1987; manager, Noise and Vibration Technology, 1987-1994; director, Noise and Vibration Technology, 1994-2000); Roush Industries, Inc., Livonia, Michigan, program manager—Powertrain NVH, 2001-2004; and Visteon Corporation, Van Buren Township, Michigan, senior manager—chassis systems NVH, 2004-2005. He leads projects in engine design, performance, and emissions development and gasoline and diesel engine NVH improvement. He has led a number of programs, including several emissions reduction projects and the clean sheet design and development of a new off-highway diesel engine. Mr. Reinhart has over 25 years of experience in diesel engine and power train design, analysis, and development, with particular expertise in noise and vibration testing and analysis. He has published 14 technical papers on a range of diesel NVH topics. He has a wide range of experience in the NVH issues of applications, ranging from trucks through agricultural equipment, construction, forestry, marine, rail, and military vehicles. Mr. Reinhart has worked with customers on a range of issues, including NVH, drivability, fuel consumption, and adaptation of engines to a wide range of applications. He holds four patents for ideas related to diesel engine NVH control. For several years he was a member of Cummins' patent review committee. His work experience also covers a wide range of development projects on gasoline and diesel engines, as well as transmissions and fuel cell vehicle power trains. He is a member of the Institute of Noise Control Engineering (INCE), the Society of Automotive Engineers, and the International Institute of Acoustics and Vibration. He is also a member of the board of directors of INCE. He has been a member of the organizing committee for the SAE Noise and Vibration Conference since 2002 and chair of the Diesel Noise session at this conference since 2003. He has a B.S. and an M.S. in mechanical engineering from Purdue University.

Aymeric P. Rousseau is manager of the Advanced Powertrain Vehicle Modeling Team at Argonne National Laboratory. He received his engineering diploma at the Industrial System Engineering School in La Rochelle, France, in 1997. After working for PSA Peugeot Citroen for several years in the hybrid electric vehicle research department, he joined Argonne National Laboratory where he is now responsible for development of the Powertrain System Analysis Toolkit (PSAT). PSAT is the primary vehicle model for all FreedomCAR and 21st Century Truck Partnership activities by the

U.S. Department of Energy (DOE) and counts several hundred users. PSAT is currently used by the DOE to support the Advanced Technology Vehicles Manufacturing Loan Program. He was awarded an R&D 100 Award in 2004, presented to the 100 most technologically significant new products and processes introduced into the market each year for the development of PSAT. PSAT is currently used by more than 130 companies worldwide with more than 750 users. He also has helped to provide direction to DOE's R&D activities with the publication of dozens of technical papers.

Charles K. Salter is retired after working 39 years with Mack Trucks, Inc./Volvo PowerTrain NA (3.5 years). His experience covers a wide range of heavy-duty diesel engine engineering and development. His most recent position was as executive director, of engine development, where he was responsible for all engine/system functions (design and analysis; emissions control/fuel economy optimization; electronics system development, performance durability testing, manufacturing, supplier, sales and service liaison). This responsibility included design and production introduction of the world's first fully electronically controlled diesel unit pumps for 12-liter, six-cylinder engines in 1990. He jointly initiated (with Detroit Diesel) and developed, with the Environmental Protections Agency (EPA) and various industry participants a urea infrastructure for targeted 2007 calendar year engine production (then delayed to 2010). He participated in industry collaborative research through the U.S. Department of Energy Diesel Crosscut Committee, which was part of the 21st Century Truck Partnership. He was a consultant to Volvo PowerTrain NA from 2005 to 2007 on an advanced large truck diesel exhaust gas recirculation cooler vibration study/amelioration and on heavy-duty truck hybrid power train duty cycle test procedure development for comparative fuel consumption (EPA/industry/HTUF). He has been a member of the Society of Automotive Engineers for 43 years; an organizer for World Congress technical sessions on heavy-duty diesel fuel injection systems for several years; and company representative to the Engine Manufacturers Association for 25 years, including 13 years on its board of directors, where he has been treasurer, vice president, and president. He holds a B.S. in mechanical engineering from Pennsylvania State University and an M.S. in engineering, solid mechanics, from the University of Maryland.

James J. Winebrake is chair of the Department of Science, Technology, and Society/Public Policy at Rochester Institute of Technology (RIT). Dr. Winebrake focuses his research on solving problems related to energy security, environmental quality, and transportation. He has published extensively in scholarly journals, coauthored a textbook on environmental modeling, and was editor and lead contributor for a book on alternative energy. He is also co-principal investigator on a recently awarded $2 million National Science Foundation grant to study the impact of greenhouse gas policies on the transportation sector. Dr. Winebrake's recent research has been on sustainable goods movement, including evaluations of greenhouse gas emissions from trucks, trains, ships, and planes. At RIT, Dr. Winebrake is co-director of the Laboratory for Environmental Computing and Decision Making and director of the University-National Park Energy Partnership Program. Dr. Winebrake received a B.S. in physics from Lafayette College, an M.S. in technology and policy from the Massachusetts Institute of Technology, and a Ph.D. in energy management and policy from the University of Pennsylvania.

John Woodrooffe heads the Transportation Safety Analysis Division, University of Michigan Transportation Institute (UMTRI). He is responsible for the Center for National Truck and Bus Statistics, which conducts nationwide surveys of Trucks Involved in Fatal Accidents and Buses Involved in Fatal Accidents, and the Statistical Analysis Group, which performs analytical modeling and conducts research to advance statistical methods for road and vehicle safety analysis. He is an international expert on policy and safety evaluation of combination vehicles. Prior to joining UMTRI, Mr. Woodrooffe founded the Road Vehicle Research Program at the National Research Council of Canada and developed it into a successful, internationally active heavy truck research laboratory. He was a consultant to Australia's National Road Transport Commission for a unique 3-year performance-based standards development project that produced a new performance-based regulatory system for large vehicle combinations. He has also served as chair of the Large Truck-Tractor Trailer working group for the 21st Century Truck Partnership through the U.S. Department of Energy. The program evaluated vehicle systems and forecasted the probable influence of emerging technologies on fuel consumption and vehicle emissions. Mr. Woodrooffe holds master's and bachelor's degrees in mechanical engineering from the University of Ottawa.

Martin B. Zimmerman is Ford Motor Company Clinical Professor of Business Administration, University of Michigan. His career has spanned academia, government, and business. He has served as chief economist and group vice president at Ford Motor Company, where he was responsible for corporate economics, governmental affairs, environmental and safety engineering, and corporate social responsibility. Prior to joining Ford he taught at the Business School of the University of Michigan and at the Sloan School of Management at MIT. He serves on the National Commission on Energy Policy and also served as a member of the Panel of Economic Advisers of the Congressional Budget Office and as a Senior Staff Economist on the President's Council of Economic Advisors. His research is concerned with energy policy, government regulation of business, and economic developments in the automotive industry. He earned an A.B. from Dartmouth College (1967) and a Ph.D. from the Massachusetts Institute of Technology (1975).

D

Abbreviations and Acronyms

21CTP	21st Century Truck Partnership
ABS	antilock brake system
ACC	adaptive cruise control
ACEA	European Automobile Manufacturers' Association
AMPT	alternative maritime power
AMT	automated manual transmission
ANL	Argonne National Laboratory
APU	auxiliary power unit
AT	automatic transmission
ATI	automatic tire inflation
BAC	battery air conditioning
BES	(Office of) Basic Energy Sciences (DOE)
BOP	balance of plant
BSFC	brake-specific fuel consumption
Btu	British thermal unit
C&S	codes and standards
CAFE	corporate average fuel economy
CAR	Cooperative Automotive Research
CARB	California Air Resources Board
CCP	coordinated cam phasers
CCPPR	capital cost per percent reduction
CCS	carbon capture and sequestration
C_d	drag coefficient
CDL	commercial driver's license
CFD	computational fluid dynamics
CFRP	carbon-fiber-reinforced polymer
CIL	component-in-the-loop
CLEERS	crosscut lean exhaust emission reduction simulation
CMEM	Comprehensive Modal Emissions Model
CNG	compressed natural gas
CO_2	carbon dioxide
CO	carbon monoxide
COE	center of excellence
CPM	cost per mile
CRC	Coordinating Research Council
C_{rr}	coefficient of rolling resistance
CVISN	Commercial Vehicle Information System and Networks Program
CVO	commercial vehicle operations
CVT	continuously variable transmission
DCP	dual cam phasers
DCT	dual clutch transmission
DDC	Detroit Diesel Corporation
Deac	cylinder deactivation
DF	durability/deterioration factor
DOC	U.S. Department of Commerce
DOC	diesel oxidation catalyst
DOE	U.S. Department of Energy
DOT	U.S. Department of Transportation
DPF	diesel particulate filter
DSRC	Dedicated Short Range Communications
DTI	Directed Technologies, Inc.
DTT	delivery technical team
DVVL	discrete variable valve lift
E85	85 percent ethanol
EC	European Commission
EERE	(Office of) Energy Efficiency and Renewable Energy
EEA	Energy and Environmental Analysis, Inc.
EGR	exhaust gas recirculation
EIA	Energy Information Administration
EISA	Energy Independence and Security Act
EPA	U.S. Environmental Protection Agency
ePTO	electric power takeoff
ePS	electric power steering
EPS	electrified parking space
ePump	electric engine oil and coolant pump
ETC	electronic toll collection
EUCAR	European Council for Automotive R&D
EU	European Union
EV	battery electric vehicle

FACE fuels for advanced combustion engines
FC fuel consumption
FCC Federal Communications Commission
FCHEV fuel cell hybrid electric vehicle
FE fuel economy
FFV flexible fuel vehicle
FMVSS Federal Motor Vehicle Safety Standard
FOH fuel-operated heater
FPITT fuel pathway integration technical team
FTP federal test procedure

GATE Graduate Automotive Technology Education
GDI gasoline direct injection
GDL gas diffusion layer
GFRP glass-fiber-reinforced plastic
gge gallons gasoline equivalent
GHG greenhouse gas
GPS Global Positioning System
GREET Greenhouse Gas, Regulated Emissions, and Energy Use in Transportation (model)
GS generator set
GVW gross vehicle weight
GVWR gross vehicle weight rating
GW gigawatt (1 billion watts)

H or H_2 hydrogen
H2A Hydrogen Technology Analysis (model)
HAMMER Hazardous Materials Management and Emergency Response (facility)
HC hydrocarbon
HCCI homogeneous charge compression ignition
heAlt high-efficiency alternator
HEV hybrid electric vehicle
HFCIT Hydrogen, Fuel Cells and Infrastructure Technologies (program)
HFCV hydrogen fuel cell vehicle
HFET highway fuel economy test
HFI Hydrogen Fuels Initiative
HHDDT Heavy Heavy-Duty Diesel Truck schedule
HHV hydraulic hybrid vehicle
HIL hardware-in-the-loop
HILS hardware-in-the-loop simulation
HLA hydraulic launch assist
HOT high-occupancy toll (lane)
HOV high-occupancy vehicle (lanes)
HSS high-strength steel
HV hybrid vehicle
HVAC heating, ventilation, and air conditioning
HyTrans Hydrogen Transition (model)

ICC International Codes Council
ICE internal combustion engine
IEA International Energy Agency
IGBT insulated gate bipolar transistor
IGCC integrated gasification combined cycle
ITS intelligent transportation system
IVT Intelligent Vehicle Technologies

kg kilogram
kW kilowatt
kWe kilowatt (electric)
kWh kilowatt-hour

Li-ion lithium ion
LCV longer combination vehicle
LCFS low-carbon fuel standard
LNC lean NO_x catalyst
LNG liquefied natural gas
LPG liquefied petroleum gas
LSFC load-specific fuel consumption
LTC low-temperature combustion

M85 85 percent methanol
MARKAL Market Analysis (model)
MATT modular automotive technology testbed
MBD Model-Based Design
MEA membrane electrode assembly
Meeh mechanical driven accessories
MHDVs medium- and heavy-duty vehicles
MOU memorandum of understanding
MPa megapascal
mpg miles per gallon
MSM MacroSystem Model
MT manual transmission
MWe megawatt (electric)

NAE National Academy of Engineering
NAS National Academy of Sciences
NE Office of Nuclear Energy (DOE)
NEMS National Energy Modeling System
NFPA National Fire Protection Association
NGNP Next-Generation Nuclear Powerplant
NHTSA National Highway Traffic Safety Administration
NiMH nickel metal hydride
NIST National Institute of Standards and Technology
NO_x nitrogen oxides
NORPASS North American Preclearance and Safety System
NPC National Petroleum Council
NPV net present value
NRC National Research Council
NREL National Renewable Energy Laboratory

O&M operation and maintenance
OBD onboard diagnostic (system)
OCTA Orange County Transit Authority
OEM original equipment manufacturer
O&M operations and maintenance

ORNL	Oak Ridge National Laboratory
PBA	(Office of) Planning, Budget, and Analysis (DOE)
PCCI	premix charge compression ignition
PCE	passenger car equivalent
PDP	product development process
PEIS	programmatic environmental impact statement
PEM	proton exchange membrane
PeMS	(California) performance measurement system
PHEV	plug-in hybrid electric vehicle
PHMSA	Pipeline and Hazardous Materials Safety Administration
PM	particulate matter
PNGV	Partnership for a New Generation of Vehicles
PNNL	Pacific Northwest National Laboratory
PRD	pressure relief device
PSAT	Powertrain Systems Analysis Toolkit
PSR	power split ratio
PTO	power takeoff
PV	photovoltaic
RCP	rapid control prototyping
RDR	rear drive ratio
RFP	request for proposal
RITA	Research and Innovative Technology Administration (DOT)
RMC	ramp modal cycle
ROI	return on investment
RPE	retail price equivalent
rpm	revolutions per minute
RSPA	Research and Special Projects Administration (DOT)
SA	selective availability (of GPS signal)
SAE	Society of Automotive Engineers
SBIR	Small Business Innovation Research
SCI	special crash investigation
SCR	selective catalytic reduction
SER	strategic environmental review
SiC	silicon carbide
SIL	software-in-the-loop
SMR	steam methane reforming
SNL	Sandia National Laboratories
SRI	Stanford Research Institute
STTR	small business technology transfer
SUV	sport utility vehicle
SwRI	Southwest Research Institute
TG	number of transmission gears
TMC	traffic management center
TP	test procedure
TPMS	tire pressure monitoring system
TRB	Transportation Research Board
TREAD	Transportation Recall Enhancement, Accountability, and Documentation
TSS	thermal storage system
UC	ultracapacitor
UDDS	Urban Dynamic Driving Schedule
USABC	U.S. Advanced Battery Consortium
USCAR	U.S. Council for Automotive Research
VDP	Vehicle Development Process
VGT	variable geometry turbine
VITT	Vehicles Integration Tag Team
VIUS	Vehicle Inventory and Use Survey
VMT	vehicle miles traveled
VVA	variable valve actuation
VVL	variable valve lift
VVT	variable valve timing
WAAS	weighted aerodynamic average speed
WBS	wide-base single (tires for low rolling resistance)

E

Fuel Economy and Fuel Consumption as Metrics to Judge the Fuel Efficiency of Vehicles

Figure E-1 shows the relationship of fuel consumption versus fuel economy. The negative slope and the shape of this relationship are both important. The slope of the fuel consumption/fuel economy (FC/FE) curve indicates the amount of change in FC relative to a change in FE. For example, when the slope magnitude in Figure E-1 is high, such as at 10 mpg, there is a large change in FC for a small change in FE. On the other hand, at 50 mpg, there is a small change in FE, since the slope magnitude is very low and approaching zero as indicated by the lower right-hand slope scale on Figure E-1.

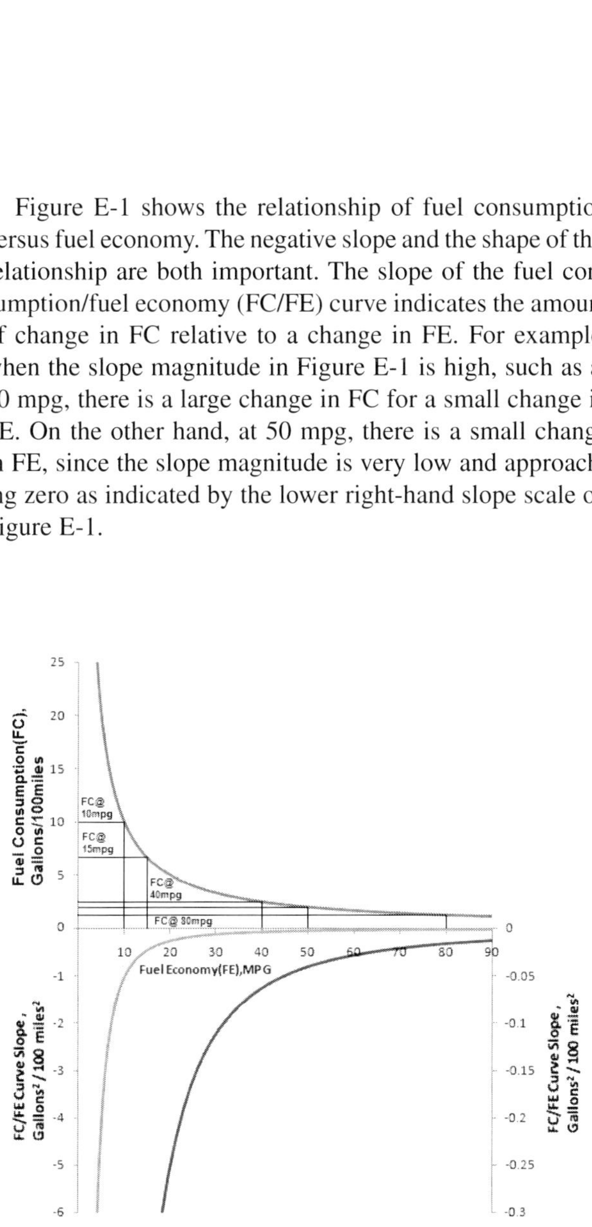

FIGURE E-1 Fuel consumption (FC) versus fuel economy (FE) (upper half of figure) and slope of FC/FE curve (lower half of figure). The light-colored lower curve matches the left-hand y-axis, while the dark curve matches the right-hand y-axis.

Fuel consumption decreases slowly after 40 mpg since the slope of the FC/FE curve approaches zero (Figure 2-1 lower curve and right-hand scale). The slope rapidly decreases past 40 mpg since it varies as the inverse of FE squared, which then results in a small decrease in FC for large FE increases. This fact is very important since fuel consumption is the metric in corporate average fuel economy (CAFE) standards for light-duty vehicles. For example, the fuel consumption is 2.5 gallons/100 miles at 40-mpg and 1.25 gallons/100 miles at 80 mpg. Thus, a 40 mpg change in fuel economy results in a change in fuel consumption of only 1.25 gallons/100 miles. In going from 8 to 9 mpg, there is a change in fuel consumption of approximately 1.39 gallons/100 miles. This means that a change from 8 to 9 mpg saves more fuel than a change from 40 to 80 mpg. This nonlinear relationship between fuel economy and fuel consumption has important meaning for regulations, where a reduction in fuel use or in greenhouse emissions is desired. Improving vehicles with high fuel consumption (low mpg) and high vehicle miles traveled (VMT) has much more effect on fuel savings than improving low-consumption (high-mpg) and low-VMT vehicles.

$$\frac{\sum_1^n N_n}{\sum_1^n N_1 \frac{1}{FE_1} + \ldots + N_n \frac{1}{FE_n}}$$

Tables E-1, E-2, and E-3 show vehicle groups and national average payload data that can serve as the basis for National Highway Traffic and Safety Administration (NHTSA) use in determining payloads to be used for testing and simulating any future medium- and heavy-duty vehicle fuel consumption procedures as related to standards. The data in Tables E-1 through E-3 and other data in the report on which the tables are based merit careful study by NHSTA before they are considered for use in a regulation.

The following equations hold for calculating FE and FC:

$$\text{Total miles traveled} = \int_0^t \frac{\text{miles}}{\text{hour}}\,dt.$$

$$\text{FC} = \frac{100}{\text{FE}}, \text{ gallons/100 miles.}$$

The equations above hold from engine on to engine off in order to capture idle time. Chapter 2 refers to an "average payload" to calculate load-specific fuel consumption (LSFC) but does not indicate how to calculate it on a trip delivering cargo. The calculation for average payload is as follows:

$$\text{Average payload} = \frac{P_1t_1 + P_2t_2 + \ldots P_xt_x}{t_1 + t_2 + \ldots t_x} = P_{ave}$$

where P_x = payload in tons carried for time x when the vehicle is moving, and t_x = time in hours carrying payload P_x. This is a "time average payload" for a vehicle operating in the field and excludes idle time—it is the integral of payload to get the average payload.

From Figure 2.6, LSFC does not decrease significantly for a payload increase as long as the payload is greater than 70 percent of the full payload.

In the equation for payload, if any P_x is zero, there is zero in the numerator for that segment, but the time is counted in the denominator, which then lowers the average payload. The FC during the no-load segment would decrease, lowering the total gallons of fuel used. If the time average payload is less than 70 percent of full load, LSFC will increase—if it is greater than 70 percent, LSFC will increase somewhat based on Figure 2.6.

Then,

$$\text{Total payload moved on trip from time 0 to } t_x = P_1 - P_x, \text{ tons}$$

and the

$$\text{Payload delivered in Segment 1} = P_1 - P_2, \text{ tons, etc.}$$

Therefore,

$$\text{Trip average LSFC} = \frac{\text{Total gallons used} \times 100}{P_{ave} \times \text{Total miles traveled}}$$

For picking up cargo, the average payload equation would need a plus payload term for each segment to account for the pickup. Therefore, $P_2 = P_1 - P_d + P_p$ where P_d = payload delivered and P_p = payload picked up after segment 1,

TABLE E-1 Gross Vehicle Weight Groups

Group	Gross Vehicle Weight (lb)
1	<6,000
2	6,001-10,000
3	10,001-14,000
4	14,001-16,000
5	16,001-19,500
6	19,501-26,000
7	26,001-33,000
8	>33,000

NOTE: Vehicle groups used for average payloads in Tables E-2 and E-3.
SOURCE: *Development of Truck Payload Equivalent Factor (TPEF)*, final report submitted to Office of Freight Management and Operations, Federal Highway Administration, Washington, D.C., by Battelle, 505 King Avenue, Columbus, Ohio 43201. June 15, 2007. Available at http://ops.fhwa.dot.gov/freight/freight_analysis/faf/faf2_reports/reports9/index.htm#toc.

TABLE E-2 Average Payload (lb) by Commodities and Gross Vehicle Weight Group VIUS—National

Commodities	Group 1	Group 2	Group 3	Group 4	Group 5	Group 6	Group 7	Group 8
Live animals and fish	-	5,055	7,638	5,424	9,472	17,200	16,345	40,022
Animal feed or products of animal origin	-	4,682	6,138	3,760	8,330	11,778	18,980	39,841
Cereal grains	-	13,348	15,234	8,690	14,334	17,640	24,208	41,922
All other agricultural products	-	10,728	6,889	5,985	7,660	11,348	26,793	34,616
Basic chemicals	-	*	*	3,386	*	11,180	14,264	38,431
Fertilizers and fertilizer materials	-	8,062	2,937	5,382	7,898	12,308	25,148	30,134
Pharmaceutical products	-	*	*	-	*	7,455	*	14,507
All other chemical products	-	2,715	3,046	4,357	6,193	9,712	17,574	36,411
Alcoholic beverages	-	-	2,670	-	*	16,177	20,142	35,758
Bakery and milled grains	-	2,000	2,407	7,083	*	3,198	27,732	31,389
Meat, seafood, and their preparation	-	*	10,402	3,646	-	8,819	10,738	40,012
Tobacco products	-	-	2,700	-	-	*	9,253	34,381
All other prepared foodstuff	-	4,354	3,607	3,617	5,486	13,240	23,736	38,894
Logs and other wood in rough	-	5,838	4,880	*	9,384	11,029	22,746	46,774
Paper and paperboard articles	-	-	*	-	6,718	8,842	18,591	37,932
Printed products	-	8,864	3,418	4,699	2,126	8,578	8,805	21,340
Pulp, newsprint, paper, or paperboards	-	*	*	*	-	10,904	15,815	41,774
Wood products	-	3,303	3,592	5,410	7,263	8,218	16,182	34,699
Articles of base metal	*	1,808	2,016	4,399	4,323	8,095	12,840	29,564
Base metal finished or semi-finished form	*	3,375	3,871	3,731	4,080	6,356	12,110	38,010
Non-metallic mineral products	-	3,737	2,088	3,438	6,652	10,527	28,977	35,962
Non-powered tools	-	2,675	3,167	4,353	5,421	6,680	9,899	14,810
Powered tools	-	3,894	3,602	4,849	8,513	7,405	12,242	25,241
Electronic and other electrical equipment	-	2,463	4,068	2,060	*	7,877	9,946	26,353
Furniture, mattresses, lamps, etc.	-	2,056	2,769	2,591	-	6,397	17,501	22,598
Machinery	-	4,271	4,277	9,265	5,020	9,958	17,598	35,754
Miscellaneous manufactured products	-	1,401	2,411	6,148	5,615	8,571	17,861	27,236
Precision instruments and appliances	-	1,455	1,373	10,095	-	4,391	*	26,195
Textile, leather, and related articles	-	2,073	2,986	*	8,701	7,599	41,925	36,656
Vehicle, including parts	-	3,751	5,506	5,896	7,333	8,173	23,554	31,945
All other transportation equipment	-	-	2,025	5,431	*	16,312	18,286	42,517
Coal	-	*	*	-	-	6,748	-	50,011
Crude petroleum	-	-	-	-	-	8,590	-	39,890
Gravel and crushed stones	-	6,544	6,931	6,276	10,122	13,770	24,305	39,130
Metallic ores and concentrates	-	-	*	-	10,000	-	-	42,272
Monumental and building stones	-	*	3,460	5,782	14,100	10,392	9,473	35,960
Natural sand	-	7,306	3,029	12,849	6,000	11,643	28,662	38,067
All other nonmetallic minerals	-	7,337	3,064	2,478	7,662	16,262	13,580	38,835
Fuel oils	-	4,484	14,811	-	*	15,422	17,525	39,634
Gasoline and aviation turbine	-	*	-	2,825	-	15,128	18,916	53,423
Plastic and rubber	-	*	2,931	3,329	*	8,113	12,548	30,379
All other coal and refined petroleum	-	4,519	4,336	*	4,874	10,326	18,672	41,027
Hazardous waste	-	*	1,500	-	-	6,854	15,517	37,856
All other waste and scrap	-	3,384	2,927	*	5,951	8,120	12,823	24,944
Recyclable products	-	3,153	4,878	3,689	*	8,425	13,743	27,532
Mail and courier parcels	*	7,976	5,559	4,608	7,342	10,884	33,344	31,628
Empty shipping containers	-	2,661	*	-	*	2,309	16,129	26,699
Passengers	-	2,264	2,501	*	*	*	*	*
Mixed freight	-	2,080	2,633	4,051	*	20,137	28,811	37,094
Multiple categories	-	3,602	3,375	4,198	5,463	8,127	17,189	31,946
Products not classified, blank, not reported or applicable	-	2,471	*	6,556	7,809	11,622	17,644	30,545

SOURCE: *Development of Truck Payload Equivalent Factor (TPEF)*, final report submitted to Office of Freight Management and Operations, Federal Highway Administration, Washington, D.C., by Battelle, 505 King Avenue, Columbus, Ohio 43201. June 15, 2007. Available at http://ops.fhwa.dot.gov/freight/freight_analysis/faf/faf2_reports/reports9/index.htm#toc.

TABLE E-3 Vehicle Groups and National Average Payload (lb)

Major Body Type	Group 1	Group 2	Group 3	Group 4	Group 5	Group 6	Group 7	Group 8
Truck or Truck + Trailer								
Bulk	-	6,249	4,798	5,004	10,831	13,152	15,707	30,966
Flatbed	*	4,027	4,767	4,858	7,836	9,465	11,405	16,693
Tank	-	6,077	4,438	7,266	7,249	12,991	15,743	26,858
Van	*	4,179	3,071	4,322	6,770	7,516	9,519	6,233
Reefer	-	4,820	3,196	4,263	*	9,161	10,983	10,414
Logging	-	15,036	*	*	*	10,787	*	42,857
Other	*	2,969	3,075	4,442	5,235	7,391	11,887	23,853
Tractor + Trailer								
Automobile	-	-	-	*	-	*	25,443	34,257
Livestock	-	-	-	*	-	44,361	27,747	42,535
Bulk	-	*	-	*	-	36,846	45,319	50,135
Flatbed	-	*	*	*	*	24,997	31,949	41,874
Tank	-	-	-	-	-	*	47,656	49,788
Van	-	-	-	-	-	23,995	28,079	38,721
Reefer	-	-	-	-	-	19,390	24,775	41,426
Logging	-	-	-	*	-	*	*	50,004
Other	-	-	-	-	-	*	31,498	31,800

SOURCE: *Development of Truck Payload Equivalent Factor (TPEF)*, final report submitted to Office of Freight Management and Operations, Federal Highway Administration, Washington, D.C., by Battelle, 505 King Avenue, Columbus, Ohio 43201. June 15, 2007. Available at http://ops.fhwa.dot.gov/freight/freight_analysis/faf/faf2_reports/reports9/index.htm#toc.

F

Details of Aerodynamic Trailer Device Technology

Tables F-1 through F-3 report results from a collection of suppliers that provided trailer aerodynamic device results in more detail for trailer skirts, trailer base devices, and trailer face devices, three of the areas identified in Figure 5-9 (Chapter 5) as prime for aerodynamic device improvement in tractor-trailer combination trucks. These data are principally those returned by nine manufacturers responding to a committee questionnaire. Those responses were supplemented by information from the Web sites of four other manufacturers.

Interestingly, these most recent data on reduction of fuel consumption received from developers/manufacturers for trailer skirts (Table F-1) substantially group around 7 percent.

TABLE F-1 Trailer Skirt Information from Manufacturers

Item	Manufacturer	Qualified for SmartWay (Y/N)	Fuel Consumption Reduction (gal/mile) (%)	Evaluation Method (provide details)	Weight to Equip 53-ft Trailer, (lb)	Retail Price Equivalent for One Trailer (US$)	Estimated Annual Maintenance Cost (USD)	Other Useful Information
1	Laydon	Y	6	J1321	300	1,900	0	Very flexible meeting systems
2	FreightWing	Y	7	J1321, 62 mph	160	1,599	$50	Impact resistant; small road clearance
3	AdamWorks	Y	7	self truck test	<200	2,400	$400	Automatically deploys to 6-inch ground clearance
4	TransTex[a]	[a]	7.4	J1321, 61 mph	[a]	[a]	[a]	[a]
5	Windyne[a]	Y	6.9	J1321	[a]	[a]	[a]	Improved handling in side winds
6	ATDynamics	Y	7.4	J1321, 60 mph	175	2,200	0	Reduced road spray, 5-year warranty
7	Wabash	Y	5.6	J1321, 65 mph	250	1,625	0	12-inch ground clearance

[a]Committee questionnaire not responded to.

TABLE F-2 Trailer Base Device Information from Manufacturers

Item	Manufacturer	Qualified for SmartWay (Y/N)	Fuel Consumption Reduction (gal/mile) (%)	Evaluation Method	Weight to Equip 53-ft Trailer (lb)	Retail Price Equivalent for One Trailer (US$)	Estimated Annual Maintenance Cost (US%)	Other Useful Information
1	ATDynamics boat tail	Y	5.1	J1321, 62 mph	175	2,800	0	Folds flat in 6 sec; improves stability
2	AeroTrailerSys[a] inflatable tail	[a]	3	[a]	[a]	[a]	[a]	Automatically deploys
3	TransTex[a] boat tail	[a]	2.9	[a]	[a]	[a]	[a]	Reduces road spray
4	AirTab vortex generators	N	2-3	Truck test, 47 mph	1	220	0	Reduces road spray

[a]Committee questionnaire not responded to.
SOURCE: Data from responses to committee questionnaire and from manufacturers' websites.

TABLE F-3 Trailer Face Device Information from Manufacturers

Item	Manufacturer	Qualified for SmartWay (Y/N)	Fuel Consumption Reduction (gal/mile) (%)	Evaluation Method	Weight to Equip 53-ft Trailer (lb)	Retail Price Equivalent for One Trailer (US$)	Estimated Annual Maintenance Cost (USD)	Other Useful Information
1	Laydon Vortex Stabilizer	N	1	J1321	40	495	0	Better performance in yaw
2	Laydon Nose Fairing	Y	2	J1321	95	795	0	No tractor interference
3	FreightWing Gap Fairing	Y	2	J1321, 65 mph	75	849	$50	Better performance with low aerodynamic tractor
4	NoseCone Eyebrow	Y?	>3	J1321?	30	—	—	For high tractor roof fairing
5	NoseCone	Y?	>4	J1321?	75	1,264	$35	No yaw effect in J1321

G

Vehicle Simulation

Vehicle simulation has been referred to several times in this report as part of the fuel consumption assessment and certification process and is described in Chapter 3 as already part of Japan's heavy vehicle fuel consumption rules. Any simulation relies on the availability of accurate submodels or good-quality test data from the components and on accurate portrayal of the physical and control linkages between the components.

Several key requirements are necessary to answer both industry needs to accelerate the introduction of advanced technologies and regulatory needs to evaluate benefits in the most cost-effective manner.

The simulation tool should provide a set of default models, processes, and postprocessing, but also allow users to integrate any legacy code. Indeed, future regulations might recommend that companies use the same assumptions but might also give the option to use legacy codes (e.g., engine and vehicle models) that have been internally developed. Using the same models regardless of the technology considered might penalize a particular company. However, if proprietary models are used, a validation process should be clearly defined to ensure their accuracy under specific operating conditions.

Due to the large number of power train configurations, which will continue to increase with hybrid electric vehicles, the tool should also be able to quickly simulate any drivetrain configurations. Finally, all the physical equations and control parameters should be open source, at least to the regulator, to ensure transparency of the process. It may be necessary to require that proprietary codes be available to the regulatory body either as soon as they are used for regulatory compliance or after some waiting period.

A review of currently available software reveals that, while the tools all provide a set of existing models, each has existing limitations. Some of the existing tools do not represent realistic vehicle behavior (e.g., ADVISOR), are not open source (e.g., AVL CRUISE, GT-DRIVE, AMESIM) or cannot be compiled to perform model-based design (MBD; e.g., AVL CRUISE), or linkage with database management is not available or incomplete.

Most of the models used throughout the industry to simulate fuel consumption are based on steady-state look-up tables representing the losses of the components. Table G-1 lists the main maps for each component. Some of the look-up tables listed can also be multidimensional (e.g., the transmission will have different maps for each gear, the electric machine losses and maximum torque might depend on voltage). The models also require additional parameters such as mass, inertia, ratios, and fuel characteristics.

Most of the parameters can be directly obtained from manufacturers' specifications. However, some, like tire losses, require specific testing. Additional testing is also required to characterize the losses of the different components. While some of the test procedures are well characterized, others remain different from one manufacturer to the next and consequently should be clearly defined.

TABLE G-1 Main Vectors for Component Models

Component	X-Axis	Y-Axis	Z-Axis
Engine	Speed	Torque	Fuel Rate
	Speed	Maximum torque	
	Speed	Closed throttle Torque	
Transmission	Speed	Torque	Efficiency
Final drive	Speed	Torque	Efficiency
Electric machine	Speed	Torque	Efficiency
	Speed	Continuous torque	
	Speed	Maximum torque	
Energy storage	State-of-charge	Open-circuit voltage	
	State-of-charge	Internal resistance	

VEHICLE SIMULATION TOOL REQUIREMENTS FOR REGULATORY USE

In a world of growing competitiveness, the role of simulation in vehicle development is constantly increasing. Because of the number of possible advanced power train architectures that can be employed, development of the next generation of vehicles requires accurate, flexible simulation tools. Such tools are necessary to quickly narrow the technology focus to those configurations and components that are best able to reduce fuel consumption and emissions.

With improvements in computer performance, many researchers started developing their own vehicle models. But often computers in simulation are used only to "crunch numbers." Moreover, model complexity is not the same as model quality. Using wrong assumptions can lead to erroneous conclusions; errors can come from modeling assumptions or from data. To answer the right questions, users need to have the right modeling tools. For instance, one common mistake is to study engine emissions by using a steady-state model or to study component transient behavior by using a backward model.

Figure G-1 summarizes the main requirements, discussed below, for vehicle simulation tools required to fulfill both needs.

Basic Requirements

Maximum Reusability

While numerous plant and control models exist throughout companies, it is critical that the work performed during a project can be reused throughout the companies for future applications. Several approaches are necessary to achieve this goal:

- Duplication of systems without duplication of models stored. For example, a wheel model should be reused numerous times without storing it several times under different names, which would make versioning management difficult.
- Location of expert models in a single site. For example, an engine system comprised of control, actuator, plant and sensor models, and initialization file, by being located under the same folder, would facilitate its transfer to another expert.
- Open source of the plant and control models (rather than compiled) to facilitate understanding of the assumptions and the modifications of equations to model new phenomena.

Maximum Flexibility

With the consistently increasing number of possible power train configurations for medium- and heavy-duty applications and the need to select the different level of modeling to properly meet different needs (i.e., fuel efficiency, emissions, drive quality), the need to quickly simulate any application is crucial. A vehicle modeling software should be able to provide the following features:

- Simulation of subsystems, systems, collections or combinations of systems and subsystems (e.g. power trains), or entire vehicles. Providing a common environment to different experts (e.g., engine and vehicle experts) will facilitate the model's reusability and ensure process consistency (e.g., validation, calibration).
- Allow any configuration (assembly of systems) to be quickly modified and built automatically. For maintenance purposes, saving hundreds of models (a number

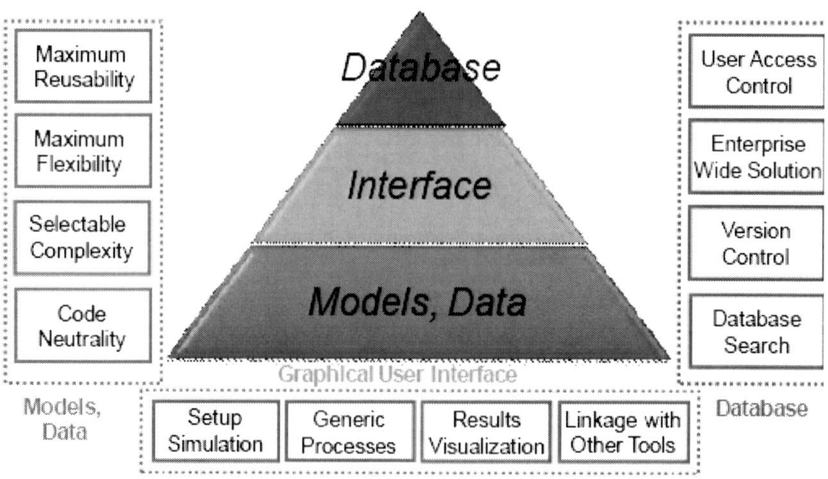

FIGURE G-1 Vehicle modeling tool requirements.

that can easily be achieved through combination of configurations and model complexity) is not feasible.

- Allow users to quickly add their own configurations.
- Allow users to implement any test data from subsystems, systems, or entire vehicles in the same environment as the models to facilitate the validation process.

Selectable Complexity

Different studies (e.g., fuel efficiency, emissions, drive quality) require different levels of modeling. Throughout a project, the level of model complexity will increase to take into accounts new physical phenomena.

- Common nomenclature, including naming convention, units. If nomenclature is not consistent, an automated process should be provided to users to easily integrate any legacy code into the agreed upon format.
- Common model organization to facilitate interactions of different expert models. For example, consistent format between controllers and plant would allow integration between both areas of expertise.
- Model compatibility check. When used in a large organization, users do not know what models are compatible with each other. For example, a particular gearbox should be used along with a specific torque converter. Using another combination could lead to a software crash—or worse—erroneous results. While the original developers are aware of the potential issue, it is necessary to enforce that when one model is used, it is in conjunction with the other one.

Code Neutrality

While most software companies claim to be able to model any particular plant with different levels of accuracy, some software packages are used mainly for specific applications. As a consequence, different experts will use different packages to model specific plants. One needs to have a plug-and-play platform that allows the user to:

- Integrate any legacy code from any software package and
- Run all models in the same environment or through co-simulation.

Graphical User Interface

Setup Simulation

The graphical user interface (GUI) should be able to allow users to quickly set up different simulations, including:

- Select architecture, model, and data
- Check model compatibilities to avoid crash or erroneous results
- Select simulation type, including component evaluation, vehicle fuel efficiency, or drive quality

Generic Processes

When evaluating specific technologies, having consistent processes is critical for proper comparison. Differences in the definitions of processes could lead to discrepancies in results, which could become a significant issue for regulatory purposes. For example, the definition of the term "validation" varies significantly from one engineer to another. In addition clear definition of generic processes (e.g., calibration, validation, tuning) for major tasks throughout a company will lead to increased productivity.

Users should have the ability to easily modify any processes or implement new ones. One could assume that specific processes would be developed and agreed upon for validation, report generation, and so forth for regulatory purposes.

Results Visualization

The GUI should allow users to quickly analyze the simulation.

- Predefined calculation. Since most tools only record efforts (e.g., torque, voltage) and flows (e.g., rotational speed, current), existing calculations should allow users to quickly calculate powers, energies, efficiencies, and so forth.
- Predefined plots should be available to quickly analyze the operating conditions of each component or control strategies.
- Energy balance information should be available.
- Reports should be automatically generated.
- All results should be saved along with the assumptions and any files required to rerun the simulation.
- Any existing calculation, plot, or report should be easily modified by users or new ones should be implemented.

Linkage with Other Tools

As discussed previously, linkage with other tools is compulsory to properly integrate detailed legacy models. While numerous tools exist, the list should include at a minimum MathWorks toolboxes, GT-Power, AMESim, TruckSim, ADAMS, and AVL DRIVE.

Database

User Access Control

The sharing and distribution of proprietary models can be achieved successfully only if their producers can trust that only the proper users will have access to them. User access control is the cornerstone of that trust.

User access control can be used in two ways:

- Intra-enterprise, to define the access at each process steps. For instance, during the design stage, only the design team can access the model. Once a version is ready, access can be granted to a larger group, such as calibration, testing, and so forth.
- Extra-enterprise, to define the access to outside users, including suppliers, regulatory committees, and so forth.

Access control should be of at least four types:

- Producer, for the people who can add and/or modify models and data on the database.
- Consumers with full access, for people who can download the models and data to run on their computers, but not modify them (or at least not upload them on the database).
- Consumers with restricted access, who can only run the models remotely on a dedicated server (no access to the models or data themselves).
- Administrator, who manages access control for everyone.

Users can also be a combination of these types. For example, some people creating models may need to access existing ones, and consumers with full access on some models may have only restricted access on others, or they can access only low-fidelity versions of some models.

Enterprise-Wide Solution

Another requirement for the sharing and distribution of proprietary models is their enterprise-wide accessibility, including for producer and consumer teams spread across the country or even the world for some global companies—for example, a control design team can have members in the United States and England, or a model calibration and validation team might be located hundreds of miles from the model design team.

Up-to-date models should be accessible to all people who have the right access, wherever they are located.

This constraint requires a unique and secure point of access for all users. However, there can be one point of access for intra-enterprise use only in each company and another global one outside, specifically for regulatory purposes.

Version Control

As models and data evolve with time owing to improved data and/or algorithms, or even issues such as new modeling software version compatibility, the need for version control is mandatory for auditing and regulatory purposes. Any study done with those models needs to specify which version was used to ensure 100 percent traceability of the results.

Moreover, version control can also be used intra-enterprise as a way to get feedback on the original designs. For example, the model producer can follow which modifications were needed to his model during the calibration and validation process, which can then be used to create a better model next time. Version control can also be used to locate the original designer to get more information about some of the model.

User access control applies on versioning as well. Some users should have access to all model versions, when some others have access only to the latest version and others can only see the history.

Database Search

To maximize the reusability of models, any user should be able to search for an existing one available to them. Search should be available on name and versions of the files, as well as specific criteria, such as, engine technology, displacement, and wheel radius.

The search should also be possible by specific vehicle or project, so that all of the models and data used for a specific application can be found together and eventually run or downloaded on the user's computer.

Only the models and data that the user has access to should be returned in the search query. As an optional functionality, the search could inform the user that other models exist but are not available and could provide the coordinates of people to contact to request their access.

SINGLE VERSUS MULTIPLE TOOLS SELECTION FOR REGULATION

Numerous tools are currently being used by companies, both internally developed and commercially available. For regulatory purposes, consistency between all approaches is critical for a fair comparison. As a result, while legacy code shall be used, a single platform is necessary to ensure proper integration of the different systems. Indeed, due to the large number of companies involved, the models used to simulate a specific application will most likely come from numerous sources. Common tool and formalism are then critical. As shown in Figure G-2, the lack of common nomenclature makes reusability of models among companies very cumbersome.

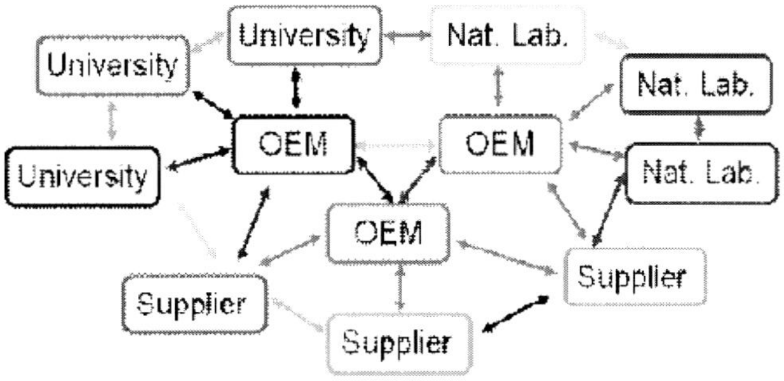

FIGURE G-2 Different nomenclatures within each company currently make model exchange very difficult.

H

Model-Based Design

KEY ELEMENTS OF MODEL-BASED DESIGN

Model-based design (MBD) is a mathematical and visual method of addressing the problems associated with designing complex control systems and is being used successfully in many motion control, industrial equipment, aerospace, and automotive applications. It provides an efficient approach for the four key elements of the development process cycle: modeling a plant (system identification), analyzing and synthesizing a controller for the plant, simulating the plant and controller, and deploying the controller—thus integrating all of these multiple phases and providing a common framework for communication throughout the entire design process.

This MBD paradigm is significantly different from the traditional design methodology. Rather than using complex structures and extensive software code, designers can now define advanced functional characteristics using continuous-time and discrete-time building blocks. These built models along with some simulation tools can lead to rapid prototyping, virtual functional verification, software testing, and validation. MBD is a process that enables faster, more cost-effective development of dynamic systems, including control systems, signal processing, and communications systems. In MBD, a system model is at the center of the development process, from requirements development, through design, implementation, and testing. The control algorithm model is an executable specification that is continually refined throughout the development process.

MBD allows efficiency to be improved by:

- Using a common design environment across project teams
- Linking designs directly to requirements
- Integrating testing with design to continuously identify and correct errors
- Refining algorithms through multidomain simulation
- Automatically generating embedded software code
- Developing and reusing test suites
- Automatically generating documentation
- Reusing designs to deploy systems across multiple processors and hardware targets

The different phases of MBD are indicated in Figure H-1. Throughout the different phases of MBD, several levels of modeling are required, both from the plant and the control points of view, in order for the functional behavior of the model to match that of the generated code. Figure H-2 shows an example of different levels used for different applications.

METHODOLOGY

This section explains different processes used as part of the MBD approach.

The first step is the simulation (Figure H-3), where neither the controller nor the plant operates in real time. This step, usually used toward the beginning of the process, allows engineers to study the performance of the system and design the control algorithm(s) in a virtual environment, by running computer simulations of the complete system, or subsystem.

Rapid control prototyping (RCP) is a process that lets the engineer quickly test and iterate control strategies on a real-time computer with real input/output devices RCP (see Figure H-4) differs from (HIL) hardware-in-the-loop in that the control strategy is simulated in realtime and the "plant," or system under control, is real. RCP is now the typical method used by engineers to develop and test their control strategies. It was first used to develop power train control strategies. The simple reason is that the control software, which is in the engine and transmission control units, is difficult and time consuming to modify. It has since been adopted industry wide in applications such as antilock braking, antiroll, vehicle stability, active cruise control, and torque distribution.

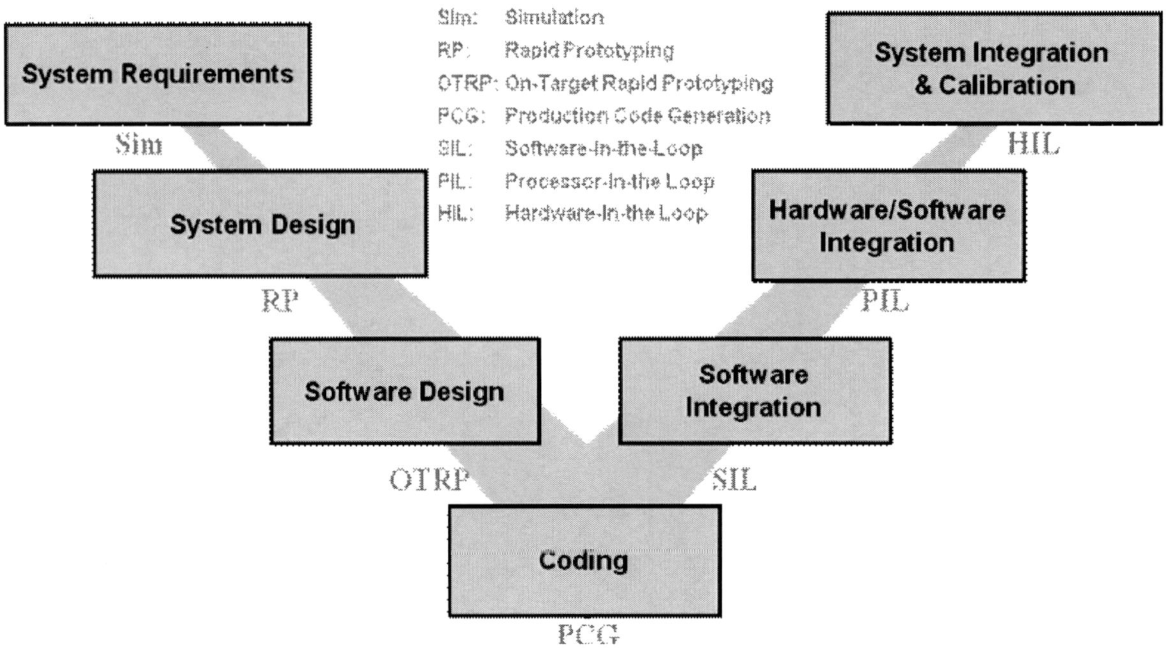

FIGURE H-1 V diagram for software development.

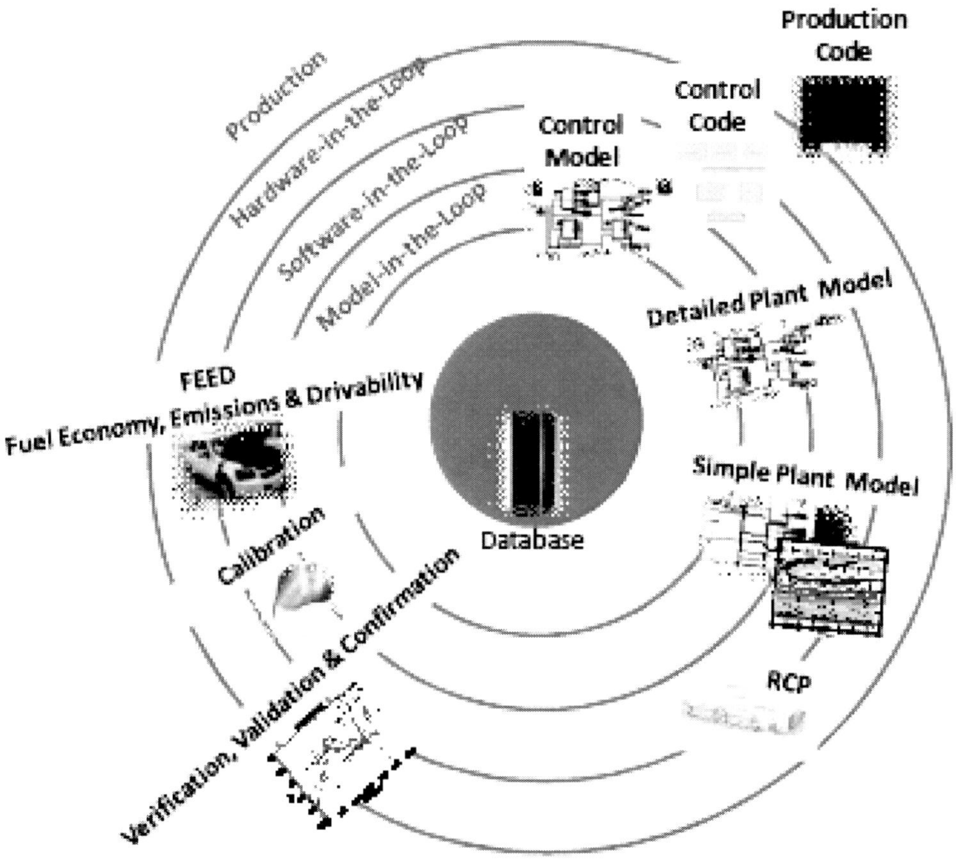

FIGURE H-2 Different levels of modeling required throughout the model-based design process.

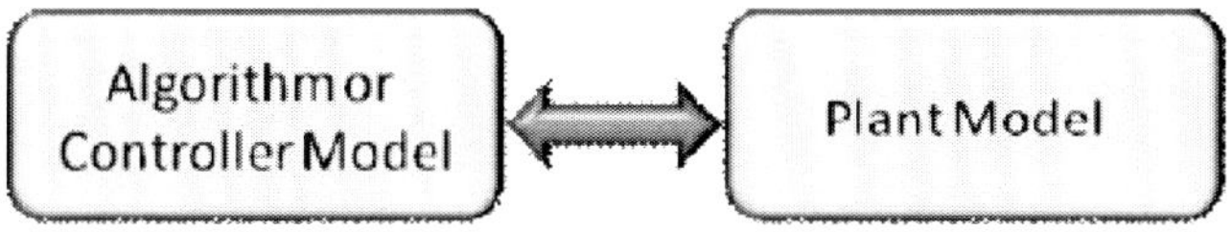

FIGURE H-3 Simulation.

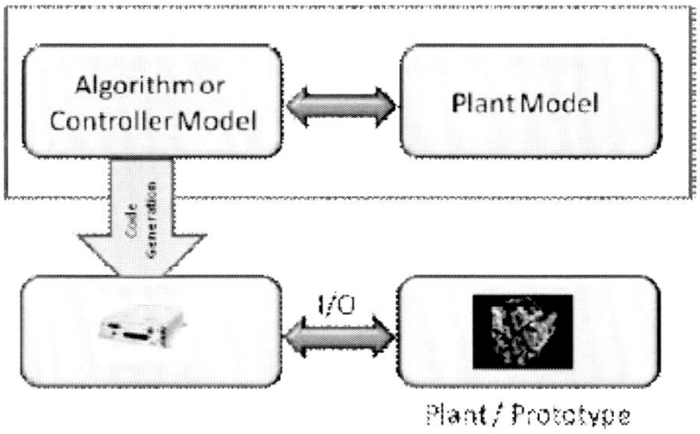

FIGURE H-4 Rapid control prototyping

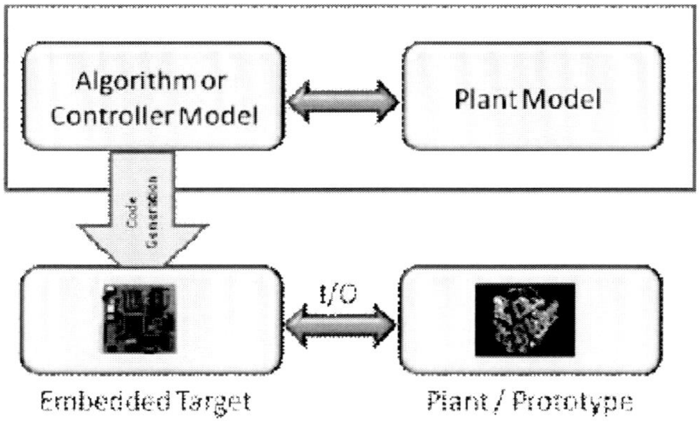

FIGURE H-5 On-target rapid prototyping.

For the on-target rapid prototyping case (see Figure H-5), new or modified functionality is added to the production code in the controller-embedded target processor to verify the additions/changes.

Once all the functions have been developed and tested, the production code is finally implemented (see Figure H-6).

In the software-in-the-loop (SIL) phase (see Figure H-7), the actual production software code is incorporated into the mathematical simulation that contains the models of the physical system. This is done to permit inclusion of software functionality for which no model(s) exists or to enable faster simulation runs.

During the processor-in-the-loop phase (see Figure H-8), the control is compiled and downloaded into an embedded target processor and communicates directly with the plant model via standard communications such as Ethernet. In this case, no input/output devices are used for the communication.

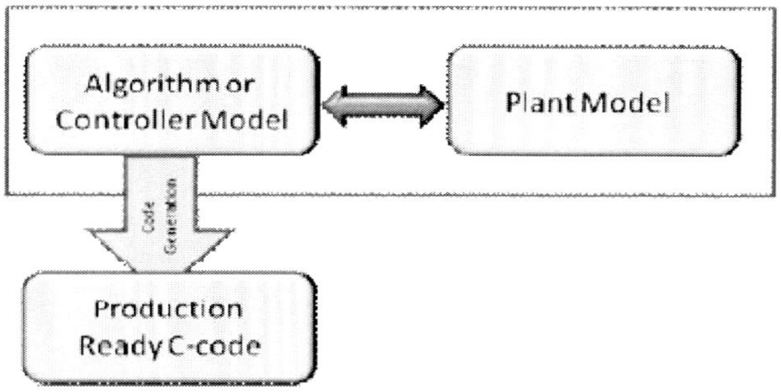

FIGURE H-6 Production code generation.

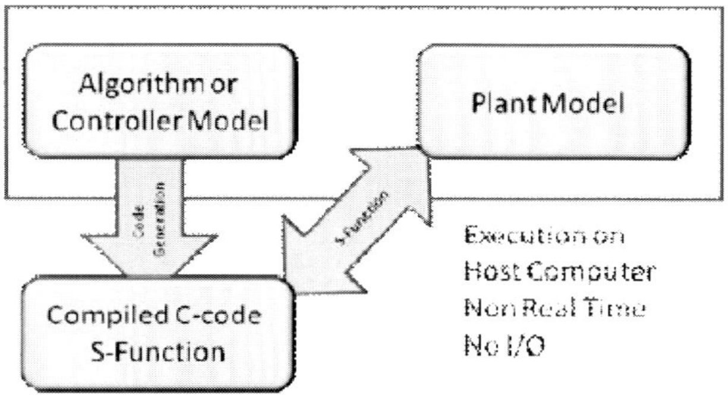

FIGURE H-7 Software-in-the-loop.

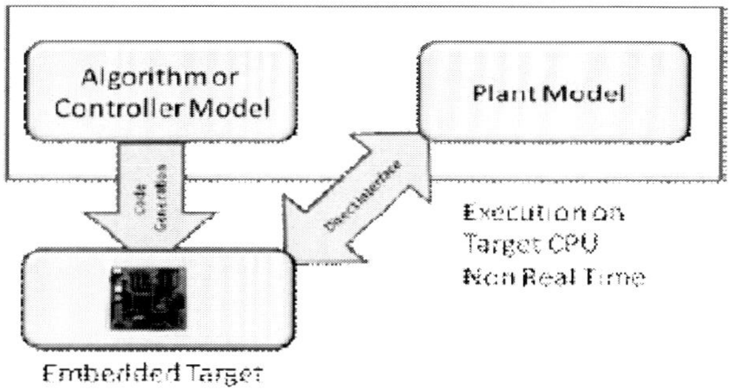

FIGURE H-8 Processor-in-the-loop

HIL (see Figure H-9) is a technique for combining a mathematical simulation model of a system with actual physical hardware, such that the hardware performs as though it were integrated into the real system. For testing and development of embedded electronic controllers, the hardware controller and associated software are connected to a mathematical simulation of the system plant, which is executed on a computer in real time. To connect the real-time model to the hardware controller, the real-time computer receives electrical signals from the controller as actuator commands

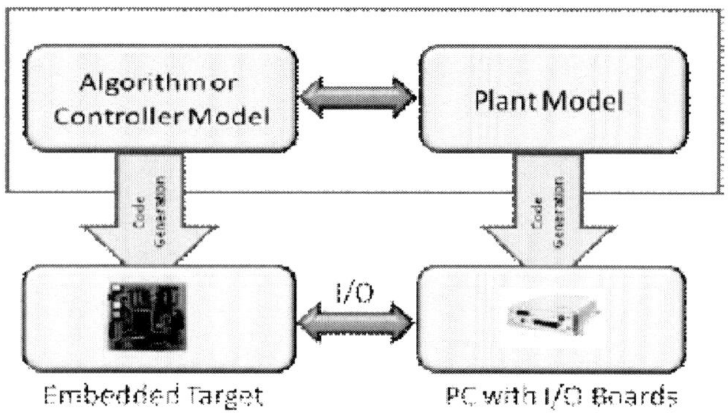

FIGURE H-9 Hardware-in-the-loop.

to drive the plant and converts these signals into the physical variables connected to the plant model. The plant model calculates the physical variables that represent the outputs of the plant, which are converted into electrical signals that represent the voltages produced by the sensors that feed the controller.

Another option to evaluate fuel consumption is component-in-the-loop (CIL), a combination of HIL and RCP. In CIL, an entire system is connected to a source emulating the rest of the vehicle. For example, Figure H-10 shows an engine and its controller connected to an AC dynamometer that would be controlled to represent the rest of the vehicle losses.

Figure H-11 shows a similar approach using a battery and a DC supply source emulating the remainder of the vehicle. In both cases the hardware component will be the one that (1) represents the new technology or (2) has not been properly validated yet or (3) cannot be accurately modeled (e.g., due to transients or thermal issues).

It should also be noted that more than one component can be hardware while some of them are still emulated. For example, both an engine and a battery could be hardware while the rest of the power train and the vehicle are emulated. One of the issues in using that approach, however, is the potential for communication-related delays since some of the signal transfer most likely has to go through the Internet.

An approach to characterize a system using several hardware components without building the entire vehicle is shown in Figures H-12 and H-13. The Modular Automotive Technology Testbed (MATT) has been developed to easily replace components by switching different plates. In the example below, a pretransmission parallel hybrid is shown. This concept allows the entire power train (or most of it) on a rolling chassis dynamometer in a controlled environment. However, like most approaches, it also shows some limitations, including lack of under-hood thermal management or the presence of a T-shaped reduction box to connect the wheels.

FIGURE: H-10 Engine on dynamometer. SOURCE: Courtesy of Cummins.

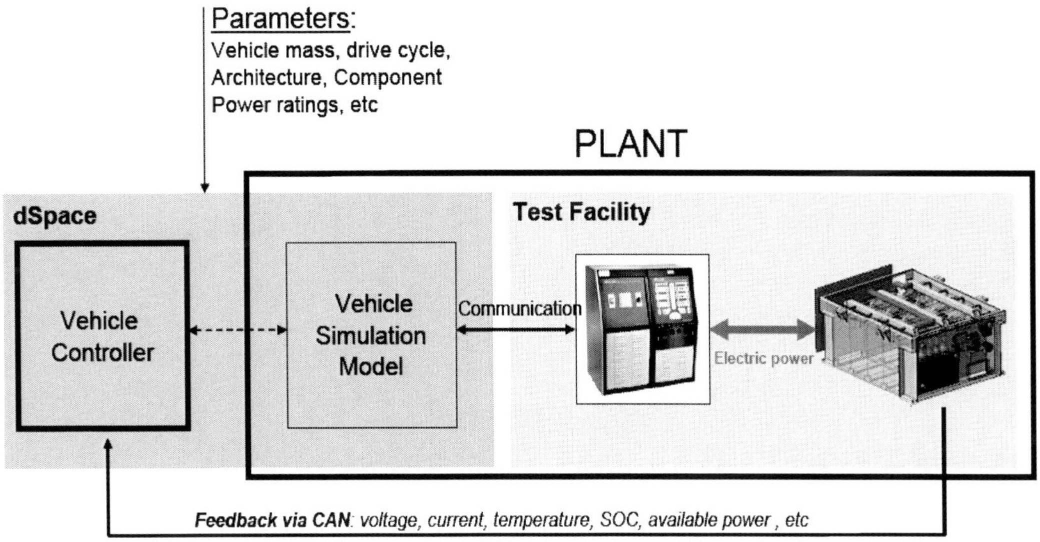

FIGURE H-11 Battery connected to a DC power source. SOURCE: Argonne National Laboratory.

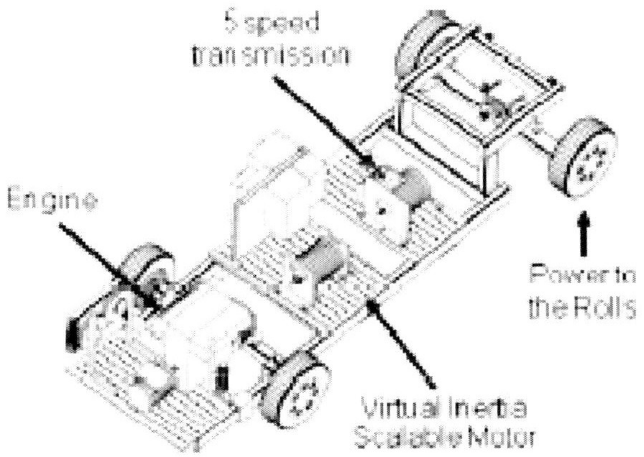

FIGURE H-12 Several components in the loop—MATT example. SOURCE: Argonne National Laboratory.

FIGURE H-13 Mixing components hardware and software—MATT example. SOURCE: Argonne National Laboratory.

PROCESS SELECTION

Different processes can be used to provide inputs for regulation depending on the technology considered and the degree of validation of the models. Ideally, if all the models have been thoroughly validated, one would like to only perform simulations to provide regulatory inputs. Realistically, since the state-of-the-art models do not yet fulfill all engineering expectations (e.g., engine emissions or cold start), a combination of hardware and software will most certainly have to be used for the foreseeable future. A couple of examples highlighting the potential use of each process are given in Figure H-14.

UNDERSTANDING UNCERTAINTIES

To select a process to properly characterize a particular technology, it is compulsory to understand and quantify the uncertainties associated with each process. Examples of questions that need to be addressed within each process and in between processes are given below.

Process Uncertainty

Uncertainty resides within each process and should be properly quantified. The following provides some examples for different processes.

- *Test facility to test facility variability.* A 2002 report from the Automotive Testing Laboratory [source CRC E-55-1 Inter-laboratory Crosscheck of Heavy-Duty Vehicle D.PDF] highlights the discrepancies between several vehicle testing facilities. Figure H-15 shows significant differences among the six laboratories. The main difference (Lab C) is mainly due to high-altitude impact, while the smaller discrepancies among the other laboratories are related to a series of reasons ranging from testing process to road load curve to driver technique. However, it should be noted that for the truck employed, particulate matter (PM) was a species that is far more sensitive to test conditions than fuel use.
- *Test-to-test variability.* While the testing conditions (e.g., temperature, humidity) are maintained constant during testing, several other factors affect dynamometer test results. One of the main factors is due to the driver, whether related to gear selection or engine on/off for hybrid vehicles. It is important to note that the driver model chosen in simulations will also affect results and is more repeatable than a human driver but must be chosen to be representative of a human driver.

While the impact can be important for conventional vehicles, especially when using manual transmissions, it is even more so for hybrid electric vehicles, due to

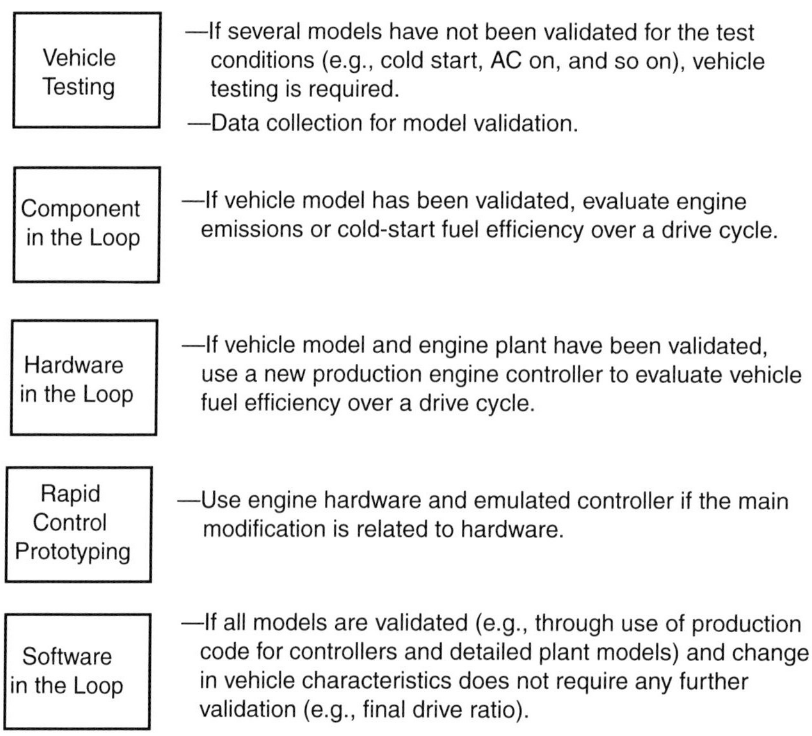

FIGURE H-14 Example of potential process use.

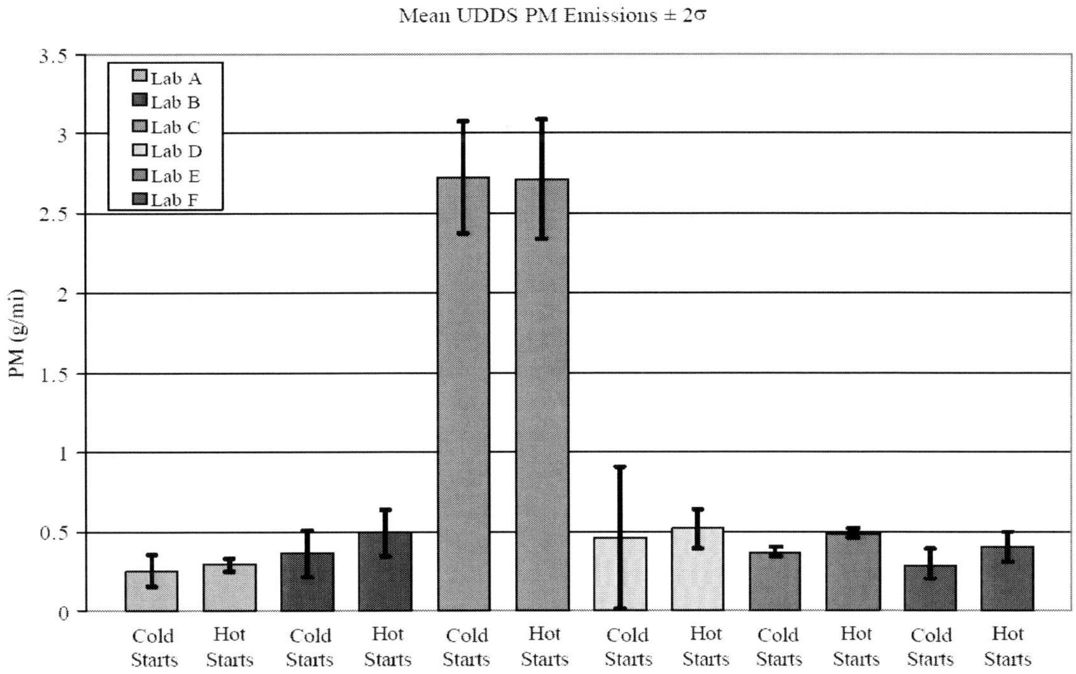

FIGURE H-15 Mean particulate matter results with two standard deviation error bars. SOURCE: Argonne National Laboratory.

the sensitivity of the engine on/off related to the pedal position.

- *Delay impact.* For several processes that include a mix of hardware and software, including CIL, SIL, or HIL, delays introduced by some hardware on the command and feedback can significantly affect the results. Dynamometer slew rate and command methodology (e.g., analog, digital, CAN) are some of the examples to be addressed. Such delays over an entire drive cycle could lead to several percentage point differences in energy, which would impact the results for both fuel efficiency and emissions. As a result, levels of acceptable delays should be defined for each process and potentially each technology to provide a low level of uncertainties.
- *Appropriate selection of level of modeling.* To simulate specific phenomena properly, an appropriate level of modeling must be selected. As such, engineers do not use the same models at the beginning of a project to compare different power train configurations as toward the end of a project when the focus is on drive quality and emissions. For regulatory purposes, the approach might be similar. The committee recommends that a study should be conducted to assess the uncertainty between different levels of modeling for specific components.
- *Data collection for model instantiation.* For the models to represent technologies properly, it is necessary to populate them with accurate sets of parameters. The conditions at which these parameters are measured along with the instrumentation will influence the uncertainty of the final simulations. As such, characterization of the parameters for several systems should be clearly defined. This would include rigorous evaluation protocols for the evaluation of coefficients for tire rolling resistance and aerodynamic drag to ensure consistency and minimize uncertainty.

In addition, even if using more detailed models (e.g., zero-dimensional engine model rather than steady state) can lead to better representation of the transients, such models do require significantly more testing and data collection to properly represent the system. As such, a trade-off analysis should be performed to evaluate the additional testing required to populate the detailed models compared to the added accuracy they provide. This accuracy evaluation may be dependent on the evolutionary stage of the technology and cannot be considered static.

In-between Processes

When selecting a process, it is important to understand the uncertainties introduced by the methodology employed. For example, using an engine on the dynamometer (CIL) versus testing the entire vehicle will lead to differences in results as they each have different uncertainty sources. While the driver will have the largest impact on the results during chassis dynamometer testing, the driver is not a factor anymore during the CIL process, where delays and model uncertainties account for most of the uncertainties. As a result, it is important to quantify each process, as one might not necessarily lead to greater uncertainties than another one.

Need for Process Standardization

As shown in Figure H-16, each process should be standardized, from data gathering to model validation and reporting of results.

- *Hardware set-up process.* For any process involving hardware, from HIL to RCP or vehicle testing, detailed test procedures should be developed to ensure consistency across organizations. While some work has been performed for vehicle testing, little to no work has been done for HIL and RCP, and more work is required to validate or improve vehicle test protocols.
- *Validation process.* From a modeling point of view, a critical need is to define what validation means and how it should or could be quantified. While all engineers claim their models are validated, the assumptions behind each one can vary significantly.

  A detailed process should be developed, describing what tests should be performed to validate specific subsystems, systems, or vehicles. A report should be provided to the regulatory agency demonstrating the process and the results of the validation. This report could be generic and automatically developed based on the list of required parameters or comparisons for the regulation.
- *Appropriate modeling level.* Using wrong assumptions can lead to erroneous conclusions; errors can come from modeling assumptions or from data. To answer the right questions, users need to have the right modeling tools. For instance, one common mistake is to study engine emissions by using a steady state model or to study component transient behavior by using a backward model. A study providing general guidance would accelerate development of the required models.
- *Regulatory report.* Since the results must be approved for regulatory purposes, a generic report should be defined so that every original equipment manufacturer provides the same information. This report or set of reports would include not only the results but also the assumptions and details of the simulations or tests for selected critical parameters to ensure validity and consistency of the results.

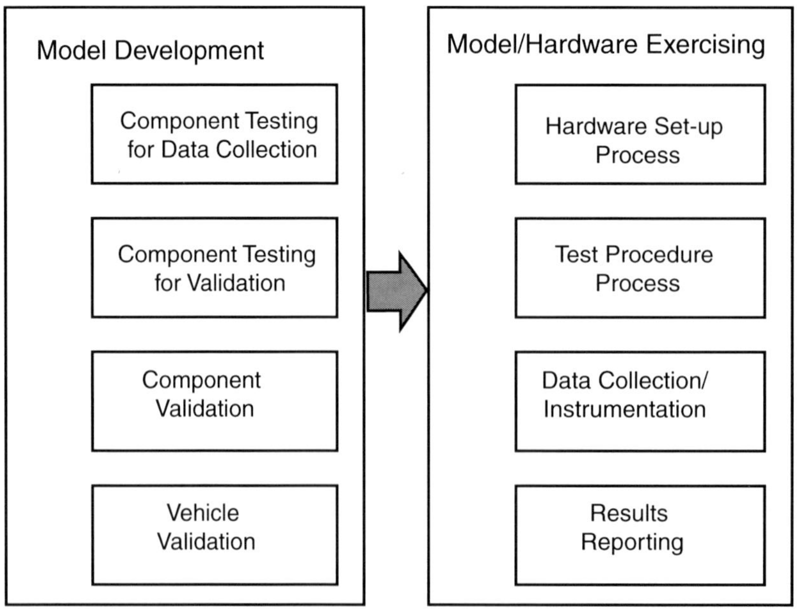

FIGURE H-16 Main phases requiring standardized processes.